Inorganic Chemistry Concepts
Volume 3

Editors

Margot Becke
Michael F. Lappert
John L. Margrave
Robert W. Parry

Christian K. Jørgensen
Stephen J. Lippard
Kurt Niedenzu
Hideo Yamatera

Philipp Gütlich
Rainer Link
Alfred Trautwein

Mössbauer Spectroscopy and Transition Metal Chemistry

With 160 Figures

Springer-Verlag
Berlin Heidelberg New York 1978

Prof. Dr. Philipp Gütlich
Dr. rer. nat. Rainer Link

Johannes Gutenberg-Universität
Institut für Anorganische und Analytische Chemie
D-6500 Mainz

Prof. Dr. Alfred X. Trautwein

Fachbereich „Angewandte Physik"
der Universität des Saarlandes
D-6600 Saarbrücken 11

ISBN 3-540-08671-4 Springer-Verlag Berlin Heidelberg New York
ISBN 0-387-08671-4 Springer-Verlag New York Heidelberg Berlin

Library of Congress Cataloging in Publication Data. Gütlich, Philipp, 1934-
Mössbauer spectroscopy and transition metal chemistry. (Inorganic chemistry
concepts; v. 3) Bibliography: p. Includes index. 1. Transition metals–Spectra.
2. Mössbauer spectroscopy. I. Link, Rainer, 1943- joint author. II. Trautwein,
Alfred, 1940- joint author. III. Title. IV. Series. QC462.T86G83 546'.6
78-2364

Typesetting: Elsner & Behrens, Oftersheim.
Printing and bookbinding: Zechnersche Buchdruckerei, Speyer.
2152/3140-543210

Preface

Two decades have passed since the original discovery of recoilless nuclear gamma resonance by Rudolf Mössbauer; the spectroscopic method based on this resonance effect — referred to as Mössbauer spectroscopy — has developed into a powerful tool in solid-state research. The users are chemists, physicists, biologists, geologists, and scientists from other disciplines, and the spectrum of problems amenable to this method has become extraordinarily broad.

In the present volume we have confined ourselves to applications of Mössbauer spectroscopy to the area of transition elements. We hope that the book will be useful not only to non-Mössbauer specialists with problem-oriented activities in the chemistry and physics of transition elements, but also to those actively working in the field of Mössbauer spectroscopy on systems (compounds as well as alloys) of transition elements.

The first five chapters are directed to introducing the reader who is not familiar with the technique to the principles of the recoilless nuclear resonance effect, the hyperfine interactions between nuclei and electronic properties such as electric and magnetic fields, some essential aspects about measurements, and the evaluation of Mössbauer spectra. Chapter 6 deals with the interpretation of Mössbauer parameters of iron compounds. Here we have placed emphasis on the information about the electronic structure, in correlation with quantum chemical methods, because of its importance for chemical bonding and magnetic properties. Some selected references to original work from the large number of publications on ^{57}Fe Mössbauer spectroscopy have been cited here; trying to be complete in this instance was neither our intention in writing this chapter, nor would it have been possible within the scope of this volume.

In Chapter 7, which deals with Mössbauer active transition elements other than iron, we have, however, strenuously tried to be complete as far as the literature is concerned, which deals with investigations of solid-state problems. It is our hope that this chapter will be particularly useful to scientists who are actively concerned with Mössbauer work on non-iron transition elements.

Chapter 8 finally is intended to give the less informed reader an impression about the various fields in which the Mössbauer effect technique has been of great importance. Here again, of course, we could not refer to all the original work. The many review articles and other selected papers we have cited here should, however, in our opinion be well suited to guide the more deeply interested reader into the relevant literature. The authors wish to express their thanks to the *Deutsche Forschungsgemeinschaft,* the *Bundesministerium für Forschung und Technologie,* and the *Fonds der Chemischen Industrie* for the financial support of their research work in the field of Mössbauer spectroscopy. They also wish to convey their particular thanks to Professor U. Gonser, Professor F. E. Harris, Professor E. Kankeleit, and Professor K. H. Lieser for stimulating and fruitful discussions over many years.

We wish to thank Mrs. G. Lehr and Miss E. Börner for preparing the manuscript and the many drawings.

Mainz and Saarbrücken, 1978   P. G.   R. L.   A. T.

Contents

1. Introduction

Almost twenty years ago, Rudolf L. Mössbauer, while he was working on his doctoral thesis under Professor Maier-Leibnitz at Heidelberg, discovered the *recoilless nuclear resonance absorption of gamma rays,* otherwise known as the *Mössbauer effect* [1.1–3]. This phenomenon developed rapidly to a new spectroscopic technique, called Mössbauer spectroscopy, of high sensitivity to energy changes in the order of 10^{-8} eV (ca. 10^{-4} cm^{-1}) and extreme sharpness of tuning (ca. 10^{-12}).

In the early stage, Mössbauer spectroscopy was somehow restricted to low-energy nuclear physics (e.g., determination of excited state lifetimes and nuclear magnetic moments). After Kistner and Sunyar's report on the observation of a "chemical shift" in the quadrupolar perturbed magnetic Mössbauer spectrum of α-Fe$_2$O$_3$ [1.4], however, one immediately realized that the new spectroscopic method could be particularly useful in solid state research, covering the interests of physicists as well as of chemists. As a matter of fact, by far the largest portion of the approximately 10,000 papers on Mössbauer spectroscopic studies published so far deals with various kinds of problems arising from or directly related to the electronic shell of Mössbauer active atoms in metals and insulators, e.g., magnetism, electronic fluctuations, relaxation processes, electronic and molecular structure, bond properties.

Up to the present time, the Mössbauer effect has been observed for nearly 100 nuclear transitions in about 80 nuclides distributed over 43 elements (cf. Fig. 1.1). Of course, as with many other spectroscopic methods, not all of these transitions are suitable for actual studies, for reasons we shall discuss below. About 15–20 elements remain for applications in Physics, Chemistry, Metallurgy, Biology, Earth and Moon Sciences, etc.

It is the purpose of the present book to scope only Mössbauer active transition elements (Fe, Ni, Zn, Tc, Ru, Hf, Ta, W, Re, Os, Ir, Pt, Au, Hg). A great deal of space will be devoted to the spectroscopy of ^{57}Fe, which is by far the most intensely investigated nuclide of all. Here we shall try to introduce the reader to the various kinds of chemical information one can extract from the electric and magnetic hyperfine interactions reflected in the Mössbauer spectra. Particular emphasis will be put on bonding and structural properties in connection with quantum chemical bonding theories.

There is neither enough room in a review like this to cover a nearly complete bibliography, nor do we think it would be wise to do this in a book which is intended to give the reader a feeling for what can be learned from the application of the Mössbauer effect in Transition Metal Chemistry. We shall restrict ourselves to typical examples from the literature and from our own

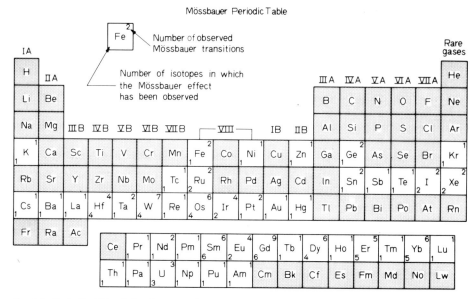

Fig. 1.1. Periodic table of the elements; those in which the Mössbauer effect has been observed are marked appropriately (From the 1974 issue of [1.6])

research which from our point of view are suitable to demonstrate the usefulness of the relatively new method. Those who are more deeply interested in or even concerned with Mössbauer spectroscopy in their research work will have to find their way to the library and consult original papers, as compiled, e.g., in special review articles and compilations [1.5, 6] and relevant books [1.7–16].

References

[1.1] Mössbauer, R. L.: Z. Physik *151*, 124 (1958)

[1.2] Mössbauer, R. L.: Naturwissenschaften *45*, 538 (1958)

[1.3] Mössbauer, R. L.: Z. Naturforsch. *14a*, 211 (1959)

[1.4] Kistner, O. C., Sunyar, A. W.: Phys. Rev. Letters *4*, 229 (1960)

[1.5] Muir Jr., A. H., Ando, K. J., Coogan, H. M.: *Mössbauer Effect Data Index, 1958–1965.* New York: Interscience 1966

[1.6] Stevens, J. G., Stevens, V. E.: *Mössbauer Effect Data Index, 1965–1975.* London: Adam Hilger

[1.7] Frauenfelder, H.: *The Mössbauer Effect.* New York: Benjamin 1962

[1.8] Wertheim, G. K.: Mössbauer Effect: *Principles and Applications.* New York: Academic Press 1964

[1.9] Wegener, H.: *Der Mössbauer-Effekt und seine Anwendung in Physik und Chemie.* Mannheim: Bibliographisches Institut 1965

[1.10] Goldanskii, V. I., Herber, R. (eds.): *Chemical Applications of Mössbauer Spectroscopy.* New York: Academic Press 1968

[1.11] May, L. (ed.): *An Introduction to Mössbauer Spectroscopy.* New York: Plenum Press 1971

[1.12] Greenwood, N. N., Gibb, T. C.: *Mössbauer Spectroscopy*. London: Chapman and Hall 1971

[1.13] Bancroft, G. M.: *Mössbauer Spectroscopy, An Introduction for Inorganic Chemists and Geochemists*. London – New York: McGraw-Hill 1973

[1.14] Gonser, U. (ed.): *Mössbauer Spectroscopy*, in Topics in Applied Physics, Vol. 5. Berlin – Heidelberg – New York: Springer 1975

[1.15] Gruverman, I. J. (ed.): *Mössbauer Effect Methodology, Vol. 1*. New York: Plenum Press 1965 and annually afterwards

[1.16] Greenwood, N. N. (ed.): In *Spectroscopic Properties of Inorganic and Organmetallic Compounds,* Chem. Soc., Specialist Periodical Report 1967 and periodically afterwards

2. Basic Physical Concepts

Before we describe the Mössbauer effect and the relevant hyperfine interactions, the reader might appreciate a brief summary on the background concepts connected with nuclear resonance phenomena. We prefer doing this by collecting formulae without deriving them, otherwise we would go far beyond the scope of this review; moreover, instructive descriptions have already been given at length in a number of introductory books [1.7–14].

2.1. Spectral Line Shape and Natural Line Width

An excited state (nuclear or electronic) of mean lifetime τ can never be assigned a sharp energy value. Instead the energy level spreads over a certain energy range of width ΔE (cf. Fig. 2.1), which correlates with the uncertainty in time Δt via the Heisenberg uncertainty relation in the form of the conjugate variables energy and time,

$$\Delta E \Delta t \geqslant \hbar \tag{2.1}$$

($h = 2\pi\hbar$ = Planck's constant). Δt, also considered as the time interval available to measure the energy E, is in the order of the mean lifetime: $\Delta t \approx \tau$. From (2.1) we see that a ground state of infinite lifetime has zero uncertainty in energy.

Nuclear transitions from an excited state (e) to the ground state (g), or vice versa, involve all possible energies within the range of ΔE. The transition probability or intensity as a function of the transition energy, I(E), therefore yields a spectral line centered around the most probable transition energy E_0 (Fig. 2.1).

Weisskopf and Wigner [2.1] have shown that in general

$$\Gamma\tau = \hbar \tag{2.2}$$

holds, if $\Gamma = \Delta E$ stands for the full width of the transition spectral line at half maximum. They also found that the spectral line has Lorentzian or Breit-Wigner form and follows the formula

$$I(E) \sim \frac{\Gamma/2\pi}{(E - E_0)^2 + (\Gamma/2)^2}. \tag{2.3}$$

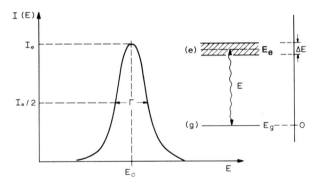

Fig. 2.1. Intensity I(E) as a function of transition energy E. $\Delta E = \Gamma = \hbar/\tau$ is the energy width of the excited state (e) with a mean lifetime τ as well as the width of the transition spectral line at half maximum (Heisenberg natural line width)

The mean lifetime τ of the excited state governs the width of the relevant transition line. Lifetimes of excited nuclear states suitable for Mössbauer spectroscopy range from $\sim 10^{-6}$s to $\sim 10^{-11}$s. Longer lifetimes produce too narrow transition lines, which would — in case of emission and absorption lines in a Mössbauer experiment — no longer overlap sufficiently because of experimental difficulties (extremely small Doppler velocities of $< \mu$m s^{-1} required). Lifetimes shorter than 10^{-11}s are connected with too broad transition lines; the resonance overlap between them would be too much smeared out and no longer distinguishable from the base line of a spectrum. As an example, the first excited state of ^{57}Fe has the mean lifetime $\tau = t_{1/2}/\ln2 = 1.43 \cdot 10^{-7}$s ($t_{1/2}$ = half-life); Γ therefore evaluates to $4.55 \cdot 10^{-9}$ eV.

In addition to the lifetime conditions there is another condition of practical importance in Mössbauer spectroscopy; this concerns the transition energy E. We shall come back to this later.

2.2. Nuclear Resonance

Suppose a nucleus in an excited state of energy E_e undergoes transition to the ground state of energy E_g by emitting a gamma quantum of energy $E_0 = E_e - E_g$. Under certain conditions which we shall discuss below, the quantum energy E_0 may be totally absorbed by a nucleus of the same kind (same number of protons Z and same number of neutrons N) in its ground state, whereby transition to the excited state of energy E_e takes place. The phenomenon, called *nuclear resonance absorption of γ-rays*, is visualized in Fig. 2.2.

Maximum resonance absorption only occurs if the spectral line for the emission process as well as the one for the absorption process appear at the same energy position, e.g., E_0. The resonance absorption cross section (shape of the absorption line) is expressed by the Breit-Wigner formula

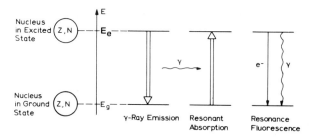

Fig. 2.2. Schematic representation of nuclear resonance absorption of γ-rays (Mössbauer effect) and nuclear resonance fluorescence

$$\sigma(E) = \sigma_0 \Gamma^2 / [\Gamma^2 + 4(E - E_0)^2], \tag{2.4}$$

where

$$\sigma_0 = \frac{\lambda^2}{2\pi} \cdot \frac{2I_e + 1}{2I_g + 1} \cdot \frac{1}{\alpha + 1} \tag{2.5}$$

is the maximum absorption cross section, I_e and I_g are nuclear spin quantum numbers of the excited and the ground state, respectively, λ is the wave length of the γ-ray, and α is the internal conversion coefficient (for ^{57}Fe $\alpha = 8.21$).

After the resonant γ-ray absorption the nucleus remains in the excited state of energy $E_0 = E_e - E_g$ for the mean lifetime τ and then undergoes transition back to the ground state by isotropic emission of a γ-ray or conversion electrons due to internal conversion (energy transfer from the nucleus to the electron shell), which in most Mössbauer active nuclei competes with γ-ray emission. Here we speak of *nuclear resonance fluorescence.*

Nuclear resonance absorption of γ-rays does not occur between nuclei of isolated atoms or molecules (in gaseous or liquid state) because of the large energy loss of the transition energy E_0 due to recoil effects.

2.3. Recoil Energy Loss, Thermal Broadening of Transition Lines

If a γ-ray (photon) is emitted from an excited nucleus of mass M and of mean energy $E_0 = E_e - E_g$, which is supposed to be at rest before the decay, a recoil is imparted to the nucleus, due to which the nucleus moves with velocity $\vec{v}$ in opposite direction to the direction of the γ-ray emission (see Fig. 2.3) and takes up the kinetic recoil energy

$$E_R = \frac{1}{2} Mv^2. \tag{2.6}$$

Momentum conservation requires that

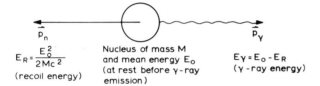

$E_R = \dfrac{E_0^2}{2Mc^2}$

(recoil energy)

Nucleus of mass M
and mean energy E_0
(at rest before γ-ray
emission)

$E_\gamma = E_0 - E_R$
(γ-ray energy)

Fig. 2.3. Recoil of momentum $\vec{p}_n$ and energy E_R imparted to an isolated nucleus upon γ-ray emission

$$p_n = -p_\gamma = -E_\gamma/c, \qquad (2.7)$$

where p_n and p_γ are the linear momenta of the nucleus and the γ-quantum, respectively, c is the velocity of light, and

$$E_\gamma = E_0 - E_R \qquad (2.8)$$

is the energy of the emitted γ-quantum. Because of the large mass of the nucleus we may write in the nonrelativistic approximation

$$E_R = p_n^2/2M = E_\gamma^2/2Mc^2. \qquad (2.9)$$

Since E_R is very small compared to E_0 it is reasonable to assume $E_\gamma \approx E_0$; then we may use the following formula for computing the recoil energy of a nucleus in an isolated atom or molecule:

$$E_R = E_0^2/2Mc^2 = 5.37 \cdot 10^{-4} \, E_0^2/A \text{ eV}, \qquad (2.10)$$

where A is the atomic number of the nucleus and E_0 is given in keV. As an example, for the Mössbauer transition between the first excited state and the ground state of ^{57}Fe ($E_0 = E_e - E_g = 14.4$ keV), E_R is evaluated to be $1.95 \cdot 10^{-3}$ eV. This is about six orders of magnitude larger than the natural width of the spectral transition line under consideration ($\Gamma = 4.55 \cdot 10^{-9}$ eV).

The recoil effect causes a displacement of the emission line from the position E_0 to smaller energies by an amount E_R (see Fig. 2.4). In the absorption process, the γ-ray to be absorbed by a nucleus requires the total energy $E_\gamma = E_0 + E_R$ to make up for the transition from the ground to the excited state and the recoil effect (for which $\vec{p}_n$ and $\vec{p}_\gamma$ have the same directions now). As shown schematically in Fig. 2.4 the transition lines for emission and absorption are separated by a distance of $2E_R$ on the energy scale, which is about 10^6 times larger than the natural line width Γ. Overlap between the two transition lines and hence nuclear resonance absorption is not possible in isolated atoms or molecules in the gaseous or liquid state.

Atoms in a gas are never at rest. If γ-ray emission takes place while the nucleus (or atom) is moving at a velocity v_n in the direction of the γ-ray propagation, the γ-photon of energy E_γ receives a Doppler energy E_D,

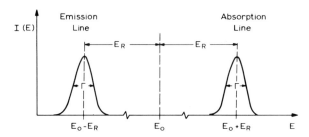

Fig. 2.4. Consequences of the recoil effect caused by γ-ray emission and absorption in isolated nuclei. The transition lines for emission and absorption are separated by $2E_R \approx 10^6 \Gamma$. There is no overlap between emission and absorption line and hence no resonant absorption possible

$$E_D = \frac{v_n}{c} E_\gamma, \tag{2.11}$$

which is to be added to E_γ

$$E_\gamma = E_0 - E_R + E_D. \tag{2.12}$$

The resulting mean Doppler broadening of the transition line can be evaluated from

$$\overline{E_D} = \overline{(2E_K M v_n^2)^{1/2}} = \overline{2(E_K E_R)^{1/2}} = E_\gamma \overline{(2E_K/Mc^2)^{1/2}}, \tag{2.13}$$

where

$$E_K = \frac{1}{2} M v_n^2 = \frac{3}{2} k_B T \tag{2.14}$$

is the mean kinetic energy of the moving nucleus (atom). The average of $\overline{E_D}$ is taken over all angles ϑ between $\vec{p}_\gamma$ and $\vec{v}_n$. $\overline{E_D}$ is in the order of E_R or larger (e.g., ^{57}Fe: $E_0 = 14.4$ keV, $E_R = 1.95 \cdot 10^{-3}$ eV, $\overline{E_D} \simeq 10^{-2}$ eV at 300 K). Therefore there is some very low probability for resonance absorption even in case of relatively large recoil loss: the Doppler broadened emission and absorption lines overlap in a small energy region.

In the solid state the situation is different. Here nuclear resonance absorption of γ-rays is possible; we shall describe briefly why.

In the solid state a Mössbauer active atom under consideration is more or less rigidly bound to the lattice. If a γ-ray is emitted from the excited Mössbauer nucleus, the concomitant recoil energy may be assumed to consist of two parts,

$$E_R = E_{tr} + \overline{E_{vib}}. \tag{2.15}$$

E_{tr} is the translational energy transferred through a linear momentum to the crystallite as a whole, which accommodates the Mössbauer nucleus under consideration. E_{tr} may be evaluated using the formula (2.9), in which M stands now for the mass of the whole crystallite. Because of the very large mass of the crystallite as compared to the mass of a single nucleus E_{tr} turns out to be many orders of magnitude smaller than Γ and can therefore be neglected.

Most of the recoil energy E_R is converted into mean lattice vibrational energy $\overline{E_{vib}}$, i.e., the recoil energy is mostly transferred to the lattice vibrational system. The (free atom) recoil energy E_R is larger than the characteristic lattice vibration (phonon) energy but smaller than the displacement energy (~ 25 eV); therefore the decaying Mössbauer atom will remain in its lattice position and will dissipate $\overline{E_{vib}}$ by heating the near lattice surroundings. In case, however, E_R is smaller than the characteristic phonon energy (which is in the order of 10^{-2} eV for solids), $\overline{E_{vib}}$ causes a change in the vibrational energy of the oscillators by integral multiples of the phonon energy $\hbar\omega_E$ (if ω_E represents the Einstein frequency), i.e., $0 \cdot \hbar\omega_E$, $\pm 1\hbar\omega_E$, $\pm 2\hbar\omega_E$, etc. The model tells us that there is a certain probability f that no lattice excitation (energy transfer of $0\hbar\omega_E$, called zero-phonon-process) takes place during the γ-emission or γ-absorption process. f is called the *recoil-free fraction* and denotes the fraction of nuclear transitions which occur without recoil. We may therefore write (for $E_R \ll \hbar\omega_E$)

$$E_R = (1 - f) \hbar\omega_E \tag{2.16}$$

and

$$f = 1 - E_R/\hbar\omega_E = 1 - k^2\langle x^2 \rangle, \tag{2.17}$$

where $\langle x^2 \rangle$ is the expectation value of the squared vibrational amplitude in the x-direction and k the propagation vector. The recoil-free fraction in Mössbauer spectroscopy is equivalent to the fraction of scattering processes of x-rays without lattice excitation. The more general expression for f is [2.2]:

$$f = \exp(-E_R/\hbar\omega_E) = \exp(-k^2 \langle x^2 \rangle). \tag{2.18}$$

From (2.18) we obtain (2.17) by taking $E_R \ll \hbar\omega_E$.

The Debye model for solids leads to the following expression for the recoil-free fraction:

$$f = \exp\left[\frac{-6E_R}{k_B\Theta_D} \left\{ \frac{1}{4} + \left(\frac{T}{\Theta_D}\right)^2 \int_0^{\Theta_D/T} \frac{x}{e^x - 1} \, dx \right\} \right], \tag{2.19}$$

which reduces to the approximations

$$f = \exp\left[-\frac{E_R}{k_B\Theta_D} \left(\frac{3}{2} + \frac{\pi^2 T^2}{\Theta_D^2} \right) \right] \text{ for } T \ll \Theta_D, \tag{2.20}$$

$$f = \exp\left(-\frac{6E_R T}{k_B\Theta_D^2} \right) \text{ for } T > \Theta_D. \tag{2.21}$$

k_B is the Boltzmann factor and $\Theta_D = \hbar\omega_D/k_B$ the Debye temperature. From these formulae we notice that

(i) f increases with decreasing recoil energy, i.e., decreasing transition energy E_γ;

(ii) f increases with decreasing temperature;

(iii) f increases with increasing Debye temperature Θ_D. Θ_D may be considered a measure for the strength of the bonds between the Mössbauer atom and the lattice. (Θ_D is usually high for metallic systems and low for metal organic complexes).

f is generally called the *Debye-Waller factor* (or *Lamb-Mössbauer factor*). Characteristic f-values are, e.g., 0.91 for the 14.4 keV gamma transition in ^{57}Fe, and 0.06 for the 129 keV gamma transition in ^{191}Ir.

For a more detailed instruction on the recoil-free fraction and lattice dynamics, the reader is referred to relevant textbooks [1.7–10].

2.4. The Mössbauer Effect

From Section 2.3 we have learned that the recoil effect in free or loosely bound atoms shifts the γ-transition line by E_R, and thermal motion broadens the transition line by twice the geometric mean of $E_R E_K$. For nuclear resonance absorption of γ-rays to be successful, one must bring the emission and absorption lines into coincidence, or at least to partially overlap, by making use of the Doppler effect to compensate for the recoil energy loss. Moon, in 1950, succeeded first by mounting the source to an ultracentrifuge and moving it at suitably high velocities toward the absorber [2.3]. Other experiments later on were also successful; they all were basically similar in that the recoil energy loss was compensated for somehow by the Doppler effect.

The real breakthrough in nuclear resonance absorption of γ-rays, however, came with Mössbauer's discovery at Heidelberg. By means of an experimental arrangement similar to the one described by Malmfors [2.4], he intended to measure the lifetime of the 129 keV state in ^{191}Ir. Nuclear resonance absorption was planned to be achieved by making the emission line and the absorption line partially overlap each other through thermal broadening. By lowering the temperature it was generally expected that the transition lines would sharpen up because of less effective Doppler broadening and consequently smaller degree of overlap. The opposite was observed by Mössbauer: the resonance effect increased by cooling both source and absorber. Mössbauer not only observed this striking experimental effect, which was not consistent with the prediction of sharpening of the γ-transition lines with decreasing temperature, he also presented an explanation which is based on zero-phonon processes in the crystal associated with emission and absorption of γ-rays. In other words, with a certain probability f, which increases with decreasing temperature, the quantum state of the lattice remains unchanged during the nuclear transition. The factor f describes the recoil-free fraction of the nuclear transition and is therefore a measure of the recoilless nuclear absorption of γ-radiation — the Mössbauer effect.

In an actual Mössbauer experiment (see Fig. 2.5) one moves the source and the absorber relative to each other (either by moving the source and keeping the absorber fixed or vice versa) in a controlled fashion and registers the transmitted γ-quanta as a

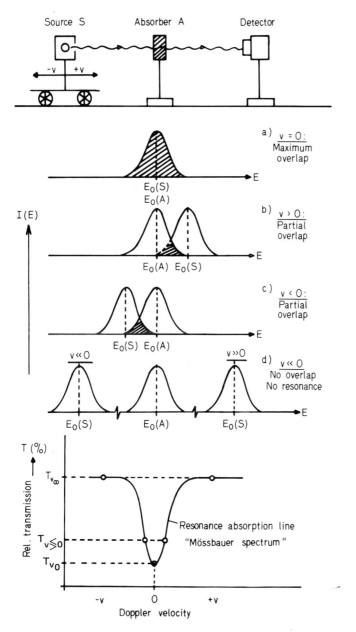

Fig. 2.5. Schematic illustration of the experimental arrangement for recoilless nuclear resonance absorption, and relative transmission of recoilless γ-quanta as a function of Doppler velocity

function of the relative velocity (Doppler-velocity). The Mössbauer spectrum, the plot of the relative transmission versus Doppler velocity, shows maximum resonance and therefore minimum relative transmission at relative velocities where emission and absorption lines overlap ideally. In Chapter 4 we shall see what a modern Mössbauer effect spectrometer looks like and how it is used.

References

[2.1] Weisskopf, V., Wigner, E.: Z. Physik *63*, 54 (1930); *65*, 18 (1930)
[2.2] Lamb Jr., W. E.: Phys. Rev. *55*, 190 (1939)
[2.3] Moon, P. B.: Proc. Phys. Soc. *63*, 1189 (1950)
[2.4] Malmfors, K. G.: Arkiv Fysik *6*, 49 (1953)

3. Hyperfine Interactions

In Chapter 2 we have, for the sake of simplicity, dealt with transitions between unperturbed energy levels of "bare" nuclei and have called the mean transition energy E_0. In reality, however, we deal with atoms and molecules, with gases, liquids, and solids. Nuclei are therefore generally embedded in electric and magnetic fields, which may be created by shell electrons and charges in the near neighborhood. We also have to keep in mind that nuclei are positively charged and may possess various kinds of nuclear moments. These generally interact with the electric and magnetic fields in the nuclear region and perturb the nuclear energy levels. The perturbation, called *nuclear hyperfine interactions,* may be such that it only shifts the nuclear energy levels, as is the case in *electric monopole interaction (e0)*, or such that it splits degenerate nuclear levels into sublevels without shifting the centroid of the multiplet, as is observed in *electric quadrupole interaction (e2)* and *magnetic dipole interaction (m1)*. Only these three kinds of interactions have to be considered in practical Mössbauer spectroscopy. Electric dipole interaction *(e1)* does not exist because of symmetry arguments (invariance of nuclear forces relative to change in sign of coordinates). Interactions of higher order *(m3, e4,* etc.) are negligible; their energy effects are so small that they cannot be resolved in a Mössbauer spectrum.

A Mössbauer spectrum, in general, reflects the nature and the strength of the hyperfine interactions. The *e0* interaction affects the position of the resonance lines on the Doppler velocity (energy) scale and gives rise to the so-called *isomer shift* (chemical shift) δ. *e2* and *m1* interactions split resonance lines originating from transitions between degenerate nuclear levels; the *electric quadrupole splitting* ΔE_Q and the *magnetic splitting* ΔE_M are the resulting Mössbauer parameters. Most valuable chemical information can be extracted from these three "Mössbauer parameters" δ, ΔE_Q, and ΔE_M.

3.1. Electric Hyperfine Interaction

The total energy of the electrostatic interaction between a nucleus with charge Ze and surrounding charges may be expressed as [1.9]:

$$E_{el} = \int \rho_n(r) \, V(r) \, d\tau \tag{3.1}$$

where $\rho_n(r)$ stands for the nuclear charge density at a point with coordinates $r = (x_1, x_2, x_3)$ and $V(r)$ is the Coulomb potential set up at a point $r = (x_1, x_2, x_3)$ by all other charges. $d\tau = dx_1 \cdot dx_2 \cdot dx_3$ represents the volume element. The center of the coordinate system coincides with the center of symmetry of the nuclear charge.

We now expand $V(r)$ in a Taylor series at the point $r = 0$,

$$V(r) = V_0 + \sum_{i=1}^{3} \left(\frac{\partial V}{\partial x_i}\right)_0 x_i + \frac{1}{2} \sum_{i,j=1}^{3} \left(\frac{\partial^2 V}{\partial x_i \partial x_j}\right)_0 x_i x_j + \dots \tag{3.2}$$

and insert (3.2) into (3.1)

$$E_{el} = V_0 \int \rho_n(r) d\tau + \sum_{i=1}^{3} \left(\frac{\partial V}{\partial x_i}\right)_0 \cdot \int \rho_n(r) x_i d\tau + \frac{1}{2} \sum_{i,j=1}^{3} \left(\frac{\partial^2 V}{\partial x_i \partial x_j}\right)_0 \cdot \int \rho_n(r) x_i x_j d\tau + \dots \tag{3.3}$$

nuclear charge   *elect. dipole interaction (does not)*

With $eZ = \int \rho_n(r) d\tau$ for the nuclear charge, the first term in (3.3) becomes eZV_0. This term represents the electrostatic interaction between the nucleus considered as a point and other charges in the material. It contributes to the potential energy of the crystal as a whole, which is of no further interest here. The second term in (3.3) expresses the electric dipole interaction, which we have stated earlier does not exist because of symmetry arguments. Higher odd-order terms do not exist for the same reason. Even-order terms higher than the third term in (3.3) are negligible because their interaction energies are too small to be resolved by Mössbauer spectroscopy. Thus, the only term in (3.3) which we are interested in is the third one.

The quantities $(\partial^2 V/\partial x_i \partial x_j)_0 = V_{ij}$ form a (3 x 3) second-rank tensor. We may choose the coordinate system such that all tensor elements V_{ij} vanish except the diagonal ones, V_{ii} (principal axes system). We may then write for the third term of (3.3) [1.9]

$$E = \frac{1}{2} \sum_{i=1}^{3} V_{ii} \cdot \int \rho_n(r) x_i^2 d\tau = \frac{1}{2} \sum_{i=1}^{3} V_{ii} \cdot \int \rho_n(r) \left(x_i^2 - \frac{r^2}{3}\right) d\tau$$

$$+ \frac{1}{6} \sum_{i=1}^{3} V_{ii} \int \rho_n(r) r^2 d\tau, \text{ with } r^2 = \sum_{i=1}^{3} x_i^2. \tag{3.4}$$

By adding and subtracting the quantity

$$\frac{1}{6} \sum_{1=i}^{3} V_{ii} \int \rho_n(r) r^2 d\tau$$

in (3.4) we introduce into the first term the definition of the *nuclear quadrupole moment,* the diagonal tensor elements of which in the principal axes system are given by

$$Q_{ii} = \int \rho_n(r) (3x_i^2 - r^2) d\tau. \tag{3.5}$$

With Laplace's differential equation

$$\Delta V + 4\pi\rho_e = 0 \tag{3.6}$$

we find at the point $r = (x_1, x_2, x_3) = 0$ (center of symmetry of the nucleus)

$$(\Delta V)_0 = \left(\sum_{i=1}^{3} V_{ii}\right)_0 = 4\pi e|\psi(o)|^2, \tag{3.7}$$

where $\rho_e = -e|\Psi(0)|^2$ is the charge density exerted by the surrounding electrons at the nucleus ($r = 0$). Inserting (3.7) into (3.4) we obtain

$$E = \frac{2}{3}\pi e|\psi(o)|^2 \int \rho_n(r)r^2 d\tau + \frac{1}{2}\sum_{i=1}^{3} V_{ii} \cdot \int \rho_n(r)\left(x_i^2 - \frac{r^2}{3}\right)d\tau \equiv E_I + E_Q. \tag{3.8}$$

The first term of (3.8) represents the *electric monopole interaction*. It causes a shift of nuclear energy levels and gives rise to the *isomer shift* δ. The second term in (3.8) represents the *electric quadrupole interaction*. It splits degenerate nuclear energy levels and yields the *quadrupole splitting* ΔE_Q.

3.1.1. Electric Monopole Interaction; Isomer Shift

By electric monopole interaction we simply mean the electrostatic Coulomb interaction between the nuclear charge, which spreads over a finite volume, and electrons inside the nuclear region. s-electrons have the ability to penetrate the nucleus and spend a fraction of their time there. Electrons with nonzero angular momentum (p-, d-, f-electrons) do not have this ability, unless we consider relativistic effects, in which case $p_{1/2}$-electrons may also spend a very small fraction of their time inside the nucleus.

We consider the first term of (3.8) and substitute

$$\int \rho_n(r)r^2 d\tau \equiv \langle r^2 \rangle \cdot Ze, \tag{3.9}$$

where $\langle r^2 \rangle$ is the expectation value of the square of the nuclear radius and $\int \rho_n(r)d\tau$ is the nuclear charge Ze. We then obtain

$$E_I = \frac{2}{3}\pi Ze^2|\psi(o)|^2 \langle r^2 \rangle \equiv \delta E \tag{3.10}$$

as the interaction energy, by which the energy level of a nuclear state is shifted due to Coulomb interactions. We put $\delta E \equiv E_I$ to express the very small change in energy ($\simeq 10^{-8}$ eV).

As the nuclear volume and hence the quantity $\langle r^2 \rangle$ is different in each state of excitation, the electrostatic shift δE will also be different for each nuclear state. This is shown in Fig. 3.1, where the nuclear levels of the excited state and the ground state

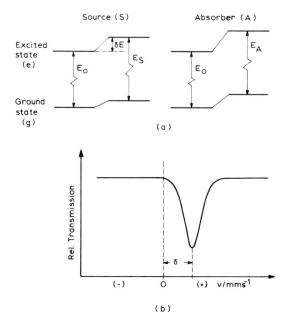

Fig. 3.1. Origin of isomer shift. (a) Electric monopole interaction between the nuclear charge and electrons at the nucleus shifts nuclear energy levels without changing the degeneracy; (b) resultant Mössbauer spectrum (schematic)

are shifted to different extents. Therefore, in a nuclear transition between the ground (g) and an excited (e) state the energy change of a γ-ray will be

$$\Delta E = E_S - E_0 = (\delta E)_e - (\delta E)_g = \frac{2}{3}\pi Z e^2 |\psi(o)|^2 [\langle r^2 \rangle_e - \langle r^2 \rangle_g]. \tag{3.11}$$

In a Mössbauer experiment, where an appropriate Doppler velocity is applied to either the source (S) or the absorber (A) in order to bring the γ-ray emission line into coincidence (optimum overlap) with the corresponding absorption line, one never observes the quantities $(\Delta E)_A$ and $(\Delta E)_S$ separately, but only the difference of the electrostatic shift (3.10) in the source and the absorber. Any difference (in electron configuration, structure, temperature, pressure, etc.), between the source and the absorber will influence the electron density at the nucleus. We therefore obtain the following general expression for the difference in the electrostatic shift between the source and the absorber:

$$\delta = (\Delta E)_A - (\Delta E)_S = \frac{2}{3}\pi Z e^2 [|\psi(o)|_A^2 - |\psi(o)|_S^2] \cdot [\langle r^2 \rangle_e - \langle r^2 \rangle_g]. \tag{3.12}$$

δ is called the *isomer shift*. If we consider atomic nuclei to be of spherical symmetry with radius R and constant charge density $\rho_n(r) = 3Ze/4\pi R^3$, we find for $\langle r^2 \rangle$

$$\langle r^2 \rangle = \frac{1}{Z_e} \int \rho_n(r) r^2 d\tau = \frac{3}{4\pi R^3} \int r^2 \cdot d\tau. \tag{3.13}$$

Replacing $d\tau$ by the volume element $r^2 dr \sin \vartheta d\vartheta d\varphi$ in spherical coordinates and integrating (3.13) yields

$$\langle r^2 \rangle = \frac{3}{4\pi R^3} \int_0^R r^4 dr \int_0^\pi \sin \vartheta d\vartheta \int_0^{2\pi} d\varphi = \frac{3}{5} R^2. \tag{3.14}$$

With (3.14) we may rewrite (3.12)

$$\delta = \frac{2}{5} \pi Z e^2 \left\{ |\psi(0)|_A^2 - |\psi(0)|_S^2 \right\} \cdot (R_e^2 - R_g^2). \tag{3.15}$$

Putting finally $R_e - R_g \equiv \delta R$ and $R_e + R_g \equiv 2R$ (R_e and R_g differ only slightly from each other) we find the frequently encountered expression

$$\delta = \frac{4}{5} \pi Z e^2 \left\{ |\psi(0)|_A^2 - |\psi(0)|_S^2 \right\} \cdot (\delta R/R) R^2. \tag{3.16}$$

The corresponding Doppler velocity v_D, which is necessary to restore resonance between source and absorber in a Mössbauer experiment, is given by

$$v_D = (4\pi c/5 E_\gamma) Z e^2 R^2 (\delta R/R) \left\{ |\psi(0)|_A^2 - |\psi(0)|_S^2 \right\} \tag{3.17}$$

and can be evaluated readily from a Mössbauer spectrum as the distance of the resonance line (or centroid of a resonance multiplet) from zero Doppler velocity (Fig. 3.1).

So far we have found expressions for the isomer shift δ in nonrelativistic form. These expressions may be applied to lighter elements, say up to iron, without causing too much of an error. In heavier elements, however, the wave function ψ is subject to considerable modification, particularly near the nucleus, by relativistic effects (remember that the spin-orbit coupling coefficient increases with Z^4!). Therefore the electron density at the nucleus $|\psi(0)|^2$ will be modified as well and the above equations for the isomer shift require relativistic correction. This has been done [3.1] in a somewhat restricted approach using Dirac wave functions[1] and first-order perturbation theory; one finds this way that the relativistic correction simply consists of a dimensionless factor $S'(Z)$, which is introduced in the above equations for δ,

$$\delta = (4\pi/5) Z e^2 S'(Z) R^2 (\delta R/R) \left\{ |\psi(0)|_A^2 - |\psi(0)|_S^2 \right\}. \tag{3.18}$$

Values of the "relativity factor" $S'(Z)$ for $Z = 1$ to 96 have been compiled by Shirley [3.1]. For example, $S'(Z) = 1.32$ for iron ($Z = 26$), 2.48 for tin ($Z = 50$), 19.4 for neptunium ($Z = 93$). The problem of relativistic corrections does not arise in Mössbauer effect studies, where one compares compounds of the same Mössbauer nuclide, because the relativity factor $S'(Z)$ is constant for all compounds of a given

---

[1] A comparison between relativistic and nonrelativistic calculations shows that the correction factor $S'(Z)$ is slightly different for 1s, 2s, and 3s electrons of iron. One finds

for $3d^5$: $S'_{1s}(Z) = 1.2619$ for $3d^6$: $S'_{1s}(Z) = 1.2619$
$S'_{2s}(Z) = 1.2998$ $S'_{2s}(Z) = 1.3002$
$S'_{3s}(Z) = 1.3079$ $S'_{3s}(Z) = 1.3077$

These small differences are of great significance in the calculation of absolute electron densities (cf. Table 3.1).

Mössbauer nuclide. Furthermore, if a standard source is used throughout in a study of a series of compounds of a particular Mössbauer nuclide, we may consider $|\psi(o)|_S^2$ = C a constant, and the δ isomer shift will be a linear function of the charge density $|\psi(o)|_A^2$ at the absorber nucleus,

$$\delta = const\, (\delta R/R) \left\{ |\psi(o)|_A^2 - C \right\}. \tag{3.19}$$

The nuclear factor $\delta R/R$, which stands for the relative change of the nuclear radius in going from the excited state to the ground state, is known, though not to any great accuracy, for many Mössbauer nuclides. It may be positive or negative. The sign of $\delta R/R$, which is of great importance in chemistry, is known for most of the Mössbauer nuclides of chemical applicability. For $\delta R/R$ positive, a positive isomer shift indicates an increase of the electron density at the nucleus in going from the source to the absorber; when $\delta R/R$ is negative, a positive isomer shift implies that the electron density at the absorber nuclei is lower than at the source nuclei. In a series of compounds of a particular Mössbauer nuclide, observed changes of δ in the positive direction (relative to the isomer shift of a reference absorber) indicates an increase (decrease) of the electron density at the absorber nuclei, if $\delta R/R$ is positive (negative).

Isomer shift values must always be reported with respect to a given reference material, because the Mössbauer spectra of a particular compound measured with different sources (e.g., in ^{57}Fe spectroscopy, ^{57}Co in metal foils of Pd, Pt, Cr, Cu, Rh, stainless steel are commonly used sources) under otherwise constant conditions will show different isomer shifts. The reason is obvious from (3.18) : $|\psi(o)|_S^2$ changes with the chemical environment of the Mössbauer nuclide. In ^{57}Fe Mössbauer spectroscopy, metallic iron and sodium nitroprusside dihydrate, $Na_2[Fe(CN)_5 NO] \cdot 2H_2O$ (SNP), are commonly used as standard reference materials in reporting isomer shifts.

The finite electron density at the nucleus is primarily due to the ability of the s-electron to penetrate the nucleus. The total s-electron density at the atomic nucleus in a compound may be considered to be composed of a contribution from filled s-orbitals of inner electron shells and a contribution from partially filled valence orbitals, where valence electrons of the Mössbauer atom as well as electrons from surrounding ligands are accommodated to form the chemical bond. The valence electron contribution to the s-electron density at the nucleus is very sensitively affected by changes in the electronic structure of the valence shell by chemical influences, such as change of oxidation state, of spin state, of bond properties by electron delocalization, etc. Such changes in the valence shell will inevitably influence the s-electron density at the nucleus, generally in two ways.

1. Directly, by altering the s-electron population in the valence shell; a change in s-electron density in the valence shell will directly change $|\psi(o)|^2$ in the same direction.

2. Indirectly, via shielding s-electrons by electrons of nonzero angular momentum; an increase (decrease) of valence electron density of p-, d-, and f-character, respectively, will cause a weaker (stronger) attraction of the s-electron cloud by the nuclear charge and thus decrease (increase) the quantity $|\psi(o)|^2$.

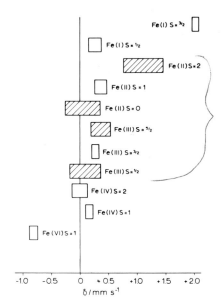

Fig. 3.2. Approximate ranges of isomer shifts observed in iron compounds (relative to metallic iron at room temperature). S refers to the spin quantum number. Shadowed ranges belong to more frequently met configurations (adapted from [3.2]). Note that for ^{57}Fe (δ R/R) < o

Correlation diagrams, in which isomer shift ranges are drawn as a function of the oxidation state, are available for many Mössbauer atoms (e.g., ^{57}Fe, ^{119}Sn, ^{127}I, ^{129}Xe, ^{99}Ru, ^{197}Au, ^{237}Np, and others). Such diagrams are very helpful in characterizing unknown compounds. A typical diagram of this kind for ^{57}Fe is shown in Fig. 3.2. This diagram [3.2] shows approximate ranges of the isomer shift observed in iron compounds with different oxidation and spin states of the central metal ion. The diagram demonstrates that for ionic (high-spin) iron compounds the δ-values become more positive with decreasing formal oxidation state, due to shielding effects by 3d electrons as has been discussed above. The δ-ranges for the different oxidation states in ionic iron compounds hardly overlap one another; it is therefore relatively easy to determine the oxidation state of iron in unknown ionic compounds. Low-spin iron (II) and iron (III) compounds, however, exhibit similar isomer shifts; it is therefore not possible to distinguish between iron (II) and iron (III) in such compounds from δ-values alone. A distinction, however, can be made by combined consideration of the isomer shift and the quadrupole splitting ΔE_Q.

The influence of the electron configuration of a free atom on the electron density at the nucleus, $|\psi(o)|^2$ can be demonstrated by results of Hartree-Fock atomic structure calculations carried out by Watson [3.3]. He found radial functions R(r) for the first transition metal ion series in convenient analytical expressions as accurate solutions to the Hartree-Fock equations. The one-electron s-functions, which may be written as

$$\psi_{ns} = R_{ns}(r)/2\sqrt{\pi},\qquad(3.20)$$

where n denotes the principal quantum number, are the only ones which directly contribute to the electron density at the nucleus (r = 0). The total electron density at

Table 3.1. s-Electron densities at the iron nucleus for various valence electron configurations of the iron atom (in atomic units: electrons per cubic Bohr radius)

	Partial electron densities $\|\psi_{ns}\|^2_{r=0} = \frac{1}{4\pi} \sum_n \|R_{ns}(r)\|^2_{r=0}$ for the configuration				
	$3d^8$	$3d^7$	$3d^6$	$3d^5$	$3d^6 4s^2$ (free atom)
From one electron in 1s	5,378.005	5,377.937	5,377.840	5,377.625	5,377.873
From one electron in 2s	493.953	493.873	493.796	493.793	493.968
From one electron in 3s	67.524	67.764	68.274	69.433	68.028
From one electron in 4s					3.042
$\|\psi(0)\|^2 = 2 \sum_n \|\psi_{ns}\|^2_{r=0}$	11,878.9	11,879.2	11,879.8	11,881.7	11,885.8

$r = 0$ is then given by

$$|\psi(o)|^2 = \frac{1}{2\pi} \sum_n \left\{ R_{ns}(r) \right\}^2_{\mid r = 0}. \tag{3.21}$$

Using Watson's wave functions, s-electron densities at the iron nucleus have been computed for different electron configurations as given in Table 3.1. We notice that, in going from one electron configuration to another, differences in $|\psi(o)|^2$ originate from 1s, 2s, and 3s contributions, where the contribution from the 3s shell dominates. We also notice that, as we have stated earlier, the removal of 3d electrons leads to an increase of $|\psi(o)|^2$, due to decreased shielding of the 3s electrons from the nuclear charge by 3d electrons. Adding 4s electrons, e.g., in going from $Fe^{2+}(3d^6)$ to metallic iron ($3d^6 4s^2$), increases markedly the electron density at the nucleus.

It is obvious from the above discussion that the Mössbauer isomer shift provides very useful information about the bond properties, valency, and oxidation state of a Mössbauer atom as well as the electro-negativity and the number of coordinated ligands.

The isomer shift values assigned to a certain oxidation state of a Mössbauer atom often form a relatively broad range; this is a direct consequence of different bond properties. The electron distribution in the molecular orbitals depends on various abilities of the ligands to take part in σ- and π-bonding to a Mössbauer atom under consideration. Satisfactory quantitative correlations of electron densities and hence isomer shifts are arrived at on the basis of molecular orbital calculations. We shall say more about this aspect in Chapter 6.

3.1.2. Electric Quadrupole Interaction; Quadrupole Splitting

In the discussion about the electric monopole interaction and isomer shift (Section 3.1.1) we have considered the nuclear charge distribution to be uniform and spherically symmetric. In this case the nuclear quadrupole moment is zero and the second term of (3.8), which represents the electric quadrupole interaction, E_Q, vanishes. E_Q also becomes zero, if the extranuclear charges (electrons and lattice charges) are arranged in cubic symmetry. Therefore, electric quadrupolar interaction only occurs if there is an observable nuclear quadrupole moment and simultaneously a nonzero electric field gradient (EFG) at the nucleus.

Nuclear Quadrupole Moment

In many nuclei the nuclear charge distribution deviates more or less from spherical symmetry; the deviation may change in going from one state of excitation to another. A measure of the deviation is given by the electric quadrupole moment eQ, which is a (3 x 3) second rank tensor with elements

$$Q_{ij} = \int \rho_n(r) \, (x_i x_j - \delta_{ij} r^2) d\tau. \tag{3.22}$$

ρ_n is the nuclear charge; x_i, x_j are cartesian coordinates of r; and δ_{ij} is the Kronecker symbol.

If we choose x_i, x_j to be the coordinates x, y, z in the principal axes system, for which the off-diagonal elements Q_{ij} are zero, and if we further choose the z-axis to be the axis of preferred orientation (axis of quantization), we may define the electric quadrupole moment for nuclear charge distribution of cylindrical symmetry as

$$Q \equiv \frac{1}{e} \int \rho_n(r) \, (3z^2 - r^2) d\tau = \int \rho_n(r) r^2 \, (3\cos^2 \theta - 1) d\tau \tag{3.23}$$

where θ is the angle between the symmetry axis and the vector $\vec{r}$; $z = r \cdot \cos \theta$ in spherical coordinates.

Q is positive for an elongated (cigar-shaped) nucleus, and negative for a flattened (pancake-shaped) one. Q is zero for nuclei with spherical charge distribution $\rho_n(r)$. Nuclear states with spin quantum number I = 0, 1/2 do not possess an observable quadrupole moment. Only nuclear states with I > 1/2 have a spectroscopic electric quadrupole moment different from zero and thus can interact with an inhomogeneous electric field, described by the electric field gradient (EFG) at the nucleus. As Q is constant for a given Mössbauer nuclide, changes in the quadrupole interaction energy observed in different compounds of a given Mössbauer nuclide under constant experimental conditions can only arise from changes in the EFG at the nucleus. All information about the molecular and electronic structure of a Mössbauer atom, which can be extracted from the measured quadrupolar interaction energy, originates from changes in the EFG. The interpretation of quadrupole splittings (quadrupole coupling constants) therefore requires knowledge about the origin of the EFG and the way it is altered by chemical and physical influences. We shall give now a brief outline on basic concepts of the EFG and its coupling with the nuclear quadrupole moment. Later on, in Section 6.4, we shall learn more about the significance and the interpretation of the EFG and the quadrupole splitting ΔE_Q.

Electric Field Gradient (EFG)

A point charge q at a distance $r = (x^2 + y^2 + z^2)^{1/2}$ from the nucleus (located at the origin of the coordinate system) may cause a potential $V(r) = q/r$ at the nucleus. The electric field $\vec{E}$ at the nucleus is the negative gradient of the potential, $-\vec{\nabla} V$, and the electric field gradient EFG is given by

$$\text{EFG} = \vec{\nabla}\vec{E} = -\vec{\nabla}\vec{\nabla}V = \begin{bmatrix} V_{xx} & V_{xy} & V_{xz} \\ V_{yx} & V_{yy} & V_{yz} \\ V_{zx} & V_{zy} & V_{zz} \end{bmatrix}, \tag{3.24}$$

where

$$V_{ij} = \frac{\partial^2 V}{\partial i \partial j} = q(3ij - r^2\delta_{ij})r^{-5}, \quad (i, j = x, y, z) \tag{3.25}$$

are the nine components of the 3 x 3 second-rank EFG tensor. Only five of these components are independent, because of the symmetric form of the tensor, i.e., $V_{ij} = V_{ji}$, and because of Laplace's equation, which requires that the EFG be a traceless tensor,

$$\sum_i V_{ii} = 0, \quad i = x, y, z. \tag{3.26}$$

In the principal axes system, the off-diagonal elements vanish. If we choose the principal axes such that the ordering

$$|V_{zz}| \geq |V_{xx}| \geq |V_{yy}| \tag{3.27}$$

holds, we can specify the EFG by two independent parameters, i.e.,
    1. V_{zz}, sometimes denoted as eq (e = proton charge),
    2. the asymmetry parameter η, defined as

$$\eta = \frac{V_{xx} - V_{yy}}{V_{zz}}. \tag{3.28}$$

With (3.27) we find that $0 \leq \eta \leq 1$. For a fourfold or threefold axis of symmetry passing through the Mössbauer nucleus as the center of symmetry, $V_{xx} = V_{yy}$, and therefore $\eta = 0$; the EFG is in this case axially symmetric. In a system with two mutually perpendicualr axes of threefold or higher symmetry the EFG becomes zero.

In principle, one can consider two sources which can contribute to the total EFG:
    1. charges on distant ions which surround the Mössbauer atom in noncubic symmetry, usually called *lattice contribution;*
    2. anisotropic electron distribution in the valence shell of the Mössbauer atom, usually called *valence electron contribution.*

The lattice contribution, denoted as $(V_{zz})_{lat}$, and the concomitant asymmetry parameter η_{lat} are easy to evaluate, if the distances r_i and the angles ϕ_i and θ_i of all n contributing ions i with respect to the Mössbauer atom are known from crystal structure studies, and if charges q_i can be assigned to the ions. We employ the formulae

$$(V_{zz})_{lat} = \sum_{i=1}^{n} q_i r_i^{-3} (3 \cos^2 \theta_i - 1),$$ (3.29)

$$\eta_{lat} = \frac{1}{(V_{zz})_{lat}} \sum_{i=1}^{n} q_i r^{-3} 3 \sin^2 \theta_i \cos 2\phi_i.$$ (3.30)

$(V_{zz})_{lat}$, however, is not the actual EFG contribution felt by the nucleus. Instead, the electron shell of a Mössbauer atom is being distorted by electrostatic interaction with the noncubic charge distribution in the surrounding crystal lattice. $(V_{zz})_{lat}$ consequently will be modified. This phenomenon has been treated by Sternheimer [3.4], who accounted for the so-called "antishielding effect" by multiplying $(V_{zz})_{lat}$ by the factor $(1-\gamma_\infty)$. This factor amplifies $(V_{zz})_{lat}$ in iron compounds by a factor of approximately 10.

The valence electron contribution to V_{zz}, $(V_{zz})_{val}$, arises from anisotropic electron populations in the molecular orbitals between the Mössbauer atom and coordinated ligands. For highly ionic compounds, the crystal field theory is generally considered a sufficiently good model for qualitative estimates. To evaluate $(V_{zz})_{val}$ in this limit of approximation we may simply take the expectation value of the quantity $-e(3 \cos^2 \theta - 1) r^{-3}$ for each electron in the valence atomic orbital $|\psi\rangle$ of the Mössbauer atom and sum over all valence electrons i,

$$(V_{zz})_{val} = - e \sum_i \langle \psi_i | (3 \cos^2 \theta - 1) r^{-3} | \psi_i \rangle.$$ (3.31)

The expectation value of $1/r^3$,

$$\langle r^{-3} \rangle = \int R(r) r^{-3} R(r) r^2 dr$$ (3.32)

may be obtained from experiments [3.5]. We can therefore factorize out $\langle r^{-3} \rangle$ in (3.31) and have

$$(V_{zz})_{val} = -e \sum_i \langle \psi_i | 3 \cos^2 \theta - 1 | \psi_i \rangle \langle r_i^{-3} \rangle.$$ (3.33)

In case of covalent compounds the crystal field theory is a rather poor model because of the extensive participation of ligand atomic orbitals in the chemical bond. MO calculations are a much better choice for the characterization of covalent bonds. Such calculations yield the degree of participation and the electron population of central metal as well as ligand atomic orbitals. With this information available, the evaluation of $(V_{zz})_{val}$ for all molecular orbitals is then analogous to what we have said above for the crystal field model except that the summation of (3.33) now includes ligand atomic orbitals as well. A more detailed description of such calculations is given in Section 6.4.

Anisotropic charge distribution in the valence shell gives rise to a deformation of the filled inner-orbital shells, similarly to the antishielding effect caused by noncubic charge distribution in the surrounding lattice. Sternheimer [3.4] has shown that in this case $(V_{zz})_{val}$ has to be corrected by multiplying with the factor $(1 - R)$. The Sternheimer shielding factor R has been estimated to be 0.25–0.35 for iron.

The evaluation of η_{val}, the asymmetry parameter due to anisotropic valence electron distribution, requires working out the diagonal EFG tensor components in the x- and y-directions, as given by

$$(V_{xx})_{val} = -e \sum_i \langle \psi_i | 3 \sin^2 \theta \cos^2 \phi - 1 | \psi_i \rangle \langle r_i^{-3} \rangle, \qquad (3.34)$$

$$(V_{yy})_{val} = -e \sum_i \langle \psi_i | 3 \sin^2 \theta \sin^2 \phi - 1 | \psi_i \rangle \langle r_i^{-3} \rangle, \qquad (3.35)$$

where ψ_i may be central metal and ligand atomic orbital functions. By use of the expressions (3.33), (3.34) and (3.35), one can calculate the magnitude of the diagonal $(EFG)_{val}$ tensor elements and η for electrons of different angular momenta as given in Table 3.2. s-electrons do not contribute to the EFG because of their accommodation in orbitals of spherical symmetry. f-electrons of the higher transition elements do not participate in chemical bonding to any significant extent; they are accommodated in orbitals much closer to the nucleus than the p-, d- and s-valence electrons of a particular complex ion and are therefore well shielded by these outer electrons from ligand influences. Therefore, the only electrons of interest are the ones of p- and d-type.

The total EFG, comprising both contributions $(V_{zz})_{lat}$ and $(V_{zz})_{val}$, may be fully described by the expressions

$$V_{zz} = (1 - \gamma_\infty)(V_{zz})_{lat} + (1 - R)(V_{zz})_{val}, \qquad (3.36)$$

$$\eta = \frac{1}{V_{zz}} [(1 - \gamma_\infty)(V_{zz})_{lat}\eta_{lat} + (1 - R)(V_{zz})_{val}\eta_{val}]. \qquad (3.37)$$

For the temperature dependence of the EFG see Section 6.4.3.

Table 3.2. Magnitude per electron of the diagonal $(EFG)_{val}$ tensor elements for p- and d-electrons

Orbital	$(V_{xx})_{val}$ $e\langle r^{-3}\rangle$	$(V_{yy})_{val}$ $e\langle r^{-3}\rangle$	$(V_{zz})_{val}$ $e\langle r^{-3}\rangle$
p-Electrons			
p_x	$-4/5$	$+2/5$	$+2/5$
p_y	$+2/5$	$-4/5$	$+2/5$
p_z	$+2/5$	$+2/5$	$-4/5$
d-Electrons			
d_{xy}	$-2/7$	$-2/7$	$+4/7$
d_{xz}	$-2/7$	$+4/7$	$-2/7$
d_{yz}	$+4/7$	$-2/7$	$-2/7$
$d_{x^2-y^2}$	$-2/7$	$-2/7$	$+4/7$
d_{z^2}	$+2/7$	$+2/7$	$-4/7$

Quadrupole Splitting

The interaction between the electric quadrupole moment of the nucleus, Q, as defined by (3.23) in the principal axes system with the z-axis as axis of quantization, and the EFG at the nucleus, described by V_{zz} and η, may be expressed by the Hamiltonian [3.6]

$$\hat{H}_Q = \frac{eQV_{zz}}{4I(2I-1)}[3\hat{I}_z^2 - \hat{I}^2 + \eta(\hat{I}_+^2 + \hat{I}_-^2)/2]. \tag{3.38}$$

I is the nuclear spin quantum number, $\hat{I}$ is the nuclear spin operator, $\hat{I}\pm = \hat{I}_x \pm i\hat{I}_y$ are the shift operators, and $\hat{I}_x$, $\hat{I}_y$, $\hat{I}_z$ are the operators of the nuclear spin projections onto the principal axes. Working out the first-order perturbation matrix, one finds the eigenvalues E_Q to the perturbation operator $\hat{H}_Q$ as

$$E_Q = \frac{eQV_{zz}}{4I(2I-1)}[3m_I^2 - I(I+1)](1 + \eta^2/3)^{1/2}, \tag{3.39}$$

where $m_I = I, I-1, \ldots, -I$ is the nuclear magnetic spin quantum number. Wegener [1.9] shows that the same expression for E_Q is obtained starting out with the second term of equation (3.8).

Electric quadrupole interaction causes a splitting of the $(2I+1)$-fold degenerate energy level of a nuclear state with spin quantum number $I > 1/2$ into substates $|I, \pm m_I >$ without shifting the baricenter of the level. The substates are characterized by the magnitude of the magnetic spin quantum number $|m_I|$, but cannot be distinguished by the sign of m_I because of the second power of m_I in (3.39). Therefore, the substates $|I, \pm m_I >$ arising from nuclear quadrupole splitting remain doubly degenerate. The twofold degeneracy can be removed by magnetic perturbation (magnetic dipole-interaction).

As an example, the effect of electric quadrupole interaction in ^{57}Fe with I = 3/2 in the 14.4 keV state and I = 1/2 in the ground state is shown in Fig. 3.3. The nuclear ground state with I = 1/2 is not split, because there is no spectroscopic quadrupole moment in nuclei with I = 0, 1/2. The excited state with I = 3/2 splits into two doubly degenerate substates $|3/2, \pm 3/2 >$ and $|3/2, \pm 1/2 >$. Using (3.39), the perturbation energies $E_Q(\pm m_I)$ for the substates in case of an axially symmetric EFG ($\eta = 0$) are as follows:

$$E_Q (\pm 3/2) = 3eQV_{zz}/12 \quad \text{for} \quad I = 3/2, m_I = \pm 3/2$$

$$\tag{3.40}$$

$$E_Q (\pm 1/2) = -3eQV_{zz}/12 \quad \text{for} \quad I = 3/2, m_I = \pm 1/2.$$

From (3.40) we see that the magnitude of the perturbation energies, $|E_Q (\pm m_I)|$, is the same for both substates, indicating that the baricenter of the I = 3/2 level is not affected by electric quadrupole interaction. The energy difference ΔE_Q between the two substates is

$$\Delta E_Q = E_Q (\pm 3/2) - E_Q (\pm 1/2) = eQV_{zz}/2. \tag{3.41}$$

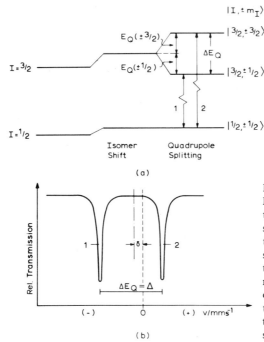

Fig. 3.3. Quadrupole splitting in ^{57}Fe with I = 3/2 in the excited state and I = 1/2 in the ground state. (a) The I = 3/2 level is split into two sublevels |I, ± m$_I$⟩ by electric quadrupole interaction. The ground state level with I = 1/2 is not split, because there is no spectroscopic quadrupole moment in a nucleus with I = 1/2. The levels of I = 3/2 and I = 1/2 are shifted by electric monopole interaction (giving rise to the isomer shift). (b) Resultant Mössbauer spectrum (schematic)

In a Mössbauer experiment with a single-line source, one would observe two resonance lines, usually called quadrupole doublet, of equal intensities (except for the occurrence of an anisotropic recoilless fraction f in a polycrystalline substance, called Goldanskii-Karyagin effect [1.10], or except for the appearance of texture; cf. Section 3.5) with the assignment as given in Fig. 3.3. The distance Δ between the two resonance lines corresponds to the energy difference ΔE_Q of (3.41) and is called *quadrupole splitting*. ΔE_Q is another Mössbauer parameter of great importance in chemical applications of the Mössbauer effect. It allows conclusions to be made about bond properties, molecular and electronic structure problems (see Section 6.4).

3.2. Magnetic Hyperfine Interaction

An atomic nucleus in the energy state E with spin quantum number I > 0 possesses a nonzero magnetic dipole moment $\vec{\mu}$ and may interact with a magnetic field $\vec{H}$ at the nucleus. The interaction is called magnetic dipole interaction or nuclear Zeeman effect and may be described by the Hamiltonian

$$\hat{H}_m = -\hat{\vec{\mu}} \cdot \hat{\vec{H}} = -g_N \beta_N \hat{\vec{I}} \cdot \hat{\vec{H}}. \tag{3.42}$$

g_N is the nuclear Landé factor, $\beta_N = e\hbar/2Mc$ (M: mass of the nucleus) is the nuclear

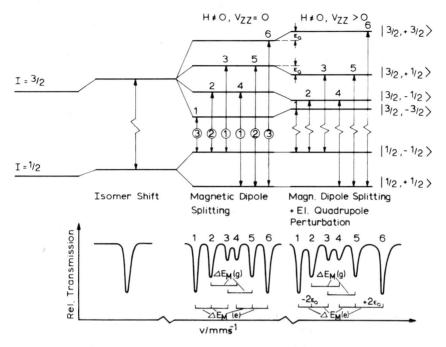

Fig. 3.4. Magnetic dipole splitting (nuclear Zeeman effect) in ^{57}Fe without (H ≠ 0, V_{zz} = 0) and with electric quadrupole perturbation (H ≠ 0, V_{zz} > 0) and resultant Mössbauer spectra (schematic). The centers of gravity of the nuclear levels are shifted by electric monopole interaction, which is always present.
$\Delta E_m(g) = g_g\beta_N H$ refers to the splitting of the ground state,
$\Delta E_m(e) = g_e\beta_N H$ to the splitting of the excited state

magneton. Diagonalizing the first-order perturbation matrix yields the eigenvalues E_M to $\hat{H}_M$ as

$$E_M(m_I) = -\mu H m_I/I = -g_N\beta_N H m_I. \qquad (3.43)$$

The nuclear Zeeman effect splits the nuclear state with spin quantum number I into 2I + 1 equally spaced and nondegenerate substates |I, m_I >, which are characterized by the sign and the magnitude of the nuclear magnetic spin quantum number m_I. Fig. 3.4 shows schematically the effect of magnetic dipole interaction in ^{57}Fe, where the I = 3/2 level is split into four substates and the ground state with I = 1/2 into two substates. The allowed gamma transitions between the sublevels of the excited state and those of the ground state are easily found following the selection rules for magnetic dipole transitions: ΔI = 1, Δm = 0, ±1. The six allowed transitions in ^{57}Fe are shown in Fig. 3.4. The circled numbers refer to the relative intensities (for isotropic orientation of the magnetic field with respect to the γ-ray propagation) as determined by the squares of the Clebsch-Gordan coefficients [1.9]. In a Mössbauer experiment with a single-line source and a magnetically ordered substance as absorber, one would usually observe a resonance sextet, the centroid of which may be shifted from zero velocity by electric monopole interaction (isomer shift).

The magnetic hyperfine splitting enables one to determine the effective magnetic field acting on the nucleus. There are various sources contributing to the effective magnetic field, the most important of which are

1. The Fermi contact field H^c, which arises from a net spin-up or spin-down s-electron density at the nucleus as a consequence of spin-polarization of inner filled s-shells by spin-polarized partially filled outer shells [3.7];

2. a contribution H^L from the orbital motion of valence electrons with the total orbital momentum quantum number L;

3. a contribution H^d, called spin-dipolar field, arising from the electron spin of the atom under consideration. These magnetic field contributions will be discussed further in Chapter 6.5.

3.3. Combined Electric and Magnetic Hyperfine Interactions

The case of pure nuclear magnetic hyperfine interaction beside the electric monopole interaction, which is always present due to $\delta R/R$ being nonzero, is rarely encountered in chemical applications of the Mössbauer effect. Metallic iron is an exception. Quite frequently a nuclear state is perturbed simultaneously by all three types of hyperfine interactions — electric monopole, magnetic dipole, and electric quadrupole interaction. In this case the nuclear energy level diagram may be constructed by a perturbation treatment for either $E_M \ll E_Q$ or $E_Q \ll E_M$. Assuming the latter case, where the electric quadrupole interaction is considered a perturbation to the magnetic dipole interaction, we find for the example in Fig. 3.4 that the sublevels $|3/2, \pm 3/2 >$ are shifted by an amount $E_Q(\pm m_I) = \Delta/2$ to higher energy and the sublevels $|3/2, \pm 1/2 >$ are shifted by E_Q to lower energy, provided V_{zz} is positive. The direction of the energy shifting by E_Q is reversed if V_{zz} is negative. The sublevels of the excited state I = 3/2 of Fig. 3.4 are no longer equally spaced. This results into an asymmetric magnetically split Mössbauer spectrum as pictured in Fig. 3.4 (H $\neq$ 0, $V_{zz} > 0$). As the sublevel spacings and thus the asymmetry of the spectrum are directly correlated with the sign of V_{zz}, one can determine the sign of the EFG of a polycrystalline sample from a magnetically split Mössbauer spectrum.

In general the axis of the principal component V_{zz} of the EFG forms an angle β to the axis of the magnetic field. Assuming $E_Q \ll E_M$ and the EFG tensor to be axially symmetric, a first-order perturbation treatment yields the general expression for the eigenvalues of the nuclear sublevels

$$E_{M,Q}(I, m_I) = - g_N \beta_N H m_I + (-1)^{|m_I|+1/2}(eQV_{zz}/8)(3 \cos^2 \beta - 1). \tag{3.44}$$

The more general case of $E_M \approx E_Q$ may be handled using appropriate computer programs.

3.4. Relative Intensities of Resonance Lines

3.4.1. Transition Probabilities

The transition probabilities or line intensities of the hyperfine components in a Mössbauer spectrum are determined by the properties of the nuclear transition. Important are the spin and the parity of the excited and the ground states of the nucleus under consideration, as well as the multipolarity of the transition and geometric configurations, i.e., the direction of the wave vector $\vec{k}$ of the emitted γ-quanta with respect to the quantization axis, e.g., the direction of the magnetic field or the principal axes system of the field gradient tensor. It will be shown that the geometric arrangement of the experimental set-up may result in additional information concerning the hyperfine interaction parameter and the anisotropy of the chemical bond (cf. Sec. 3.4.2).

For the sake of simplicity and a more instructive description, we shall restrict ourselves to the case of unpolarized single line sources of $I = 3/2 \longleftrightarrow I = 1/2$ magnetic dipole transitions (M1) as for example in ^{57}Fe, which has only a negligible electric quadrupole (E2) admixture. It will be easy to extend the relations to arbitrary nuclear spins and multipole transitions. A more rigorous treatment has been given in [3.8–11, 17, 19]. The probability P for a nuclear transition of multipolarity M1 from a state $|I_1, m_1 >$ to a state $|I_2, m_2 >$ is equal to

$$P(3/2m_{3/2}, 1m|1/2m_{1/2}, \theta, \phi) = |< 3/2m_{3/2}, 1m|1/2m_{1/2} >|^2 F_{1m_{1/2}}^{1m_{3/2}}(\theta, \phi) \cdot$$

$$< I_1 ||1||I_2 >|^2 \tag{3.45}$$

where θ, ϕ are the polar and azimuthal angles of the z-direction (defined by, e.g., the direction of the magnetic field) and the direction of the γ-ray emission (Fig. 3.5). $< I_1 m_1, Lm|I_2 m_2 >$ are the Clebsch-Gordan coefficients [3.11] coupling together the three vectors $\vec{I}_1, \vec{L}, \vec{I}_2$, and $< I_1 ||1||I_2 >$ is the reduced matrix element, which does not depend on the magnetic quantum numbers. For M1 (L = 1) and E2 (L = 2) transitions and $I_1 = 3/2, I_2 = 1/2$, the Clebsch-Gordan coefficients are tabulated in

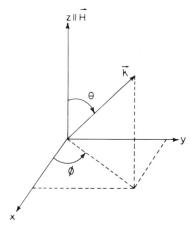

Fig. 3.5. Definition of the polar angles θ, ϕ. $\vec{k}$ is the wave vector of the emitted γ-ray. The z-axis may be defined by the direction of a magnetic field

Table 3.3. Values of the Clebsch-Gordan coefficients $\langle 3/2m_{3/2}, Lm|1/2m_{1/2}\rangle$ in case of a $I_e = 3/2$, $I_g = 1/2$ magnetic dipole transition (L = 1) and electric quadrupole transition (L = 2) (see, e.g., [3.8, 10, 12, 18])

$m_{3/2}$	$m_{1/2}$	m	$\langle 3/2m_{3/2}, 1m|1/2m_{1/2}\rangle$	$\langle 3/2m_{3/2}, 2m|1/2m_{1/2}\rangle$
3/2	1/2	−1	$\sqrt{3/6}$	$\sqrt{1/10}$
1/2	1/2	0	$-\sqrt{2/6}$	$-\sqrt{2/10}$
−1/2	1/2	1	$+\sqrt{1/6}$	$+\sqrt{3/10}$
−3/2	1/2	2	0	$-\sqrt{4/10}$
3/2	−1/2	−2	0	$+\sqrt{4/10}$
1/2	−1/2	−1	$+\sqrt{1/6}$	$-\sqrt{3/10}$
−1/2	−1/2	0	$-\sqrt{2/6}$	$+\sqrt{2/10}$
−3/2	−1/2	−1	$+\sqrt{3/6}$	$-\sqrt{1/10}$

Table 3.3. The angular dependent terms $F_{Lm}^{L'm'}(\theta, \phi)$ do not depend on quantum numbers I, m_I, and are of more general validity.

The angular functions $F_{Lm}^{L'm'}$ are given in Table 3.4. Let us first consider the case of a magnetically split spectrum from a powder sample. Here we have pure m_I-states $|Im_I>$ with the z-axis, which is parallel to the direction of the internal magnetic field, being randomly distributed. Therefore, we have to integrate over the polar (θ) and azimuthal (ϕ) angles,

$$P(3/2m_{3/2}, 1m|1/2m_{1/2}) = \int_0^\pi \int_0^{2\pi} P(3/2m_{3/2}, 1m|1/2m_{1/2}; \theta, \phi) \sin\theta d\theta d\phi \qquad (3.46)$$

or

$$P(3/2m_{3/2}, 1m|1/2m_{1/2}) \propto |<3/2m_{3/2}, 1m|1/2m_{1/2}>|^2. \qquad (3.47)$$

From (3.47) and Table 3.3 we derive the relative intensity ratios $3:2:1:1:2:3$ for the hyperfine components of a Zeeman pattern of a powder sample. It is easy to calculate the transition probablity for the case of the polar angle $\theta = \theta_0$ by integrating (3.45) only over the azimuthal angle ϕ. One obtains a factor $(1 + \cos^2\theta_0)/2$ and $\sin^2\theta_0$ for m = ±1 and m = 0, respectively, the square of the Clebsch-Gordan coefficients have to be multiplied with. As a consequence of the angular correlation of the transition probabilities, the second and fifth hyperfine components (Fig. 3.6) disappear, if the direction $\vec{k}$ of the γ-rays and the magnetic field $\vec{H}$ are parallel ($\theta_0 = 0$).

Let us now consider a pure quadrupole spectrum, which leads to the hyperfine splitting as shown schematically in Fig. 3.7. For the intensity ratio I_2/I_1 we obtain from (3.45) and Table 3.4 for the case of an axially symmetric field gradient tensor (integration over ϕ)

Table 3.4. Direction dependent terms $F_{Lm}^{L'm'}(\theta)$ for M1 and E2 nuclear γ-transitions (see, e.g., [3.9, 12])

a) $F_{1m}^{1m'}\,/\,3e^{-i(m'-m)\phi}$

m \ m'	1	0	−1
1	$\tfrac{1}{2}(1+\cos^2\theta)$	$\tfrac{1}{4}\sqrt{2}\sin 2\theta$	$\tfrac{1}{2}\sin^2\theta$
0	$\tfrac{1}{4}\sqrt{2}\sin 2\theta$	$\sin^2\theta$	$\tfrac{1}{4}\sqrt{2}\sin 2\theta$
−1	$\tfrac{1}{2}\sin^2\theta$	$\tfrac{1}{4}\sqrt{2}\sin 2\theta$	$\tfrac{1}{2}(1+\cos^2\theta)$

b) $F_{2m}^{2m'}\,/\,5e^{-i(m'-m)\phi}$

m \ m'	2	1	0	−1	−2
2	$\tfrac{1}{2}(\sin^2\theta+\tfrac{1}{4}\sin^2 2\theta)$	$-\tfrac{1}{4}(\sin 2\theta+\tfrac{1}{2}\sin 4\theta)$	$-\tfrac{1}{4}\sqrt{3/2}\,\sin^2 2\theta$	$-\tfrac{1}{4}(\sin 2\theta-\tfrac{1}{2}\sin 4\theta)$	$-\tfrac{1}{2}(\sin^2\theta-\tfrac{1}{4}\sin^2 2\theta)$
1	$-\tfrac{1}{4}(\sin 2\theta+\tfrac{1}{2}\sin 4\theta)$	$\tfrac{1}{2}(\cos^2 2\theta+\tfrac{1}{2}\sin^2 2\theta)$	$\tfrac{1}{4}\sqrt{3/2}\,\sin 4\theta$	$\tfrac{1}{2}(\cos^2\theta-\cos^2 2\theta)$	$\tfrac{1}{4}(\sin 2\theta-\tfrac{1}{2}\sin 4\theta)$
0	$-\tfrac{1}{4}\sqrt{3/2}\,\sin^2 2\theta$	$\tfrac{1}{4}\sqrt{3/2}\,\sin 4\theta$	$\tfrac{1}{4}\sin^2 2\theta$	$-\tfrac{1}{4}\sqrt{3/2}\,\sin 4\theta$	$-\tfrac{1}{4}\sqrt{3/2}\,\sin^2 2\theta$
−1	$-\tfrac{1}{4}(\sin 2\theta-\tfrac{1}{2}\sin 4\theta)$	$\tfrac{1}{2}(\cos^2\theta-\cos^2 2\theta)$	$-\tfrac{1}{4}\sqrt{3/2}\,\sin 4\theta$	$\tfrac{1}{2}(\cos^2\theta+\cos^2 2\theta)$	$\tfrac{1}{4}(\sin 2\theta-\tfrac{1}{2}\sin 4\theta)$
−2	$-\tfrac{1}{2}(\sin^2\theta-\tfrac{1}{4}\sin^2 2\theta)$	$\tfrac{1}{4}(\sin 2\theta-\tfrac{1}{2}\sin 4\theta)$	$-\tfrac{1}{4}\sqrt{3/2}\,\sin^2 2\theta$	$\tfrac{1}{4}(\sin 2\theta-\tfrac{1}{2}\sin 4\theta)$	$\tfrac{1}{2}(\sin^2\theta+\tfrac{1}{2}\sin^2 2\theta)$

c) $F_{1m}^{2m'}\,/\,\sqrt{15}\cdot 2\cos(m'-m)\phi$

m \ m'	2	1	0	−1	−2
1	$-\tfrac{1}{2}\sin 2\theta$	$\tfrac{1}{2}(\cos^2\theta+\cos 2\theta)$	$\tfrac{1}{2}\sqrt{3/2}\,\sin 2\theta$	$\tfrac{1}{2}(\cos^2\theta-\cos 2\theta)$	0
0	$-\tfrac{1}{2}\sqrt{2}\,\sin^2\theta$	$\tfrac{1}{4}\sqrt{2}\,\sin 2\theta$	0	$\tfrac{1}{4}\sqrt{2}\,\sin 2\theta$	$\tfrac{1}{2}\sqrt{2}\,\sin 2\theta$
−1	0	$-\tfrac{1}{2}(\cos^2\theta-\cos 2\theta)$	$\tfrac{1}{2}\sqrt{3/2}\,\sin 2\theta$	$-\tfrac{1}{2}(\cos^2\theta+\cos 2\theta)$	$-\tfrac{1}{2}\sin 2\theta$

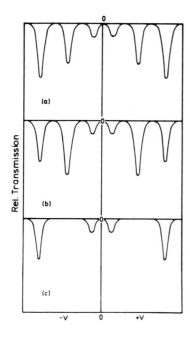

Fig. 3.6. Magnetic hyperfine pattern of a powder sample with randomly distributed internal magnetic field (a), and with (b) an applied magnetic field ($\theta_0 = 90°$), and (c) an applied magnetic field ($\theta_0 = 0°$)

$$I_2/I_1 = \frac{\int\limits_0^\pi 3(1 + \cos^2 \theta)h(\theta)\,\sin\theta d\theta}{\int\limits_0^\pi (5 - 3\cos^2 \theta)h(\theta)\,\sin\theta d\theta}.\tag{3.48}$$

This relation is only valid for a crystal with isotropic f-factor. The effect of crystal anisotropy will be treated in Section 3.4.2. The function $h(\theta)$ describes the probability of finding an angle θ between the direction of the z-axis and the γ-ray propagation. In a powder sample there is a random distribution of the principal axes system

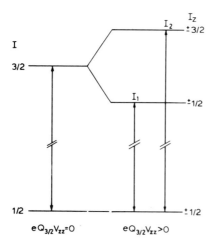

Fig. 3.7. Quadrupole hyperfine splitting for ^{57}Fe with $I_e = 3/2$ and $I_g = 1/2$; the quadrupole interaction parameter $e Q_{3/2} V_{zz}$ is assumed to be positive

of the EFG and with $h(\theta) = 1$ we expect the intensity ratio to be $I_2/I_1 = 1$, i.e., a symmetric Mössbauer spectrum. In this case it is not possible to determine the sign of the quadrupole coupling constant eQV_{zz}. For a single crystal, where $h(\theta) = \frac{1}{\sin\theta}\delta(\theta - \theta_0)$ (δ: delta-function), the intensity ratio takes the form

$$I_2/I_1 = \frac{3(1 + \cos^2\theta_0)}{5 - 3\cos^2\theta_0} \tag{3.49}$$

with values ranging from 3 for $\theta_0 = 0$ to 0.6 for $\theta_0 = 90°$. Therefore in studies using single crystals the sign of the quadrupole coupling constant eQV_{zz} can be determined. This in turn yields the sign of V_{zz} (EFG) if the sign of the electric quadrupole moment is known (which is the case for ^{57}Fe, viz., $Q(^{57}\text{Fe}) > 0$).

The situation becomes somewhat more complicated if both magnetic dipole and electric quadrupole interaction are present. Then the states are no longer pure m_I states $|I, m_I >$, but linear combinations of these, e.g., for ^{57}Fe

$$|\Psi_{3/2i}> = \sum_{m_{3/2}=-3/2}^{+3/2} C_{m3/2}^{3/2i}|3/2m_{3/2}>, \quad i = 1, 2, 3, 4,$$

$$\tag{3.50}$$

$$|\Psi_{1/2j}> = \sum_{m_{1/2}=-1/2}^{+1/2} C_{m1/2}^{1/2j}|1/2m_{1/2}>, \quad j = 1, 2.$$

The coefficients $C_{m3/2}^{3/2i}$, $C_{m1/2}^{1/2j}$ depend on the strength of the magnetic dipole and electric quadrupole interactions and are calculated by diagonalizing the appropriate Hamiltonian $\hat{H} = \hat{H}_M + \hat{H}_Q$. The transition probability will then be given by [3.12, 17]

$$P(3/2i, 1/2j; \theta, \phi) \propto \sum_{m_{3/2}} \sum_{m'_{3/2}} \sum_{m_{1/2}} \sum_{m'_{1/2}} C_{m1/2}^{1/2j} C_{m3/2}^{3/2i} C_{m'3/2}^{3/2i} C_{m'1/2}^{1/2j} \times$$

$$<3/2m_{3/2}, L = 1m|1/2m_{1/2}><3/2m'_{3/2}, L = 1m'|1/2m'_{1/2}> F_{1m}^{1m'}(\theta, \phi). \tag{3.51}$$

In those cases where we are dealing with nuclear transitions which are a mixture of multipolarity M1 and E2 with a mixing parameter δ defined by $\delta = \frac{<I_1\|E2\|I_2>}{<I_1\|M1\|I_2>}$ (positive or negative), one obtains the extended relation

$$P(3/2i, 1/2j; \theta; \phi) \propto (3.51) + \delta^2 \cdot [(3.51) \text{ with } L = 2] -$$

$$\delta \sum_{m_{3/2}} \sum_{m'_{3/2}} \sum_{m_{1/2}} \sum_{m'_{1/2}} C_{m1/2}^{1/2j} C_{m3/2}^{3/2i} C_{m'3/2}^{3/2i} C_{m'1/2}^{1/2j} \times \tag{3.52}$$

$$<3/2m_{3/2}, L = 1m|1/2m_{1/2}><3/2m'_{3/2}, L = 2m'|1/2m'_{1/2}> F_{1m}^{2m'}(\theta, \phi).$$

For a polycrystalline absorber, where the z-axis is randomly distributed, (3.52) reduces to

$$P(3/2i, 1/2j) \propto \left[\sum_{m\,3/2} \sum_{m'3/2} \sum_{m_{1/2}} \sum_{m'_{1/2}} C^{1/2j}_{m_{1/2}} C^{3/2i}_{m_{3/2}} C^{3/2i}_{m'3/2} C^{1/2j}_{m'_{1/2}} \times \right.$$

$$\left. <3/2m_{3/2}, L = 1m|1/2m_{1/2}><3/2m'_{3/2}, L = 1m'|1/2m_{1/2}> \right] + \qquad (3.53)$$

$$\delta^2 \cdot [\text{with } L = 2].$$

3.4.2. Effect of Crystal Anisotropy on the Relative Intensities of Hyperfine Splitting Components

In anisotropic crystals the amplitudes of the atomic vibrations are essentially a function of the vibrational direction. As has been shown theoretically by Karyagin [3.13] and proved experimentally by Goldanskii et al. [3.14], this is accompanied by an anisotropic Lamb-Mössbauer factor f, which in turn causes an asymmetry in quadrupole split Mössbauer spectra, e.g., in the case of $I_e = 3/2 \longleftrightarrow I_g = 1/2$ nuclear transitions in polycrystalline absorbers. A detailed description of this phenomenon, called the Goldanskii-Karyagin effect, is given, e.g., in [3.15].
The Lamb-Mössbauer factor is given by

$$f = \exp\left(-\langle(\vec{k}\vec{r})^2\rangle\right); \qquad (3.54)$$

$\vec{k}$: wave vector of the emitted γ-quantum,
$\vec{r}$: radius vector with the origin in the center of the vibrating atom.
According to Fig. 3.5, $\langle(\vec{k}\vec{r})^2\rangle$ can be calculated as a function of the angles θ and ϕ

$$\langle(kr)^2\rangle = k^2[(\langle r_x^2\rangle \cos^2\phi + \langle r_y^2\rangle \sin^2\phi)\sin^2\theta + \langle r_z^2\rangle \cos^2\theta]. \qquad (3.55)$$

For an axially symmetric crystal $\langle r_x^2\rangle = \langle r_y^2\rangle = \langle r_\perp^2\rangle$ and $\langle r_z^2\rangle = \langle r_\parallel^2\rangle$. Using this notation one obtains the following expression for f:

$$f(\theta) = \exp\left(-k^2\langle r_\perp^2\rangle\right) \exp\left(-k^2(\langle r_\parallel^2 - r_\perp^2\rangle)\cos^2\theta\right). \qquad (3.56)$$

Inserting (3.56) into (3.48) we find a modified relation for the relative intensity ratio I_2/I_1 for crystals with anisotropic (but axially symmetric) Lamb-Mössbauer factor f

$$I_2/I_1 = \frac{\int_0^\pi 3(1 + \cos^2\theta)\, h(\theta)\, f(\theta)\, \sin\theta\, d\theta}{\int_0^\pi (5 - 3\cos^2\theta)\, h(\theta)\, f(\theta)\, \sin\theta\, d\theta} \qquad \text{or}$$

$$I_2/I_1 = \frac{\int_0^\pi 3(1 + \cos^2\theta)\, h(\theta)\, \exp\left(-N\cos^2\theta\right) \sin\theta\, d\theta}{\int_0^\pi (5 - 3\cos^2\theta)\, h(\theta)\, \exp\left(-N\cos^2\theta\right) \sin\theta\, d\theta}, \qquad (3.57)$$

$$N = k^2(\langle r_\parallel^2\rangle - \langle r_\perp^2\rangle).$$

With $h(\theta) = \dfrac{1}{\sin\theta}\ \delta(\theta - \theta_0)$ one obtains the same result as given by (3.49), which implies that the anisotropy of the f factor cannot be derived from the intensity ratio of the two hyperfine components in the case of a single crystal. Of course, it can be evaluated from the absolute f value of each hyperfine component. However, for a polycrystalline absorber $(h(\theta) = 1)$, (3.57) leads to an asymmetry in the quadrupole split Mössbauer spectrum. The ratio of I_2/I_1, as a function of the difference of the mean square amplitudes of the atomic vibration parallel and perpendicular to the γ-ray propagation, is given in Fig. 3.8.

The anisotropic f factor may also manifest itself in the relative line intensities of Zeeman split hyperfine spectra in a polycrystalline absorber. Expanding $f(\theta)$ in a power series

$$f(\theta) = \sum_i a_i P_i(\theta), \quad (P_i(\theta): \text{Legendre polynomials}), \tag{3.58}$$

Cohen et al. [3.16] obtained the following relations for the transition intensities $I_{m_{3/2} - m_{1/2}}$:

$$I_{\pm 3/2 - \pm 1/2} = \frac{3(120 - 48\ N)}{120 - 40\ N},$$

$$I_{\pm 1/2 - \pm 1/2} = \frac{2(120 - 24\ N)}{120 - 40\ N}, \tag{3.59}$$

$$I_{\pm 1/2 - \pm 1/2} = \frac{120 - 48\ N}{120 - 40\ N}.$$

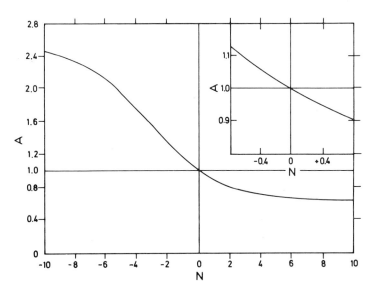

Fig. 3.8. The ratio of $A = I_2/I_1$ as function of the difference of the mean square of the vibrating amplitudes, $N = k^2(\langle r_\parallel^2 \rangle - \langle r_\perp^2 \rangle)$ (from [3.15])

In [3.16, 17] the line intensities for electric quadrupole and Zeeman (magnetic dipole) splitting and including the anisotropy of the f-factor are also given for $I_e = 2 \leftrightarrow I_g = 0$ transitions (even-even isotopes, e.g., in the rare earth region or in W, Os).

We wish to conclude this section with a warning. Not each asymmetry in the intensities of observed quadrupole split spectra (3/2–1/2) is caused by an anisotropic Lamb-Mössbauer factor. There are other effects which may give rise to an asymmetric spectrum, for example the line broadening due to the cosine smearing effect for spectra with nonzero isomer shift (cf. Sec. 4.4), and texture effects, i.e., preferred orientation of crystallites in the absorber introduced during the preparation (packing) of the absorber, which cause $h(\theta)$ not to be isotropic. One has to measure the Mössbauer spectra as a function of the angle between the normal of the absorber plane and the direction of the γ-ray propagation as well as the temperature dependence of the asymmetry, which should increase with increasing temperature, in order to evaluate the effect of crystal anharmonicities on the Lamb-Mössbauer factor and finally to find the correct source of line asymmetry.

3.5. Experimental Line Shape and Width of Resonant Absorption

The relative intensity of a γ-ray with energy E emitted by a Mössbauer source, which is moved with velocity v, is given by [3.20]

$$I_{em}(v, E) = \frac{f_s \, \Gamma/(2\pi)}{[E - E_0(1 + \frac{v}{c})]^2 + (\Gamma/2)^2} \tag{3.60}$$

f_s is the fraction of resonantly emitted γ-quanta in the source, Γ is the natural line width, and E_0 is the transition energy for a source kept at $v = 0$. In a Mössbauer experiment we are interested in the transmission probability T(E) of a γ-quantum of energy E through a resonance absorber. T(E) depends on the thickness d and on the particle density ρ of the absorber,

$$T(E) = \exp{(-d\rho\sigma(E))}, \tag{3.61}$$

where $\sigma(E)$ is the absorption cross section per atom. $\sigma(E)$ can be divided into one part which is nearly energy independent, at least in the energy region definded by the natural line width, and one part which describes the resonance absorption,

$$\sigma(E) = \bar{\sigma} + \sigma_r(E). \tag{3.62}$$

$\bar{\sigma}$ is mainly determined by the cross section for the photoeffect and for nonresonant scattering. By time reversal arguments the cross section for resonant absorption should be proportional to the emission probability. It turns out that $\sigma_r(E)$ is given by [3.20]

$$\sigma_r(E) = f_{abs} \cdot \beta \cdot \sigma_0 \frac{(\Gamma/2)^2}{(E - E_0)^2 + (\Gamma/2)^2}, \quad \text{with } \sigma_0 = \frac{2\pi}{k^2} \frac{1}{1 + \alpha} \frac{2I_a + 1}{2I_g + 1}. \tag{3.63}$$

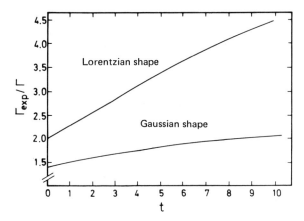

Fig. 3.9. Dependence of the experimental line width Γ_{exp} on the effective absorber thickness t (from [3.21]) for Gaussian and Lorentzian line shape

β is the relative abundance of the Mössbauer isotope. σ_0 is the maximum resonant cross section $(E = E_0)$ and α is the internal conversion coefficient. In the case of ^{57}Fe we obtain for example $\sigma_0 = 2.56 \cdot 10^{-22} m^2$. In a Mössbauer transmission experiment (cf. Chap. 4), the absorber is placed between the source and the detector. The count rate in the detector is then given by the transmission integral

$$C(v) = (1 - f_s) \exp(-d\rho\bar{\sigma}) + \int I_{em}(v, E) T(E) \, dE. \tag{3.64}$$

$(1 - f_s)$ is the nonresonant fraction of γ-rays emitted by the source. The count rate off resonance is given by

$$C(\infty) = \exp(-d\rho\bar{\sigma}). \tag{3.65}$$

In case $dn\sigma_r(E)$ is small, an analytical expression can be obtained for the count rate in the detector

$$\frac{C(\infty) - C(v)}{C(\infty)} = \frac{\rho\beta d f_s f_{abs}\sigma_0}{2} \frac{\Gamma^2}{(E_0 \frac{v}{c})^2 + \Gamma^2}. \tag{3.66}$$

We see that the line shape is of Lorentzian form with a half width twice the natural line width, assuming equal width Γ for source and absorber. However, if the effective absorber thickness $t = f_{abs}N\sigma_0 = f_{abs}\sigma_0 N_A \beta\rho\Delta x / A$ (N_A, Avogadro number; A, atomic weight; $\rho\Delta x$ in g/cm^2 of resonant nuclei) is not small compared to 1, one can no longer expect a Lorentzian. However, as shown by Margulies and Ehrman [3.21], the Mössbauer line shape still can be represented by a Lorentzian form, the width being a function of the effective absorber thickness. Fig. 3.9 shows the dependence of the observed experimental line width Γ_{exp} as function of t. The curve can be approximated by

$$\Gamma_{exp}/2 \simeq (1 + 0.135t)\Gamma_{nat}, \qquad \text{for } t \leq 4$$

and $\Gamma_{exp}/2 \simeq (1.01 + 0.145t - 0.0025t^2)$ for $t > 4$.

With the availability of large and fast computers it is no problem to calculate the transmission integral (3.64) numerically following a procedure described by Cranshaw [3.22] and Shenoy et al. [3.23].

The line broadening due to the absorber thickness t is in many cases much smaller than the broadening from inhomogeneous distribution of hyperfine fields and unresolved hyperfine interaction in the source or absorber. In this case the line shape can become more similar to a Gaussian form, which may be treated mathematically by a convolution of a Gaussian and Lorentzian function.

Line broadening may also occur by Doppler broadening of an unproperly working Mössbauer drive system (cf. Sec. 4.1) or because of inappropriate geometric arrangements which lead to the so-called cosine smearing effect.

Additional line width changes may be caused by special physical processes. For instance in delayed coincidence experiments, where the Mössbauer spectrum is accumulated at selected times of the order of the lifetime of the excited Mössbauer level after the nuclear transformation, e.g., the decay of $^{57}Co(EC, \gamma)^{57}Fe$, it is even possible to observe line widths smaller than the natural line width according to the Heisenberg uncertainty relation $\Delta E_{exp} \Delta t_{exp} \approx \hbar$ [3.24].

As a consequence of the nuclear transformation, a thermal spike may occur and create long-lived excited phonon states, which eventually decay during the lifetime of the Mössbauer level in ^{57}Fe. As a result, the Lamb-Mössbauer factor of the source increases with the time the excited nuclear level has existed. This means that the γ-rays contributing to the resonance lines come from excited nuclei with an effective lifetime longer than their mean lifetime. Again as a consequence of the Heisenberg uncertainty relation one would observe a smaller than natural line width.

A physical process which is of interest for chemists as well is the diffusion broadening of Mössbauer lines [3.25, 28]. A survey of this subject has been given by Janot [3.26]. Depending on the model, the continuous diffusion or sudden jump approximation, the change in line width is given by

$$\Delta\Gamma = 2k^2 D_0 \quad \text{or} \quad \Delta\Gamma = \frac{12}{r_0^2 f_c} D_0 (1 - \alpha_g), \tag{3.67}$$

respectively (k = wave vector of emitted γ-ray, r_0 is the jump distance, α_g is a geometric factor and f_c describes the direction correlation between successive jumps of one atom). In Fig. 3.10 the diffusion coefficient is plotted versus 1/T for Fe-3wt%Si [3.27a, b]. By adjusting Arrhenius' law to the experimental data points, one finds the diffusion coefficient to be $D_0 \approx 2 \cdot 10^{-9}$ cm^2/sec and the activation enthalpy $\Delta H \approx 49.6 \pm 3.7$ kcal/mole.

Finally we want to mention an interference effect which may influence the line shape of resonance absorption. The final states of resonant absorption and reemission of internal conversion electrons on the one hand and the photoeffect on the other hand are indistinguishable and therefore their quantum mechanical transition amplitudes

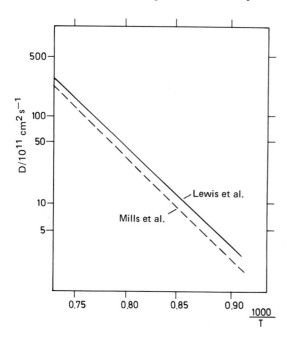

Fig. 3.10. Diffusion coefficient D_0 versus 1/T for Fe-3% Si evaluated from the measured broadening of the Mössbauer lines (Lewis et al. [3.27a], Mills et al. [3.27b])

(and not probabilities) have to be summed up and squared, giving rise to an additional dispersion term in the absorption cross section. This is due to interference of resonant scattering and Rayleigh scattering of γ-rays. These effects have been treated theoretically by Kagan, Afanas'ev and Voitovetskii [3.29–30] as well as by Trammell and Hannon [3.31] and proven experimentally, e.g., by Gorobchenko et al. [3.32] and Henning et al. [3.33] for E1 nuclear transitions. The interference effect in cases of M1 and E2/M1 transitions has been calculated, e.g., by Peregudov [3.34] or [3.30] and experimentally established by Wagner et al. [3.35]. In the latter case the effect is much smaller and normally can be neglected here (especially for energies ⩽ 100 keV). The dispersion term in the line shape is proportional to

$$\frac{E - E_0(1 + v/c)}{(E - E_0(1 + v/c))^2 + (\Gamma/2)^2};$$

the proportionality constant can by derived from theory. It is evident that this term leads to an asymmetrical Mössbauer line shape and to corrections of the isomer shift if not taken into account in the fitting procedure.

References

[3.1] Shirley, D. A.: Rev. Mod. Phys. *36*, 339 (1964)

[3.2] Greenwood, N. N., Gibb, T. C.: *Mössbauer Spectroscopy*, 1st ed. London: Chapman and Hall 1971

[3.3] Watson, R. E.: Phys. Rev. *118*, 1036 (1960); *119*, 1934 (1960); Techn. Rep. *12*, M.I.T., Solid State and Molecular Theory Group, Cambridge, Mass. (1959)

[3.4] Sternheimer, R. M.: Phys. Rev. *80*, 102 (1950); 84, 244 (1951); *130*, 1423 (1963)
       Foley, H. M., Sternheimer, R. M., Tycko, D.: Phys. Rev. *93*, 734 (1954)
       Sternheimer, R. M., Foley, H. M.: Phys. Rev. *102*, 731 (1956)

[3.5] Barnes, R. G., Smith, W. V.: Phys. Rev. *93*, 95 (1954)

[3.6] Abragam, A.: *The Principles of Nuclear Magnetism*. London – New York: Oxford University Press, Clarendon 1961, p. 161

[3.7] Freeman, A. J., Watson, R. E.: Phys. Rev. 131, 2566 (1963)

]3.8] Wegener, H.: *Der Mößbauer-Effekt und seine Anwendung in Physik und Chemie*. Mannheim: Bibliographisches Institut 1966 p. 122

[3.9] Brink, D. M., Satchler, G. R.: *Angular Momentum*. Oxford Library of the Physical Sciences (1968)

[3.10] Condon, E. U., Shortley, G. H.: *The Theory of Atomic Spectra*. Cambridge University Press (1935)

[3.11] Rose, M. E.: *Multipole Fields*. London: Chapman and Hall 1955

[3.12] Viegers, T.: Thesis 1976, University of Nijmegen, Netherlands

[3.13] Karyagin, S. V.: Dokl. Akad. Nauk. SSSR *148*, 1102 (1963)

[3.14] Goldanskii, V. I., Makarov, E. F., Khrapov, V. V.: Zh. Eksperim. Theor. Fiz. *44*, 752 (1963); Phys. Lett. *3*, 344 (1963)

[3.15] Goldanskii, V. I., Makarov, E. F.: *Chemical Applications of Mössbauer Spectroscopy*. New York – London: Academic Press 1968, p. 102

[3.16] Cohen, S. G., Gielen, P., Kaplow, R.: Phys. Rev. *141*, 423 (1966)

[3.17] Gedikli, A., Winkler, H., Gerdau, E.: Z. Phys. *267*, 61 (1974)

[3.18] Edmonds, A. R.: *Drehimpulse in der Quantenmechanik*. Mannheim: Bibliographisches Institut 1964

[3.19] Housley, R. M., Grant, R. W., Gonser, U.: Phys. Rev. *178*, 514 (1969)

[3.20] Mössbauer, R. L.: Z. Phys. *151*, 124 (1958)

[3.21] Margulies, S., Ehrman, J. R.: Nucl. Instr. Meth. *12*, 131 (1961)

[3.22] Cranshaw, T. E.: J. Phys. E *7*, 122 (1974); *7*, 497 (1974)

[3.23] Shenoy, G. K., Friedt, J. M., Maletta, H., Ruby, S. L.: in *Mössbauer Effect Methodology*, Vol. *9* (1974)

[3.24] Lynch, F. J., Holland, R. E., Hamermesh, M.: Phys. Rev. *120*, 513 (1960)

[3.25] Singwi, K. S., Sjölander, A.: Phys. Rev. *120*, 1093 (1960)

[3.26] Janot, C.: in *Mössbauer Spectroscopy and its Applications*, Proceedings. Vienna: IAEA 1972, p. 109

[3.27a] Lewis, S. J., Flinn, P. A.: Appl. Phys. Lett. *15*, 331 (1969)

[3.27b] Mills, B., Walker, G. K., Leak, G. M.: Phil. Mag. *12*, 939 (1965)

[3.28] Muller, J. G., Knauer, R. C.: in *Mössbauer Effect Methodology*, Vol. 5, 197 (1970). New York: Plenum Press

[3.29] Kagan, Yu. M., Afanas'ev, A. M., Voitovetskii, V. K.: JETP Lett. *9*, 91 (1969)

[3.30] Afans'ev, A. M., Kagan, Yu. M.: Phys. Lett. *31A*, 38 (1970)

[3.31] Trammel, G. T., Hannon, J. P.: Phys. Rev. *180*, 337 (1969)

[3.32] Gorobchenko, V. D., Lukashevich, I. I., Sklyarevskii, V. V., Filippor, N. I.: JETP Lett. *9*, 139 (1969)

[3.33] Henning, W., Baehre, G., Kienle, P.: Phys. Lett. *31B*, 203 (1970)

[3.34] Peregudov, V. N.: Hyp. Int. *3*, 353 (1977)

[3.35] Wagner, F. E., Dunlap, B. D., Kalvius, G. M., Schaller, H., Felchers, R., Spieler, H.: Phys. Rev. Lett. *28*, 530 (1972)

4. Experimental

A Mössbauer apparatus may be subdivided into three parts according to the schematic drawing of Fig. 4.1. The source, which is produced by a nuclear reaction, decays to the Mössbauer nuclide and is moved relative to the absorber using a drive system which imparts an additional Doppler energy to the emitted γ-quantum. Sometimes the source is kept fixed and the absorber is moved relative to the source. The γ-rays penetrate the absorber and may thereby be resonantly absorbed by Mössbauer nuclides of the same kind as in the source. Afterwards deexcitation in the absorber may take place by emission of γ-rays, x-rays, or electrons. An appropriate detector counts either the number of unaffected transmitted γ-rays in a transmission experiment or the reemitted γ-rays, x-rays or electrons in a scattering experiment as a function of the Doppler velocity. A block diagram illustrating the principle of a modern Mössbauer set-up is shown in Fig. 4.2. The main components are: the velocity transducer, the wave form generator and synchronizer, the multichannel analyzer, the γ-ray detection system or nuclear channel, a cryostat for low temperature and temperature dependent measurements, a velocity calibration device, the source and the absorber. It is beyond the scope of this article to discuss the entire equipment in detail and therefore we refer to the excellent review articles by Kalvius and Kankeleit [4.3a] and Cohen and Wertheim [4.3b]. We wish, however, to describe briefly the principle of a Mössbauer spectrometer and mention new developments of Mössbauer effect methodology.

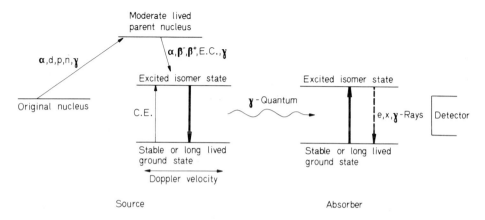

Fig. 4.1. Nuclear transitions involved in a Mössbauer effect experiment (from [4.1])

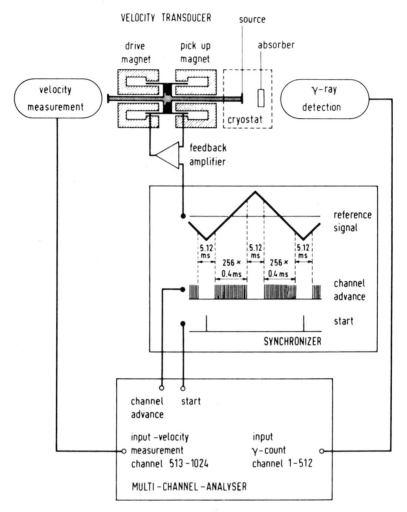

Fig. 4.2. Block diagram illustrating the principle of a Mössbauer spectrometer (from [4.2])

4.1. The Mössbauer Spectrometer

4.1.1. General Concept and Mode of Operation

The velocity transducer in nearly all modern Mössbauer spectrometers is a loudspeaker system, e.g., of the Kankeleit type [4.4]. It consists of rigidly connected driving and pickup coils moving in a homogeneous magnetic field. The amplified voltage difference between the reference signal and the induced pickup signal — the latter is proportional to the velocity of the coils — is applied to the driving coil. The negative feedback system minimizes the difference voltage such that maximum deviations smaller than $1\,^0/_{00}$ of the actual velocity from the reference waveform can be achieved.

Depending on the reference voltage waveform, the transducer may move with constant velocity, sinusoidally, in a symmetric sawtooth, or any other more complicated way. The amplitude of the reference voltage then determines the maximum velocities achieved in one period of the reference voltage. Recent improvements in the design of Mössbauer drive units have been reported by Kankeleit at the Cracow Mössbauer conference [4.5].

The detected counts are stored in a multichannel analyzer containing typically 400 or 512 channels. Synchronization of the channel number in the multichannel analyzer with the velocity increment between v and v + dv is performed by advancing the address of the memory one by one through an external clock, which divides the period of the waveform into 400 or 512 pulses according to the number of available channels. A start impulse, which coincides with the beginning of the waveform, enables the multichannel analyzer to start, with channel No. 1 being advanced by the clock pulses. This way the synchronization is accomplished in each period. The dwell time per channel is in the order of 50 to 100 μsec. An unfolded Mössbauer spectrum

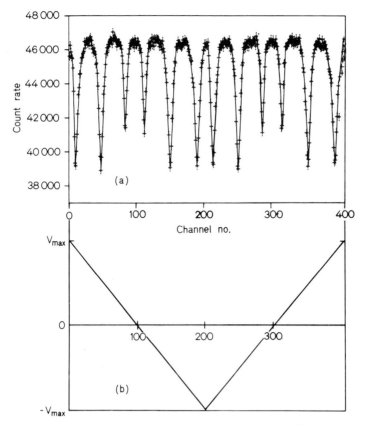

Fig. 4.3. a) Mössbauer spectrum of metallic iron taken with a ^{57}Co/Rh source. The count rate is plotted as function of the channel number. The solid line drawn through the data points represents the least squares fit of a magnetic hyperfine pattern to the experimental data points. b) Velocity as function of the channel number

of a ^{57}Co/Rh source with an iron metal absorber using a symmetric sawtooth wave-form is given in Fig. 4.3a. The scattering of points in the baseline reflects the statistical errors of the counts. The solid line drawn through the data points represents the theoretical spectrum obtained by a least squares fitting of a magnetic hyperfine pattern to the experimental data (cf. Chap. 5). In Fig. 4.3b the sawtooth velocity is plotted as a function of the channel number, from which it is evident that the Mössbauer spectrum is measured twice in one period of the sawtooth; the two spectra will appear as mirror images arranged symmetrically about the center of the channel series.

4.1.2. Calibration

The calibration of the Mössbauer spectrometer is a most essential procedure like in all other spectroscopical methods. Particularly in Mössbauer spectroscopy, where very often several kinds of sources for one particular Mössbauer nucleus are employed, e.g., ^{57}Co/Cu, ^{57}Co/Pt, ^{57}Co/Rh in case of ^{57}Fe spectroscopy, it is necessary to have reliable calibration methods at hand.

The simplest and cheapest way is to use a standard absorber with an accurately known magnetic hyperfine splitting. Iron metal, for instance, shows at room temperature an accurately measured splitting of 10.167 mm s^{-1} for the two outermost lines of the magnetic six-line-pattern. One uses this splitting as a calibration basis in measurements of hyperfine interactions with strengths corresponding to a Doppler velocity region of 0 to ± 10 mm s^{-1} (e.g. Mössbauer effect in ^{57}Fe, ^{61}Ni, ^{197}Au, ^{193}Ir and others). For nuclei such as ^{161}Dy, ^{129}I, ^{121}Sb, however, a velocity region of some cm s^{-1} is required. An extrapolation assuming linearity of the transducer to high velocities is not accurate enough and very questionable. In this case the velocity can be measured with very high accuracy using a Michelson interferometer with a He/Ne-laser as reported by Fritz and Schulze [4.6], Cranshaw [4.7], and Viegers [4.8]. The

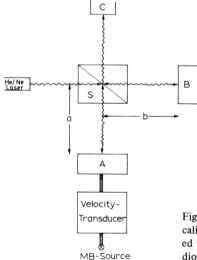

Fig. 4.4. Principle of a laser interferometer for absolute calibration of the transducer velocity. A, mirror connected to the moving transducer; B, fixed mirror; C, photodiode; S, beam splitter; light path difference x = 2 (a−b)

principle of the interferometer is shown in Fig. 4.4. A laser beam is split (S) into one half which is reflected by mirror (A) moved by the transducer, and another half which travels to a fixed mirror (B). The reflected beam from B undergoes interference with the first beam from A at a photodiode (C), the current of which is proportional to the light intensity I.

If x is the difference between the two light paths, then $I \propto \sin^2 2\pi \frac{x}{1/2\lambda} \cdot \lambda$ is the wavelength of the laser photons ($\lambda = 6338.\ 198$Å). Each time the mirror (A) has travelled by $1/4\ \lambda$ the intensity of the light registered by the photodiode (C) changes from light to dark. The intensity as a function of the channel number can easily be recorded simultaneously by routing into the second half of an 800 channel analyzer, e.g., from channel 401 to 800. In the evaluation of the spectra this absolute (not linearized) velocity can be attributed to the corresponding channel of the Mössbauer spectrum.

4.1.3. Detectors

The nuclear channel of a Mössbauer system consists of the detector, preamplifier, amplifier, discriminator (single channel analyzer), and the multichannel analyzer. All of these components are available commercially. The properties of the commonly used detectors are given in Table 4.1. It is obvious that the desired and sometimes required high resolution is in conflict with the greatly increased costs of the semiconductor detectors. In principle there is no technical limitation in improving a Mössbauer system with respect to resolution and efficiency of the γ-ray detectors. However, the maximum count rates in single pulse techniques, where it is possible to discriminate the Mössbauer γ-lines energetically from other nonresonant nuclear transitions or x-rays associated with the nuclear transformation in the source or with the photoeffect in the source and the absorber, are limited. In many cases, e.g., ^{61}Ni, ^{197}Au, ^{199}Hg Mössbauer spectroscopy, it is desirable and possible to pro-

Table 4.1. Properties of commonly used detectors in Mössbauer spectroscopy

Detector	Resolution	Efficiency	Maximum count rate in single pulse technique	Costs (approximate)
Proportional counter	2–4 keV at 14 keV	80% at 14 keV	50 kHz	$ 200
NaI	8–10 keV at 60 keV	100% at 60 keV	100–200 kHz	$ 1,500
Si (Li)	~400 eV at 10 keV	100% at 10 keV	10–40 kHz	$ 6,000
Ge (Li) Ge (pure)	~600 eV at 100 keV	80–100%	10–40 kHz	$ 8,000 –10,000

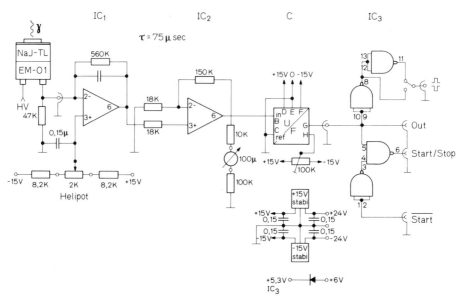

Fig. 4.5. Electrical network of the current integration device in connection with a NaI crystal

duce high source strengths (500–25,000 mCi) to make up for the small recoilless fraction resulting from high energetic nuclear transitions, for the low Debye temperatures in either source and absorber, or for low densities of Mössbauer active nuclei in the absorber. With such sources of extremely high radioactivity, conventional electronic equipment for single pulse counting would be "overloaded". This difficulty can be overcome by use of the integration counting technique by giving up the possibility of energy discrimination [4.5, 9]. At very high count rates in the detector it becomes impossible to discriminate single pulses, but it is still possible to measure the detector current. This current is converted into a voltage and fed into a voltage-to-frequency converter (UFC), the output signals of which are transferred to the input of the multichannel analyzer as in the normal multiscaling mode. If absorber and source are in resonance, the detector current and consequently the frequency of the UFC decreases, leading to the normal noninverted Mössbauer spectrum. Fig. 4.5 shows a block diagram of the current integration counting facility for use in connection with a NaI crystal[2]. This method has successfully been applied to ^{197}Au spectroscopy by Viegers and Trooster [4.8, 9] using a NaI-detector. As shown by Kankeleit [4.5], it is not possible to use NaI for source activities higher than 500 mCi because of afterlightning of the crystals. Wurtinger [4.10] has used a Ge(Li) crystal to observe the Mössbauer effect (10^{-3}% at 4.2 K) of the 158 keV transition in ^{199}Hg.

So far, we have discussed only the detection of γ- and x-rays. As indicated in Fig. 4.1, the excited absorber nuclei will reach the nuclear ground state partly by

[2]   Designed by K.-H. Häuser, Institut für Anorganische und Analytische Chemie, Johannes-Gutenberg-Universität Mainz.

internal conversion, i.e., by emission of mainly s-electrons. These or the subsequently emitted Auger electrons may be detected in Mössbauer scattering experiments. This technique has gained increasing importance in the study of surface layers, surface reactions, or implantation reactions, because of the small range of low energy electrons in solids (about 1000 Å in iron metal for 6 keV Auger electrons of Fe). To date, two different approaches are published for the detection of the emitted electrons in a Mössbauer experiment:

(i) normal electric or magnetic electron spectrometer

(ii) He/CH$_4$ filled proportional counters.

In case (i) only very few measurements concerning chemical applications have been reported [4.11, 12]. This is due to the fact that this type of spectrometer is rather expensive ($\sim$ \$ 100,000) and complicated in construction and mode of operation. The advantage is that it allows an energy discrimination to be made of the electrons of $\sim$ 1% with an efficiency of about 15% [4.11], which enables one to measure depth-selective Mössbauer spectra.

It is much cheaper ($\sim$ \$ 400) to use the second type of conversion electron detection system. The resonant γ-rays from a moved Mössbauer source enter a proportional counter, filled with a He/CH$_4$ gas mixture, and penetrate the Mössbauer absorber which forms the back-wall of the counter. The backward emitted electrons are detected in a 2π geometry; the background of the 6 keV x-rays is less than 1%. A descrip-

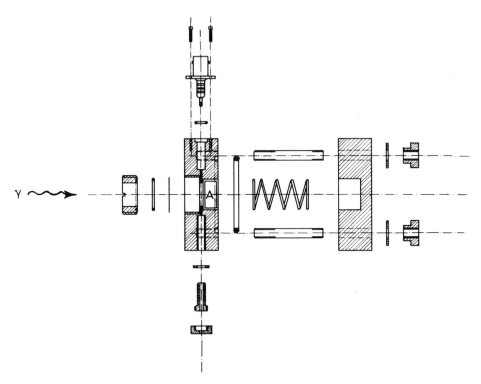

Fig. 4.6. Drawing o n improved version of a Swanson and Spijkerman type conversion electron detector. A: absorb

tion of this type of detector was first published by Swanson and Spijkerman [4.13]; it is currently widely employed in the study of corrosion and surface reactions, cf. Section 8.3. Fig. 4.6 shows a schematic diagram of an improved conversion electron detector built in our laboratory[3]. The energy resolution of this detector is so poor that it is not possible to measure depth-differential Mössbauer spectra.

A common and serious disadvantage of both electron detection systems is that one is not able to measure the temperature dependence of the Mössbauer spectra down to liquid He temperature or even lower. In the proportional counter the quench-gas will freeze out, and no satisfying spectrum [4.14] can be obtained within reason-able recording times because of the concomitant reduction of the count rate. The main technical problem in the case of a magnetic (or electric) spectrometer for low energy electrons ($\lesssim 14$ keV) in cooling the absorber is that the vacuum of the latter has to be separated from the poor spectrometer vacuum by differential pumping or by use of thin foils to prevent the condensation of gases like water vapour at the sample surface.

4.1.4. Cryostats

A very important accessory component in a Mössbauer spectrometer is a cryostat for low temperature and temperature dependent measurements. This may be necessary for various reasons like small Debye-Waller factors of the source and/or the absorber, identification of a chemical species, or determination of electronic states and molec-ular symmetry of a chemical compound under study. The temperature range varies from 1.2 to 300 K in a simple He-bath and from 6 to 300 K in He-flow cryostats. In cases with relatively high Mössbauer transition energies (e.g., ^{61}Ni, ^{197}Au), it is desirable to vary independently the temperature of the source and the absorber. All of these kinds of cryostats are commercially available. Very low temperatures (down to ~ 0.03 K) can be obtained by ^{3}He/^{4}He refrigerators, which are, however, more complicated and more expensive than the relatively simple cryostats mentioned above. With respect to cryostats we refer the reader to references [4.3a, b; 15].

4.2. Preparation of Mössbauer Sources and Absorbers

There are a number of criteria to be taken into account in the preparation of the source of a special Mössbauer isotope. First of all, if there are several different nuclear transformations leading to the excited nuclear level of interest, one should preferen-tially choose the one which leads to the highest intensity of Mössbauer quanta and which has the longest half-life of the precursor nuclide. The energy spectrum of the source, which may as well depend on the host material (Compton or x-ray back-ground), should be as simple as possible, especially if one plans to use the current integration technique (see Sec. 4.1.3). Furthermore, the chemical composition of the

---

[3]   Designed by J. Ensling and J. Grübler, Institut für Anorganische Chemie und Analytische Chemie, Universität Mainz.

source material should be such as to produce possibly single emission lines as intense and narrow as possible. Any electric quadrupole or magnetic hyperfine perturbation would split or at least broaden the emission lines, which in turn reduces the resolution of the spectra and renders the evaluation cumbersome. The Debye-Waller factor should be as large as possible in order to obtain highest possible resonance absorption. The source material should be chemically inert during the lifetime of the source and resistent against autoradiolysis. Of course, often one has to accept a compromise if not all ideal source requirements can be met. Various methods of source preparation will be mentioned in Chapter 7.

The preparation of an absorber for Mössbauer measurements is determined essentially by three factors, viz., fractional absorption, photo effect induced attenuation, and thickness broadening. The fractional absorption

$$\epsilon = (1 - e^{-t/2} J_0(i\tfrac{t}{2})) \tag{4.1}$$

($J_0(i\tfrac{t}{2})$: zeroth Bessel function) should be as large as possible, which according to Fig. 4.7 would imply a thickest possible absorber. This, however, is in contradiction to the fact that the count rate in the detector and therefore the relative statistical accuracy $\Delta_{rel} = \sqrt{1/(I/I_0)}$ decreases exponentially with the absorber thickness as a consequence of the attenuation of the incoming γ-beam due to the photoeffect:

$$I/I_0 = \exp(-\rho x \bar{\mu}) = \exp(-t/(f_a N_A \sigma_0 \beta) \bar{\mu} M). \tag{4.2}$$

$\bar{\mu} = \dfrac{1}{M} \sum_i M_i \mu_i$ (M_i : atomic weight of the ith atom in a molecule of mass M; μ_i: the absorption coefficient of the atom i for the Mössbauer γ-quanta of energy E; other symbols: see definition of t in Sec. 3.5). The absorption coefficients $\mu_i(E)$ are tabulated for example in [4.17]. In addition, the line width will increase ($\Gamma = 2\Gamma_{nat}(1 + 0.135t)$) and consequently the resolution will decrease with the absorber thickness. Therefore one has to optimize the function $g(t) = \epsilon/(\Delta_{rel}\Gamma)$

$$g(t) = \frac{(1 - e^{-t/2} J_0(i\tfrac{t}{2})) \; \sqrt{I/I_0}}{2 + 0.135\, t} = \frac{(1 - e^{-t/2} J_0(i\tfrac{t}{2})) \exp\left(\dfrac{-t\bar{\mu} M}{2 f_a N_A \sigma_0 \beta}\right)}{2 + 0.135\, t} \tag{4.3}$$

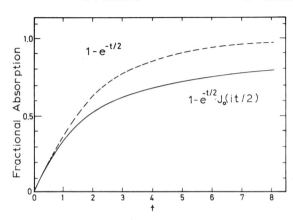

Fig. 4.7. Fractional absorption in a Mössbauer experiment as function of the effective absorber thickness $t = f_{abs}\sigma_0 N_A \beta \rho \Delta x/M$ (cf. Sec. 3.5); $J_0(it/2)$ zeroth order Bessel function

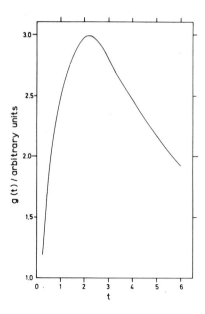

Fig. 4.8. g(t) as function of t for stainless steel using $f(298\ K) = 0.70$ and $\bar{\mu} = 58.8\ cm^2/g$. From this curve an optimal value of $t = 2.2$ can be derived

in order to obtain the optimal absorber thickness. In Fig. 4.8 we have plotted g(t) as function of t for iron in stainless steel using $\bar{\mu} = 58.8\ cm^2/g$ and $f_a(298K) = 0.70$. From this curve we derive an optimal absorber thickness $t = 2.2$ at $T = 298\ K$, which corresponds to $6.23\ mg/cm^2$ stainless steel.

4.3. Geometrical Considerations

In a transmission experiment one tends to keep the source-detector distance as short as possible and increases the solid angle of acceptance as far as possible to increase the count rate in the detector and consequently the statistical accuracy. The geometric arrangement in a Mössbauer experiment, however, may influence the shape of a spectrum considerably and has therefore to be considered with care.

Due to the relative motion between the source and the absorber, the source-detector distance will change periodically and consequently the solid angle and the count rate will be a function of the velocity as indicated in Fig. 4.9. If the two Mössbauer spectra taken during one period of the velocity sweep are folded with respect to their mirror image, this geometric effect will disappear. The situation becomes worse when the source will "see" the uncovered (by the absorber) detector directly at some critical distance R_{crit}. The former geometric effect can be almost neglected at moderate source strengths if the absorber rather than the source is moved. Here only small angle Compton scattering may lead to a geometric effect superimposed on the hyperfine pattern observed in a Mössbauer spectrum.

A very serious geometric effect is due to the so-called cosine smearing of the velocity. When the wave vector of an emitted photon forms an angle θ with the direc-

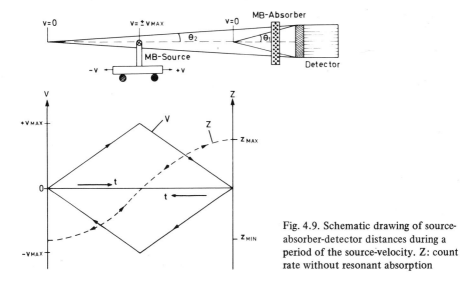

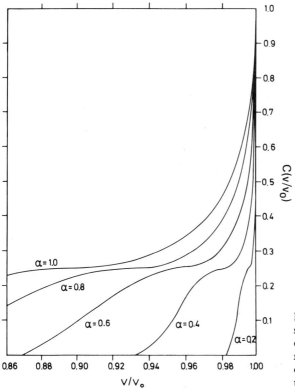

Fig. 4.9. Schematic drawing of source-absorber-detector distances during a period of the source-velocity. Z: count rate without resonant absorption

Fig. 4.10. Illustration of the cosine smearing effect in the case of identical source and detector radius. α is the aperture, the ratio of detector radius to source-detector separation (from [4.16])

tion of the relative velocity v_0 between source and absorber, the Doppler energy shift is given by $\Delta E = E_0 \cdot \frac{v_0}{c} \cos \theta$, where E_0 is the photon energy. The smearing due to this effect is shown in Fig. 4.10. As is obvious, this asymmetric broadening of the Mössbauer lines may influence the isomer shift values obtained by adjusting a symmetric Lorentzian to the experimental data points in an inappropriate geometric arrangement. Of course, one may take care of this effect as well by including it in the curve fitting procedure.

In Mössbauer scattering experiments the above arguments hold, if the source-detector distance is replaced by the source-absorber separation.

References

[4.1] Gonser, U.: Topics in Appl. Phys. *5*, 1 (1975)

[4.2] Viegers, M. P. A.: Doctoral Thesis, University of Nijmegen, 1976, p. 5

[4.3a] Kalvius, G. M., Kankeleit, E.: Proceedings of a Panel *Mössbauer Sepctroscopy and its Applications*, IAEA, Vienna (1972) p. 9

[4.3b] Cohen, R. L., Wertheim, G. K.: in *Methods of Experimental Physics 11*, 307 (1974)

[4.4] Kankeleit, E.: Rev. Sci. Instr. *35*, 194 (1964)

[4.5] Kankeleit, E.: Proceedings Int. Conf on Mössbauer Spectroscopy, Cracow (Poland), Vol. *2*, 43 (1975)

[4.6] Fritz, R., Schulze, D.: Nucl. Instr. Meth. *62*, 317 (1968)

[4.7] Cranshaw, T. E.: J. Phys. E. *6*, 1 (1973)

[4.8] Viegers, M. P. A.: Doctoral Thesis, University of Nijmegen 1976, p. 13

[4.9] Viegers, M. P. A.: Trooster, J. M.: Nucl. Instr. Meth. *118*, 257 (1974)

[4.10] Wurtinger, W.: J. Phys. Paris *C6*, 697 (1976)

[4.11] Bokemeyer, H., Wohlfahrt, K., Kankeleit, E., Eckhardt, D.: Z. Phys. *A 274*, 305 (1975)

[4.12] Schunck, J. P., Friedt, J. M., Llabador, Y.: Revue Phys. Appl. *10*, 121 (1975)

[4.13] Swanson, K. R., Spijkerman, J. J.: J. Appl. Phys. *41*, 3155 (1970)

[4.14] Sawicki, J. A., Sawicka, B. D., Stanek, J.: Nucl. Instr. Meth. *138*, 565 (1976)

[4.15] Kalvius, G. M., Katila, T. E., Lounasmaa, O. V.: in *Mössbauer Effect Methodology 5*. New York: Plenum Press 1970, p. 231

[4.16] Spijkerman, J. J., Ruegg, F. C., de Voe, J. R.: in *Mössbauer Effect Methodology 1*. New York: Plenum Press 1965, p. 119

[4.17] Storm, E., Gilbert, E., Israel, H.: *Gamma-Ray Absorption Coefficients for Elements 1 through 100*, Los Alamos Scientific Laboratory LA-2237, University of California, New Mexico (1958)

5. Mathematical Evaluation of Mössbauer Spectra

As in many other spectroscopic methods, it is necessary in Mössbauer spectroscopy to determine the positions, intensities or areas of the absorption and emission lines as accurately as possible. In the case of poorly resolved or complex hyperfine spectra due to different chemical species or several lattice sites in the sample it is not possible to evaluate the Mössbauer spectra just by hand. In addition, often one is faced with the problem of taking into account cosine smearing effects (Sec. 4.2), thick absorbers associated with the calculation of the transmission integral (Sec. 3.5), magnetic or electric relaxation (Sec. 6.7), or delayed coincidence Mössbauer measurements (Sec. 3.5), which all lead to deviations from Lorentzian and to complex line shapes. In all these cases a least-squares fitting procedure using a computer is necessary to evaluate properly the measured spectra. The weighted mean square deviation χ^2 between the experimental data points Y_i^{exp} and the corresponding theoretical values Y_i^{theo} has to be minimized,

$$\chi^2 = \sum_{i=1}^{N} \frac{1}{\sigma_i^2} (Y_i^{theo} - Y_i^{exp})^2 = \text{Minimum!} \tag{5.1}$$

$\sigma_i^2 = Y_i^{exp}$ is the square of the statistical variation of the count rate in the ith channel of the multichannel analyzer corresponding to a velocity increment v_i to $v_i + \Delta v_i$. In the most simple case Y_i^{theo} is the sum of k uncorrelated Lorentzian lines with different intensities and half widths

$$Y_i^{theo} = Y_{base\ line}^{theo} - \sum_{j=1}^{k} \frac{W_k}{(E_k - \frac{v_i}{c} E_0)^2 + \left(\frac{\Gamma_k}{2}\right)^2} . \tag{5.2}$$

$Y_{base\ line}^{theo}$ is the count rate "off resonance". Due to geometry effects like the periodical change of the solid angle during the movement of the transducer, this quantity may well be a function of the velocity (channel i). Least-squares fitting computer codes with standard theories like the sum of uncorrelated Lorentzian lines, magnetic and electric quadrupole hyperfine interaction as individual or combined [5.1] perturbation for the most frequently used Mössbauer isotopes are available in all Mössbauer laboratories to date and may be requested from there, e.g., [5.2, 3]. For the minimization procedures of χ^2, several methods are applicable. We do not want to go into detail here, but highly recommend the computer code Minuit [5.4] developed by James and Roos at the Cern (Geneva) computer center. Here a parabola fit [5.5], a Monte Carlo [5.6], and a gradient minimization [5.7] of χ^2 are included. The only

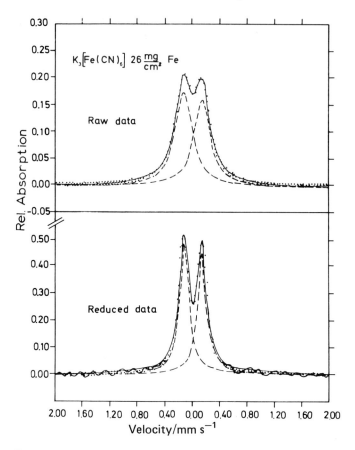

Fig. 5.1. Application of the Fourier transform method in a Mössbauer measurement to obtain the resonance absorption cross section $\sigma_r(v)$ (from [5.8])

task for the experimentalist to do is to program the theoretical values Y_i^{theo} for all velocities v_i in a subroutine (called FCN in MINUIT).

Concluding this section we want to mention the fast Fourier transformation method employed by Ure and Flinn [5.8] to get rid of the background and source emission line shape and to obtain the pure resonance absorption cross section $\sigma_r(E)$ (cf. Sec. 3.5) which in most cases is the interesting subject in Mössbauer spectroscopy. A deconvolution of (3.64) is possible after background subtraction, Fourier transformation, division by the Fourier transformed source emission line shape and retransformation. Taking now the logarithm, one obtains the resonance absorption cross section $\sigma(v)$ multiplied by the absorber thickness t. One disadvantage of this procedure, however, are the typical "Fourier transform wiggles" which are superimposed on the line shape as shown in Fig. 5.1 for the example of a $K_3[Fe(CN)_6]$ absorber (26 mg/cm²Fe).

References

[5.1] Kündig, W.: Nucl. Instr. Meth. *48*, 219 (1967)

[5.2] Bokemeyer, H., Meyer, R., Wohlfahrt, K., Wurtinger, W.: Laborbericht No. 49. Institut für Kernphysik der Technischen Hochschule, D-6100 Darmstadt (1971)

[5.3] Grimm, R., Müller, W.: Laborbericht 1/76, Institut für Anorganische Chemie und Analytische Chemie der Johannes-Gutenberg-Universität, Mainz (1976)

[5.4] James, F., Roos, M.: *Minuit* – computer code, CERN, Geneva, programme D-506

[5.5] Davidon, W. C.: Comp. J., *10*, 406 (1968)

[5.6] James, F.: Proc. of 1968 Hercey-Novi Schoool, London (1968)

[5.7] Rosenbrock, H. H.: Comp. J., *3*, 175 (1960)

[5.8] Ure, M. C. D., Flinn, P. A.: in *Mössbauer Effect Methodology 7.* New York, London: Plenum Press 1970, p. 245

6. Interpretation of Mössbauer Parameters of Iron Compounds

6.1. General Aspects

The task of this section is to show that for the interpretation of Mössbauer parameters – in the present case of electron densities, field gradients and magnetic hyperfine fields – bonding effects can play an important role, and free-ion or crystal field approaches may be a poor approximation.

In Fig. 6.1 we show the dependence of relativistic charge density $\rho(o)$ at the iron nucleus, as derived from fully relativistic configuration Dirac-Fock calculations for the free-ion, on the iron 3d and iron 4s population [6.1]. Taking the isomer shift calibration constant α in $\delta = \alpha\Delta\rho(o)$ as $\alpha \simeq -0.25\ a_0^3$ mm s^{-1} [6.1–3] we find changes in δ on going from one electronic configuration to the other as given in Fig. 6.1. In this free-ion picture there is no explanation at all for the dramatic increase in isomer shift δ on going from ferrous low-spin to ferrous high-spin compounds as illustrated in Fig. 6.2 [6.4]. Comparing further Figs. 6.1 and 6.2 it seems surprising that δ would increase by 0.75 mm s^{-1} in going from the ferrous to the ferric state, at least among low-spin compounds.

So we learn from Figs. 6.1 and 6.2 that screening of charge density at the iron nucleus by iron 3d electrons is significant but not the only effect which influences $\rho(o)$. We shall come back to this point and discuss in more detail the various contributions to $\rho(o)$.

Let us turn now for a moment to field gradients and quadrupole splittings. In a simple crystal field picture we would expect the iron 3d populations and approximate

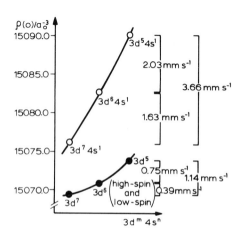

Fig. 6.1. Electron densities $\rho(o)$ (in a_0^{-3}) for iron as derived from fully relativistic mixed configuration Dirac-Fock calculation (taken from [6.1]), and for comparison, changes in isomer shift as derived from $\Delta\delta = \alpha\Delta\rho(o)$ with $\alpha = -0.25$ mm s$^{-1} a_0^3$

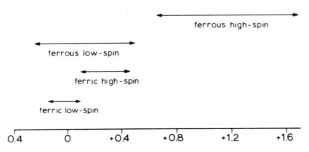

Fig. 6.2. Range of isomer shifts δ (in mm s^{-1} vs metallic iron) in iron compounds with various valencies and spin states (taken from [6.4])

quadrupole splittings, ΔE_Q, for the ferrous high-spin, ferrous low-spin, ferric high-spin, and ferric low-spin states of quasi-octahedral iron compounds as given in Fig. 6.3a. Symmetry, spin-orbit coupling, and temperature may somehow influence the estimated ΔE_Q values. In Fig. 6.3b we give examples for severe deviations from the ideal behaviour of Fig. 6.3a. Note that the ferrous low-spin and ferric high-spin compounds in Fig. 6.3b show considerably higher quadrupole splittings than the ferrous high-spin and ferric low-spin compounds, in contrast to Fig. 6.3a.

An explanation of this interesting result and a discussion of the various contributions to $\rho(o)$ will be given now on the basis of molecular orbital (MO) calculations.

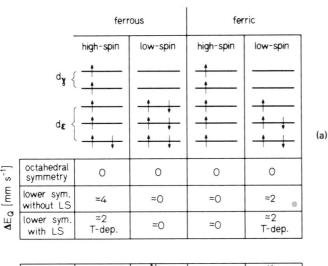

(a)

	octahedral symmetry	O	O	O	O
ΔE_Q [mm s^{-1}]	lower sym. without LS	≈4	≈0	≈0	≈2
	lower sym. with LS	≈2 T-dep.	≈0	≈0	≈2 T-dep.

compound	BaFeSiO$_4$	Na$_2$ [Fe(CN)$_5$NO] ·H$_2$O	BaFe$_{12}$O$_{19}$	K$_3$ [Fe(CN)$_5$NH$_3$]
ΔE_Q [mm s^{-1}] / T [K]	0.56 / 80	1.82 / 77	2.2 / 80	0.65 / 77
reference	6.37	6.38	6.39	6.40

(b)

ΔE_Q of the four compounds nearly temp. indep.

Fig. 6.3. Quadrupole splitting $\Delta E_Q = 1/2$ eQV$_{zz}$ for ferrous high-spin, ferrous low-spin, and ferric high-spin, ferric low-spin of quasi-octahedral iron compounds

6.2. Molecular Cluster Approach

Let us assume that we have made a molecular orbital calculation for a molecular cluster and derived a many-electron wave function Ψ describing the electronic ground state of this cluster. Since we are interested in Mössbauer parameters, we further assume that the molecular cluster under study contains ^{57}Fe or any other Mössbauer isotope. In order to calculate electron densities $\rho(r)$ or electric field gradients V_{pq} $(p, q = x, y, z)$ we simply compute the expectation values of relevant operators $\hat{O}$

$$\langle \Psi | \hat{O} | \Psi \rangle.$$

Both the electron density operator and the electric field gradient tensor operator are the sum of single-electron operators $\hat{o}_i$, each acting on the ith electron only

$$\hat{O} = \sum_i \hat{o}_i.$$

The expectation values $\langle \Psi | \hat{O} | \Psi \rangle$ can be determined easily if we use the following conditions:

(i) The many-electron wave functions Ψ are determinantal wave functions or, in general, linear combinations of those

$$\Psi = \sum_k b_k \Psi_k^S. \tag{6.1}$$

The Ψ^S (S denotes the spin) satisfy the Pauli exclusion principle, and they are eigenfunctions to $\hat{S}^2$ and $\hat{S}_z$, since we apply the antisymmetrizing operator $\hat{A}$ and the spin projection operator $\hat{O}_S$ to products of single-electron molecular spin orbitals $\widetilde{\phi}_i(\vec{r}, \vec{S})$ [6.5]

$$\Psi^S = N^S \hat{O}^S \hat{A} \prod_i \widetilde{\phi}_i(\vec{r}, \vec{S}). \tag{6.2}$$

The normalization constant N^S follows from the orthonormality condition

$$\langle \Psi^S | \Psi^S \rangle = 1.$$

(ii) The second condition is the following: the single-electron molecular spin orbitals $\widetilde{\phi}_i(\vec{r}, \vec{S})$ must be orthogonal. Since $\widetilde{\phi}_i(\vec{r}, \vec{S})$ is a product of an orbital part and a spin part,

$$\widetilde{\phi}_i(\vec{r}, \vec{S}) = \phi_i(\vec{r}) \chi_i(\vec{S}), \tag{6.3}$$

with $\chi_i(\vec{S}) = \alpha$ for $m_S = \frac{1}{2}$, and $\chi_i(\vec{S}) = \beta$ for $m_S = -\frac{1}{2}$, the orthogonality condition is only due to single-electron MO wave functions $\phi_i(\vec{r})$ and $\phi_j(\vec{r})$

$$\langle \phi_i | \phi_j \rangle = \delta_{ij}. \tag{6.4}$$

With these two conditions — Ψ being a determinantal wave function and $\widetilde{\phi}_i$ and $\widetilde{\phi}_j$ being orthogonal to each other — we have [6.6–8]

$$\langle \Psi | \hat{O} | \Psi \rangle = \sum_{k,\,l} b_k b_l \langle \Psi_k^S | \hat{O} | \Psi_l^S \rangle, \qquad (6.5\,a)$$

with

$$\langle \Psi_k^S | \hat{O} | \Psi_l^S \rangle = \sum_i \langle \widetilde{\phi}_i | \hat{o} | \widetilde{\phi}_i \rangle = \sum n_i \langle \phi_i | \hat{o} | \phi_i \rangle \qquad (6.5\,b)$$

if the same $\widetilde{\phi}_i$ contributes to Ψ_k^S and Ψ_l^S,

with

$$\langle \Psi_k^S | \hat{O} | \Psi_l^S \rangle = \langle \widetilde{\phi}_i | \hat{o} | \widetilde{\phi}_j \rangle \qquad (6.5\,c)$$

if Ψ_k^S and Ψ_l^S are different by one $\widetilde{\phi}_i$ only,

and with

$$\langle \Psi_k^S | \hat{O} | \Psi_l^S \rangle = 0 \qquad (6.5\,d)$$

if Ψ_k^S and Ψ_l^S differ by two or more $\widetilde{\phi}_i$'s.
n_i is the population number (1 or 2).

Taking now the single-electron MO's ϕ_i as linear combinations of atomic orbitals (AO), ψ_μ,

$$\phi_i = \sum_\mu a_i \psi_\mu, \qquad (6.6)$$

we finally obtain an expression for $\langle \Psi | \hat{O} | \Psi \rangle$ in terms of single-electron AO expectation values $\langle \Psi_\mu | \hat{o} | \psi_\nu \rangle$

$$\langle \Psi | O | \Psi \rangle = \sum_{\mu\nu} C_{\mu\nu}(S, m_S, n_i, b_k, b_l, a_{i\mu}, a_{i\nu}) \cdot \langle \psi_\mu | \hat{o} | \psi_\nu \rangle. \qquad (6.7)$$

All coefficients $C_{\mu\nu}$ are known from MO calculations. If we distinguish between iron AO's μ and μ' and ligand AO's μ'' and μ''' in (6.7) we obtain pure iron contributions $\langle \psi_\mu | \hat{o} | \psi_{\mu'} \rangle$, iron-ligand contributions $\langle \psi_{\mu'} | \hat{o} | \psi_{\mu''} \rangle$, and pure ligand contributions $\langle \psi_{\mu''} | \hat{o} | \psi_{\mu'''} \rangle$.

6.3. Electron Densities

6.3.1. Ab Initio Calculations

The numerical value we calculate for $\langle \Psi | \hat{O} | \Psi \rangle$ depends considerably on the atomic basis set $\psi_1, \psi_2, \ldots, \psi_\mu$, which we use for the molecular cluster under study. For ab initio SCF MO calculations of electron densities in the octahedral clusters $[FeF_6]^{3-}$, $[FeF_6]^{4-}$, $[Fe(CN)_6]^{3-}$ and $[Fe(CN)_6]^{4-}$, *Post* et al. [6.3] have used a large enough basis set, including all iron core s-AO's in order to compute

$$\rho_{unrelat.}(o) = \sum_{\mu\nu} C_{\mu\nu}\, \psi_\mu(o)\, \psi_\nu(o) \tag{6.8}$$

directly, without further corrections except a proper relativistic scaling [6.9]

$$\rho(o) = 1.3\, \rho_{unrelat.}(o).$$

Fig. 6.4 shows the linear relation between experimental δ values and calculated $\rho(o)$ values with a mean slope of $\alpha = -0.239 \pm 0.023\ a_0^3\ mm\ s^{-1}$.

6.3.2. Approximate MO Calculations

For systems considerably larger than $[FeF_6]^{-3,\,-4}$ or $[Fe(CN)_6]^{-3,\,-4}$ such as the porphyrin group in heme proteins, the computation of $\rho(o)$ on the basis of ab initio calculations with the inclusion of all AO's would be extremely time consuming. Therefore, in the case of MO calculations for heme proteins [6.10, 11] the basis set can be restricted to valence AO's, e.g. to 3d, 4s, and 4p orbitals of iron, to 2s and 2p orbitals of oxygen, nitrogen and carbon, and to the 1s orbital of hydrogen. In this case (6.8) gives only the *valence contribution* to $\rho(o)$

$$\rho_{val}(o) = \sum_{\mu\nu} C_{\mu\nu}\psi_\mu(o)\psi_\nu(o). \tag{6.9}$$

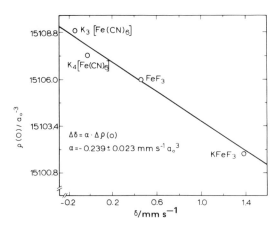

Fig. 6.4. Least-squares fit of calculated relativistic charge densities at the iron nucleus versus measured isomer shifts (taken from [6.3])

Because of the relatively high probability of finding iron 1s, 2s, and 3s electrons at the iron nucleus, we have to include as well their contributions to the total electron density at the iron nucleus, $\rho(o)$. In general, however, the orthogonality condition of (6.4) (which has to be fulfilled for calculating $\rho(o)$ from a summation of single-electron contributions) is no longer valid among MO's ϕ_i and iron core s-AO's ψ_μ

$$\langle \phi_i | \psi_\mu \rangle \neq \delta_{i\mu}.$$

Therefore, we have first to find iron core s-orbitals ϕ_{ns}, n = 1, 2, 3, which are orthogonal to the MO's ϕ_i, i = 1, ... m,

$$\langle \phi_i | \phi_{ns} \rangle = \delta_{i\,ns}.$$

Carrying out the orthogonalizations in the order 3s, 2s, 1s we find [6.1]

$$\phi_{3s}(o) = N_{3s}\left(\psi_{3s}(o) - \sum_{i=1}^{m} \langle \phi_i | \psi_{3s} \rangle \phi_i(o)\right),$$

$$\phi_{2s}(o) = N_{2s}\left(\psi_{2s}(o) - \sum_{i=1}^{m} \langle \phi_i | \psi_{2s} \rangle \phi_i(o) - \langle \phi_{3s} | \psi_{2s} \rangle \phi_{3s}(o)\right),$$

(6.10)

$$\phi_{1s}(o) = N_{1s}\left(\psi_{1s}(o) - \sum_{i=1}^{m} \langle \phi_i | \psi_{1s} \rangle \phi_i(o) - \langle \phi_{3s} | \psi_{1s} \rangle \phi_{3s}(o) - \langle \phi_{2s} | \psi_{1s} \rangle \phi_{2s}(o)\right),$$

where ψ_{3s}, ψ_{2s}, ψ_{1s} are the original (Hartree-Fock, or in the relativistic case, Dirac-Fock) iron core s-orbitals, and N_{3s}, N_{2s}, and N_{1s} are normalization factors. We could continue, obtaining $\phi_{3p}(o)$ – and $\phi_{2p}(o)$ – contributions to $\rho(o)$ from relativistic effects upon iron 3p and 2p orbitals. While these contributions to $\rho(o)$ are individually significant [6.1] (of the order of 0.5 a_0^{-3} and 3.5 a_0^{-3}, respectively), their differences are so small that they would affect $\Delta\rho(o)$ in (6.23) in the third decimal place only.

The orthogonalization procedure described by (6.10) influences the original iron core s-orbitals ψ_{ns} by overlap contributions from MO wave functions ϕ_i; therefore we call the change from core s-contributions to $\rho(o)$ of the free ion,

$$\rho_{core}^{free\ ion}(o) = 2|\psi_{3s}(o)|^2 + 2|\psi_{2s}(o)|^2 + 2|\psi_{1s}(o)|^2,$$

(6.11a)

to the new core s-contribution of the iron cluster,

$$\rho_{core}(o) = 2|\phi_{3s}(o)|^2 + 2|\phi_{2s}(o)|^2 + 2|\phi_{1s}(o)|^2,$$

(6.11b)

the *overlap distortion effect* [6.12–14]

$$\Delta\rho_{ov}(o) = \rho_{core}(o) - \rho_{core}^{free\ ion}(o).$$

(6.12)

It should be noted that such calculations of overlap distortion effects lead to an overestimation because of the neglect of subsequent *rearrangement effects*. An estimate of rearrangement effects was given by Sawatzky [6.15]. We carry out this estimate here in a slightly different way and find it less important than *Sawatzky* et al. The increase in electronic charge density at the iron nucleus due to overlap distortion effectively shields the nuclear charge so that the iron electrons are less attracted by the nucleus and tend to move outwards a little, thereby decreasing the charge density at the nucleus. Since we are interested in charge densities at the nucleus we consider only s-electrons, and since only 1s Slater type orbitals (STO's) contribute to $|\psi_{ns}(o)|^2$ we investigate the screening constant σ of the iron 1s STO

$$\psi_{1s} = C_1 e^{-(Z + \sigma) r/a_0}.$$

$Z = 26$ is the total nuclear charge number of iron, and the screening constant

$$\sigma = - \int_0^{R_0} dr^3 \rho_{1s}(r), \qquad R_0 = \text{radius of the nucleus} \tag{6.13}$$

is estimated to be -0.3 [6.16]. If we make the approximation that $\rho_{1s}(r)$ is constant up to a certain value of $r = R_0$ (Fig. 6.5), we would have

$$\sigma = \rho(o) \frac{4\pi}{3} R_0^3 \tag{6.14}$$

instead of (6.13). A change in the screening constant, $\Delta\sigma$, due to a change in the electron density at the nucleus, $\Delta\rho(o)$, then would be given by

$$\Delta\sigma = -\Delta\rho(o) \frac{4\pi}{3} R_0^3. \tag{6.15}$$

Comparing (6.14) with (6.15) we obtain

$$\Delta\sigma = -\sigma \frac{\Delta\rho(o)}{\rho(o)}. \tag{6.16}$$

Due to the modified iron 1s STO

$$\psi'_{1s} = \psi_{1s} + \frac{\partial \psi_{1s}}{\partial (Z + \sigma)} \Delta\sigma \tag{6.17}$$

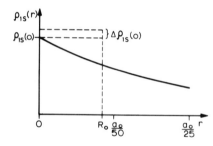

Fig. 6.5. Radial dependence of $\rho_{1s}(r) = \rho_{1s}(o)$ $\exp(-25.7r/a_0)$, which is used for an estimate of rearrangement effects (see text)

we find a change in the charge density at the nucleus due to rearrangement, $\Delta\rho_{re}(o)$, to first order in $\Delta\sigma$[4]

$$\Delta\rho_{re}(o) = 3\rho(o)\left(\frac{\Delta\sigma}{Z + \sigma}\right). \qquad (6.18)$$

With

$$\Delta\rho(o) = \Delta\rho_{ov}(o) + \Delta\rho_{re}(o) \qquad (6.19)$$

(6.18) becomes

$$\Delta\rho(o) = \Delta\rho_{ov}(o) + 3\rho(o)\left(\frac{\Delta\sigma}{Z + \sigma}\right), \qquad (6.20)$$

and with (6.16) substituting $\Delta\sigma$, (6.20) changes to

$$\Delta\rho(o) = \Delta\rho_{ov}(o) - \frac{3\sigma\Delta\rho(o)}{(Z + \sigma)}. \qquad (6.21)$$

This finally leads to

$$\Delta\rho(o) = \Delta\rho_{ov}(o) / \left(1 + \frac{3\sigma}{(Z + \sigma)}\right) = 0.96\ \Delta\rho_{ov}(o). \qquad (6.22)$$

---

[4] The total charge density in the iron 1s AO is given by

$$\psi'^2_{1s} = \psi^2_{1s} + 2\psi_{1s}\frac{\partial\psi_{1s}}{\partial(Z + \sigma)}\Delta\sigma + \ldots$$

With the iron 1s STO

$$\psi_{1s} = 2(Z + \sigma)^{3/2}\exp\left[-(Z + \sigma)r/a_0\right]$$

we derive for $\dfrac{\partial\psi_{1s}}{\partial(Z + \sigma)}$:

$$\frac{\partial\psi_{1s}}{\partial(Z + \sigma)} = 3(Z + \sigma)^{1/2}\exp\left[-(Z + \sigma)r/a_0\right] + 2(Z + \sigma)^{3/2}(-r/a_0)\exp\left[-(Z + \sigma)r/a_0\right].$$

The second term in $\dfrac{\partial\psi_{1s}}{\partial(Z + \sigma)}$ goes to zero for $r \to 0$.

Thus for $\Delta\rho_{1s, re}(0) = 2\psi_{1s}\dfrac{\partial\psi_{1s}}{\partial(Z + \sigma)}$ we have

$$\Delta\rho_{1s, re}(0) = 3\psi_{1s}^2(0)\left(\frac{\Delta\sigma}{Z + \sigma}\right).$$

Since the only contributions to $\rho(0)$ come from STO 1s AO's (a Clementi 3s AO for example contributes to $\rho(0)$ only through STO 1s AO-terms from its linear combination of STO's) we finally have

$$\Delta\rho_{re}(0) = \Delta\rho_{1s, re}(0).$$

This result clearly shows that rearrangement effects are nearly negligible compared to the pure overlap distortion effect.

Another type of rearrangement effect results from charges in the iron 3d and 4s population; this influences significantly the magnitude of $\rho(o)$. This effect, also known as *potential distortion effect,* has already been shown in Fig. 6.1, where we find different $\rho(o)$ values for iron configurations $3d^m 4s^n$ and $3d^{m'} 4s^{n'}$, respectively.

The calculational method described in this section to compute the various contributions to $\rho(o)$ — valence contributions, overlap distortion effect upon core s-contributions, rearrangement effect, potential distortion effect — has been applied to several clusters. In Table 6.1 we list the individual $\rho_{ns}(o)$ values for six selected compounds; in addition, we give the experimental values of δ, and the electronic configurations found for iron from MO calculations. The main use we make of these results is to compare similar systems, i.e., $[FeO_6]^{9-}$ (ferric high-spin) with $[FeO_6]^{10-}$ (ferrous high-spin) [6.17], $[FeF_6]^{3-}$ (ferric high-spin) with $[FeF_6]^{4-}$ (ferrous high-spin) [6.14, 6.18], and CO-hemoglobin (ferrous low-spin) with deoxy-hemoglobin (ferrous high-spin) [6.19], since the calculational procedure will be least likely affected by systematic uncertainties in $\rho(o)$ if differences $\Delta\delta$ and $\Delta\rho(o)$ between similar systems are related according to

$$\Delta\delta = \alpha \, \Delta\rho(o). \tag{6.23}$$

From inspection of Table 6.1 we come to the following conclusions:

1. Potential distortion effects contribute to $\rho(o)$ in these cases with the sign opposite compared with the overlap distortion contribution. (The same result was found by *Walch* et al. [6.20] from a systematic study of overlap distortion for ^{57}Fe embedded in an argon lattice).

2. The iron 4s contributions to $\rho(o)$ are nearly equal within similar compounds.

3. The $\rho_{2s}(o)$ and $\rho_{3s}(o)$ contributions to $\rho(o)$ are much more influenced than $\rho_{1s}(o)$ contributions by going from one cluster to the other. However, note that an increase in $\rho_{1s}(o)$ and $\rho_{3s}(o)$ is accompanied by a decrease in $\rho_{2s}(o)$, and vice versa; this is due to the specific radial dependence of Hartree-Fock core s-orbitals (zero nodes for 1s, one node for 2s, and two nodes for 3s wave functions).

4. In connection with Fig. 6.1, one realizes that for iron-oxygen and iron-fluorine clusters, $\rho(o)$ is influenced by overlap distortion *and* potential distortion effects by going from one cluster to the other.

5. The drastic increase of $\rho(o)$ during the reaction of deoxy-hemoglobin (Hb) with CO, however, is nearly exclusively due to the increased overlap distortion of the iron core s-orbitals in CO-hemoglobin (HbCO) with respect to deoxy-hemoglobin. The valence contribution and the potential distortion effect upon $\rho(o)$ are nearly the same for both, HbCO and Hb. (Notice that the 3d and the 4s populations are nearly the same in both components.) From this we see that due to stronger interaction between ligands and iron in HbCO compared to Hb, the iron core s-contribution $\phi_{ns}(o)$ of (6.10) to $\rho(o)$ is considerably higher in HbCO than in Hb, and thus δ_{HbCO} is significantly smaller than δ_{Hb}.

6. Comparing differences $\Delta\delta$ and $\Delta\rho(o)$ between similar systems according to (6.23) we find for the isomer shift calibration constant α the following values:

Table 6.1. Experimental isomer shifts δ, calculated electronic configurations for iron, and calculated relativistic electron densities ρ_{ns} (o) at the iron nucleus for various iron containing clusters

Cluster	FeO_6^{-9}	FeO_6^{-10}	FeF_6^{-3}	FeF_6^{-4}	$HbCO^e$	Hb^e
δ^{exp}(mm s^{-1})	0.50 ± 0.05^a	$1.10\pm0,02^a$	0.77 ± 0.02^b	1.42 ± 0.02^b	0.27 ± 0.05^c	0.93 ± 0.05^c
configuration^d	$3d^{5.70}4s^{0.169}$	$3d^{6.27}4s^{0.157}$	$3d^{6.41}4s^{0.13}$	$3d^{6.62}4s^{0.14}$	$3d^{6.36}4s^{0.18}$	$3d^{6.25}4s^{0.22}$
ρ_{1s}(o) (a$_0^{-3}$)	13,602.64	13,601.63	13,601.84	13,600.07	13,600.9	13,599.8
ρ_{2s}(o) (a$_0^{-3}$)	1,275.98	1,279.09	1,280.21	1,283.24	1,280.9	1,283.2
ρ_{3s}(o) (a$_0^{-3}$)	187.85	183.62	184.18	180.44	184.6	180.46
ρ_{4s}(o) (a$_0^{-3}$)	1.08	0.77	0.59	0.56	0.82	1.16

[a] Taken from [6.17]. Values of δ at 300 K (vs sodium ferrocyanide at 300 K) are derived from a sample of MgO doped with ^{57}Fe and irradiated with ^{60}Co γ-rays

[b] Taken from [6.14]. Values of δ at 300 K (vs sodium ferrocyanide at 300 K) are derived from a sample of KMgF$_3$ doped with ^{57}Fe and irradiated with ^{60}Co γ-rays

[c] Taken from [6.19]. Values of δ (vs metallic iron at 300 K) are mean values taken at 4.2 K from HbCO and Hb of various species

[d] Configurations 3d^{m}4s^n as derived from MO calculations [6.1, 11, 18]

[e] HbCO: CO-hemoglobin; Hb: deoxy-hemoglobin

$\alpha = -0.246 \pm 0.029\ a_0^3$ mm s^{-1} from the iron-oxygen clusters,

$\alpha = -0.259 \pm 0.016\ a_0^3$ mm s^{-1} from the iron-fluorine clusters, and

$\alpha = -0.242 \pm 0.039\ a_0^3$ mm s^{-1} from hemoglobin compounds.

These values are in reasonable agreement with the most recent work of other investigators, namely $-0.239 \pm 0.023\ a_0^3$ mm s^{-1} found by Post et al. [6.3] for the complexes shown in Fig. 6.4, and $-0.250 \pm 0.030\ a_0^3$ mm s^{-1} found by Regnard et al. [6.2]. (A discussion of the considerable differences between other published estimates of α is given in Duff's work [6.21].) Recently the limitation in the calculation of electron densities on the basis of approximate MO calculations has been discussed by Reschke et al. [6.75].

6.3.3. Isomer Shift and Electronegativity

An interesting question for chemists is how isomer shift and electronegativity are related to each other. Fig. 6.6 (taken from [6.15]) shows for divalent iron halides FeX_2 (X = F, Cl, Br, I) the nearly linear relation between δ and the electronegativity of the halide anion. Since the overlap integrals $\langle \psi_{\text{lig}-ms} | \psi_{\text{Fe}-ns} \rangle$ and $\langle \psi_{\text{lig}-mp} | \psi_{\text{Fe}-ns} \rangle$ (with m = 2 for F, m = 3 for Cl, m = 4 for Br, m = 5 for I, and n = 1, 2, 3, for Fe) were found to be nearly equal [6.15] — the increase of Fe-X distance from FeF_2 to FeI_2 is balanced by an increased spreading of the ligand AO's — it was assumed that the overlap distortion effect remains unaffected within the four FeX_2 compounds. From (6.10) to (6.12), however, it is clear that the bond order matrix elements

$$P_{\text{Fe}-4s,\text{lig}-ms} = \sum_{i=1}^{M} n_i a_{i,\text{Fe}-4s} a_{i,\text{lig}-ms}$$

and

$$P_{\text{Fe}-4s,\text{lig}-mp} = \sum_{i=1}^{M} n_i a_{i,\text{Fe}-4s} a_{i,\text{lig}-mp}$$

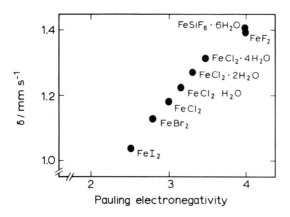

Fig. 6.6. Plot of isomer shifts δ vs the electronegativity (Pauling) of the anions for divalent iron halides (taken from [6.15])

(with MO-population n_i of ϕ_i being 0, 1, or 2, and M being the number of occupied ϕ_i's) play an important role concerning the evaluation of $\Delta\rho_{ov}(o)$. MO calculations on FeX_6^{4-} clusters indeed showed [6.22] that the overlap distortion effect is considerably different for the four iron halides. The change in $\Delta\rho_{ov}(o)$ from FeF_2 to FeI_2, however, does not account for the change in total electron density at the iron nucleus as derived from experimental isomer shifts through (6.23), since $\Delta\rho_{ov}(o)$ even goes in the wrong direction. Thus additional effects (valence contribution and potential distortion effect) have to be considered.

The decrease of electronegativity from fluorine to iodine causes a decrease in ionic character of the iron-ligand bond. In agreement with this, an increase of the iron valence shell population was found from MO cluster calculations [6.22] by going from FeF_2 to FeI_2. Consistent with Hartree-Fock results [6.23] for different $3d^n 4s^m$ configurations of the free iron atom, which show that for $n = 6, 7$ and $m = 0, 1, 2$ a $3d^n 4s^{m+1}$ configuration is energetically more stable than a $3d^{n+1} 4s^m$ configuration, the decrease of electronegativity from fluorine to iodine finally causes an increase of the iron 4s population in the MO cluster calculations. Thus the result of these calculations is that the change of valence contribution, $\rho_{val}(o)$ — which is mainly due to iron 4s population; see (6.9) — by going from FeF_2 to FeI_2 explains the change of isomer shift correctly. This change of $\rho_{val}(o)$ is pronounced enough to overcompensate the change $\Delta\rho_{ov}(o)$, which goes in the wrong direction. The potential distortion contribution was found to be practically the same for all four iron halides, indicating that charge is transferred into the iron 4s AO substituting F with Cl, Br, and I in FeX_2.

6.3.4. Second-Order Doppler Shift

To complete the discussion of energy shifts in Mössbauer spectroscopy we also should pay attention to the second-order Doppler shift δ_{SOD}, which adds to the isomer shift δ to give the measured energy shift δ_{total}

$$\delta_{total} = \delta + \delta_{SOD}. \tag{6.24}$$

In order to elucidate the physical reason for the presence of δ_{SOD}, we consider the Mössbauer nucleus ^{57}Fe with mass M executing simple harmonic motion [6.24]. The equation of motion under isotropic and harmonic approximations can be written as

$$M\ddot{r} = -Kr, \tag{6.25}$$

where K is the force constant and r is the displacement of the ^{57}Fe atoms from their mean position at any instant. The acceleration experienced by the ^{57}Fe atom at a particular instant is given by

$$\ddot{r} = -\frac{K}{M} r. \tag{6.26}$$

According to the "principle of equivalence" [6.25], this acceleration creates a gravitational field whose potential is defined as the work done by a unit mass in moving from the point under consideration to a point free from this force field and may be given by

$$\phi = \int_{r}^{0} \left(-\frac{K}{M} r\right)(-dr) = -\frac{K}{2M} r^2.$$

(6.27)

Since the period of vibration in a crystal ($\approx 10^{-13}$ s) is much smaller than the lifetime of the excited state of the Mössbauer nucleus ($\approx 10^{-8}$ s), the vibrating nucleus only sees an average value of the potential

$$\langle \phi \rangle = -\frac{K}{2M} \langle r^2 \rangle,$$

(6.28)

which causes a change [6.26] ΔE in the energy E_γ of the γ-ray emitted or absorbed by the vibrating nucleus: $\Delta E_\gamma = m_\gamma \langle \phi \rangle$. This (together with $E_\gamma = \hbar \omega_\gamma = m_\gamma c^2$) gives rise to a fractional energy shift

$$\delta_{SOD} = \frac{\Delta E}{E} = \frac{\langle \phi \rangle}{c^2},$$

(6.29a)

$$\delta_{SOD} = -\frac{K}{2Mc^2} \langle r^2 \rangle,$$

(6.29b)

which is called *second-order Doppler shift*. Since the mean square displacement $\langle r^2 \rangle$ of the vibrating nucleus is temperature dependent, δ_{SOD} also is called *temperature shift*.

Comparing changes in isomer shift, $\Delta \delta$, with changes in electron densities, $\Delta \rho(o)$, between two clusters as described in Sections 6.3.1 and 6.3.2 we, of course, have to correct the measured energy shifts δ_{total} in order to find the isomer shift δ. Since in most cases the second-order Doppler shift contributions to δ_{total} are negligibly small, very often the measured isomer shift δ is taken as the total isomer shift δ_{total}. The δ-values in Table 6.1, however, contain the proper corrections due to δ_{SOD}-contributions.

From a chemical point of view the second-order Doppler shift is very interesting with respect to its simple relation connecting δ_{SOD}, the recoil-free fraction f, and the force contant K experienced by the Mössbauer nucleus within a cluster. This relation follows from a combination of (6.29b) with (2.18) [6.24]

$$\delta_{SOD} = \frac{3\hbar^2}{2ME_\gamma^2} K \ln f.$$

(6.30)

Since it is difficult to derive δ_{SOD} directly from Mössbauer measurements we substitute (6.30) into (6.24) and obtain

$$\delta_{total} = \frac{3\hbar^2}{2ME_\gamma^2} \, K \, lnf + \delta. \tag{6.31}$$

Because of the smallness of the variation of K and δ with temperature in comparison to that of lnf, the differentiation of (6.31) with respect to temperature gives

$$\frac{\partial \delta_{total}}{\partial T} = \left(\frac{3\hbar^2}{2ME_\gamma^2} \, K\right) \frac{\partial \, lnf}{\partial T} \tag{6.32a}$$

or

$$\left(\frac{\Delta \delta_{total}}{\Delta \, lnf}\right)_T = \frac{3\hbar^2}{2ME_\gamma^2} \, K. \tag{6.32b}$$

$\Delta\delta_{total}$ and Δlnf are the changes of the total energy shift δ_{total} and of lnf, respectively, corresponding to the variation in temperature T.

Taylor et al. [6.27] have studied $\Delta\delta_{total}$ and Δlnf for ^{57}Fe in various host lattices over a wide temperature range. Taking the experimental values $\Delta\delta_{total}/\Delta lnf$ from their work, the force constants K experienced by the ^{57}Fe atoms in the different host lattices have been calculated by Gupta et al. [6.24]. These calculations indicate that $K(^{57}Fe)$ significantly depends on bonding properties and the chemical nature of the ligands which are coordinated to iron.

6.4. Quadrupole Splitting

6.4.1. The Electric Field Gradient Tensor

For the discussion of quadrupole splittings, ΔE_Q, we go back to (6.7) in order to study the electric field gradient tensor

$$(V_{pq}) = \langle\Psi||\hat{V}_{pq}||\Psi\rangle. \tag{6.33}$$

The tensor operator $(\hat{V}_{pq})$ in (6.33) is the sum of single-electron operators $\hat{v}_{pq,i}$ of the ith electron, which have the form [6.28]

$$\hat{v}_{pq} = [1 - \gamma(r)] \, Ze \, \frac{3\hat{p}\hat{q} - r^2\delta_{pq}}{r^5}, \tag{6.34}$$

where Ze is the charge of an electron (Z = 1) or of a ligand, and r follows from $r^2 = x^2 + y^2 + z^2$. The function $\gamma(r)$ is the quadrupole polarizability of the core electrons of the Mössbauer atom, if we limit our basis set in (6.7) to valence electrons only. Results of $\gamma(r)$ are usually represented in terms of Sternheimer factors [6.28, 29]

$$\gamma_\infty = \lim_{r\to\infty} \gamma(r), \quad \text{and} \quad R = \langle\gamma(r)r^{-3}\rangle/\langle r^{-3}\rangle. \tag{6.35}$$

If we distinguish between iron AO's and ligand AO's in (6.7), we are left to discuss the following types of atomic orbital matrix elements $\langle \psi_\mu | (\hat{v}_{pq}) | \psi_\nu \rangle$:

1. Matrix elements within an atomic shell of iron (i.e., 3d, 4s and 4p) can be obtained by the operator equivalence [6.30]

$$\hat{v}_{pq} = \langle ||EFG|| \rangle \, e(1 - R) \, \langle r^{-3} \rangle \left[\frac{1}{2} (\hat{l}_p \hat{l}_q + \hat{l}_q \hat{l}_p) - \frac{l(l+1)}{3} \delta_{pq} \right] \qquad (6.36)$$

The reduced matrix element $\langle ||EFG|| \rangle$ takes the value $\frac{2}{7}$ for d-electrons, $\frac{6}{5}$ for p-electrons, and 0 for s-electrons. The quantity e represents the (positive!) elementary charge. $\langle r^{-3} \rangle_{3d}$ is the radial factor, which stands for the integral $\int \psi (r^{-3}) \psi r^2 dr$, with values 4.49 a.u. for iron $3d^7$, 5.09 a.u. for iron $3d^6$, and 5.73 a.u. for iron $3d^5$ [6.31]. The Sternheimer shielding correction [6.29] is taken as $(1 - R)_{3d} = 0.68$. For iron 4p electrons the quantity $(1 - R)_{4p} \langle r^{-3} \rangle_{4p}$ is taken to be 1/3 of the corresponding quantity for 3d electrons [6.32, 33].

2. Off-diagonal elements between different shells of iron. Since the electric field gradient tensor operator is time invariant like the many-electron terms Ψ, and since its angular part transforms like the spherical harmonics Y_{2m}, matrix elements occur only between atomic orbitals of the same orbital angular momentum l and between l and l ± 2. In the present case of iron chemistry we have used 3d, 4s, and 4p AO's as basis orbitals for MO's. Thus the only nonvanishing matrix elements within this approximation are those between iron 3d and iron 4s AO's. Because the extent of mixing of iron 3d and iron 4s terms has been found to be so small in all cases studied so far, we neglect this contribution to the EFG tensor (V_{pq}).

3. Matrix elements of ligand AO's. If the radius of the ligand ion is small compared to the distance between the ligand center and the iron nucleus, V_{pq} can be approximately calculated from the charge q_{ligand} and the coordinates R_x, R_y, R_z of the ligand ion:

$$V_{pq}^{lat} = (1 - \gamma_\infty) q_{ligand} \frac{3R_p R_q - R^2 \delta_{pq}}{R^5} . \qquad (6.37)$$

For the Sternheimer factor $(1 - \gamma_\infty)$ which represents the antishielding of the core electrons due to the charge outside the iron we use the value 10.1 [6.29]. The charge q_{ligand} may be obtained from MO calculations.

4. Overlap part between iron and ligand AO's. If the overlap of the iron AO's and the ligand AO's at the ion A is limited to a small region in which r may be taken as constant $(r = \bar{R})$ the overlap contribution to V_{pq} may be estimated using the expression

$$V_{pq}^{ov} = e(1 - \gamma(\bar{R})) \frac{3R_p R_q - \bar{R}^2 \delta_{pq}}{\bar{R}^5} \sum_{\substack{\mu(Fe) \\ \nu(A)}} (C_{\nu\mu} + C_{\mu\nu}) S_{\nu\mu}. \qquad (6.38a)$$

The Sternheimer factor $(1 - \gamma(\bar{R}))$ is smaller than $(1 - \gamma(R))$, with R being the distance between iron and atom A, because of decreasing antishielding with decreasing distance. The r-dependence of $1/r^3$ goes in the opposite direction. Since we do not know the correct radial dependence of $\gamma(r)$ we are within the approximation of 3. if we estimate the overlap contribution by

$$V_{pq}^{ov} \approx e(1 - \gamma_\infty)\frac{3R_p R_q - R^2 \delta_{pq}}{R^5} \sum_{\substack{\mu(Fe) \\ \nu(A)}} (C_{\nu\mu} + C_{\mu\nu}) S_{\nu\mu}. \tag{6.38b}$$

5. Following (6.37) and (6.38b) in calculating $V_{pq}^{lat} + V_{pq}^{ov}$ might have the disadvantage that overlap contributions are overestimated, since q_{ligand} in (6.37) is estimated in many MO programs on the basis of a Mulliken population analysis[5] or on more sophisticated procedures [6.34] to divide overlap charge q_{ab} into parts q_a and q_b, which are added onto charges situated at centers a and b, in order to get effective charges. Thus, q_{ligand} in (6.37) might already contain overlap charge to some extent. To avoid an overestimation of overlap contributions to V_{pq} one either has to make sure that q_{ligand} does not contain overlap charges or has to follow an alternative estimate of $V_{pq}^{lat} + V_{pq}^{ov}$, which has been applied to ferrocene and related compounds [6.35]:

$$V_{pq}^{lat} + V_{pq}^{ov} = (1 - \gamma_\infty) \sum_{a,b} \frac{3\overline{R_p^{ab}}\,\overline{R_q^{ab}} - \overline{R_{ab}^2}\,\delta_{pq}}{\overline{R_{ab}^5}} q_{ab}. \tag{6.39}$$

The summation is over all atoms a and b. In the cases, where $a \neq b$, q_{ab} represents the overlap charge between atoms a and b, and where $a = b$, q_{ab} represents the charge of atom a. The charges q_{ab} are calculated from bond order matrix elements, $P_{\mu\nu}$, and overlap integrals, $S_{\mu\nu}$

$$q_{ab} = e(Z_a\delta_{ab} - \sum_{\mu,\nu} P_{\mu\nu} S_{\mu\nu}). \tag{6.40}$$

e is the positive elementary charge. The summation over μ includes atomic orbitals (AO) ψ_μ on center a, and that of ν includes AO's ψ_ν centered on atom b. eZ_a represents the core charge of atom a. The bond order matrix elements are defined by

$$P_{\mu\nu} = \sum_{i=1}^{M} n_i c_{i\mu} c_{i\nu}.$$

M is the number of occupied molecular orbitals (MO) $\phi_i = \sum_\mu c_{i\mu} \psi_\mu$, and $n_i = 1$ or 2 is the occupation number of MO ϕ_i. The cartesian coordinates $\overline{R_p^{ab}}$ and the distance $\overline{R_{ab}}$ between iron and the various overlap charges q_{ab} are chosen as if the q_{ab} were

---

[5]   Overlap charge q_{ab} is divided into two equal parts, one added to ligand a and the other to ligand b.

situated at the maximum of product $\psi_\mu \psi_\nu$. Since the so-defined overlap charges have a distance $\overline{R_{ab}}$ to iron larger than approximately 1.5 Å in most compounds, the use of $(1 - \gamma_\infty) = 10.1$ is adequate [6.36][6]

6.4.2. Temperature Independent Quadrupole Splitting

The compounds [6.37–40] listed in Fig. 6.3 and the ferric high-spin cluster $[FeO_4Cl_2]^{-7}$ originating from FeOCl [6.41] exhibit nearly temperature independent quadrupole splittings over a wide temperature range. This behaviour is reflected by the result from MO calculations on these clusters that excited electronic states remain unpopulated within the temperature range under study. This has the simple consequence that we are concerned with an energetically isolated electronic ground state and concerning V_{pq} only with diagonal matrix elements $\langle \Psi | (\hat{V}_{pq}) | \Psi \rangle$ as described by (6.7) and (6.33). Following the procedure described in Section 6.4.1, quadrupole splittings are derived for the five compounds mentioned above. In Table 6.2 we give first-order density matrix elements (bond order matrix elements) $P_{\mu\mu}$, atomic charges q_a of iron and its ligands, experimental and calculated ΔE_Q values. The MO interpretation of ΔE_Q values of these compounds may be summarized as follows:

1. The compounds under study are by far *not* characterized by iron $3d^5$ or iron $3d^6$ configurations but more likely by $3d^m 4s^n 4p^o$ with considerable deviations from iron charge states +3 or +2.

2. The quadrupole splittings are substantially influenced by varying population of *all* five iron 3d AO's, even in the case of ferric and ferrous low-spin compounds.

3. The quadrupole splittings are also influenced by departures of the electronic distribution from axial symmetry (Fe-NO axis for example is found to be bent and fixed in a certain position [6.42], as has also been found in MO calculations by Dorn [6.43]).

4. Asymmetric electron populations in the iron 4p AO's contribute to ΔE_Q.

5. In addition, net charges from surrounding ligands may have considerable effects upon ΔE_Q. In the case of $BaFeSiO_4$ the ligand contribution (in the literature often called the lattice contribution) to ΔE_Q is of the order of 3 mm s^{-1} [6.37, 39]; the smallness of the resulting total quadrupole splitting for this compound arises from nearly mutual cancellation of iron valence electron and lattice contributions.

6. The compound FeOCl was experimentally investigated in single-crystalline form [6.41]. Single-crystal studies have the advantage that they provide us with additional experimental results [6.44, 45], which are in the present case [6.41] (i) $\eta^{exp} = 0.32 \pm 0.03$, (ii) $V_{zz}^{exp} < 0$, and (iii) the direction of the z axis of the principal axes system of the EFG with respect to the crystallographic frame. The computational analysis of the V_{pq} tensor yields [6.46] (i) $\eta^{calc} = 0.23$, (ii) $V_{zz}^{calc} < 0$, and (iii) the z axis of the principal axes system coincides with the experimentally determined principal axes system.

---

[6]   Mathematical procedures based on numerical integration of $(1 - \gamma(r)) \langle \Psi | \hat{V}_{pq} | \Psi \rangle$ show promising results (Freeman et al. [6.95], Reschke and Trautwein, Theoret. Chim. Acta *47*, 85 (1978)

Table 6.2. Experimental and calculated quadrupole splittings, bond order matrix elements $P_{\mu\mu}$ and atomic charges q_a of iron and its next-nearest ligands as derived from MO calculations

Compound	BaFeSiO$_4$	Na$_2$Fe(CN)$_5$NO · H$_2$O	BaFe$_{12}$O$_{19}$	K$_3$Fe(CN)$_5$NH$_3$	FeOCl
ΔE_Q^{exp} (mms^{-1})	0.56± 0.002[a]	1.82±0.08[b]	1.95±0.05[c]	0.65±0.03[d]	0.916± 0.001[e]
T^{exp} (K)	80	77	120	77	300
ΔE_Q^{calc} (mms^{-1})	0.35[f]	1.84[g]	1.80[f]	0.64[h]	0.83[i]
$P_{\mu\mu}$: $d_{x^2-y^2}$	1.20	1.03	1.08	0.95	1.07
$d_{3z^2-r^2}$	1.90	0.85	0.72	0.76	1.16
d_{xz}	1.01	1.51	1.05	1.79	1.08
d_{yz}	1.01	1.51	1.05	1.79	1.11
d_{xy}	1.04	1.77	1.08	1.09	1.27
s	0.20	0.11	0.19	0.14	0.19
p_z	0.25	0.20	0.25	0.20	0.23
p_x	0.15	0.22	0.20	0.23	0.24
p_y	0.15	0.22	0.20	0.23	0.23
q_a: Fe	0.64	0.10	1.57	0	0.77
+x	−1.70	−0.22	−1.50[l]	−0.27	−0.40[m]
−x	−1.70	−0.22	−1.50[l]	−0.27	−0.40[m]
+y	−1.70	−0.22	−1.50[l]	−0.27	−1.77[m]
−y	−1.70	−0.22	–	−0.27	−1.77[m]
+z	–	−0.14[k]	−1.60	−0.03	−1.72
−z	–	−0.21	−1.60	−0.25	−1.72

[a] Taken from [6.32]; coordinates described in [6.39]
[b] Taken from [6.33]; coordinates described in [6.37]
[c] Taken from [6.39]; coordinates described in [6.39] (bipyramidal lattice site)
[d] Taken from [6.35]; coordinates described in [6.41]
[e] Taken from [6.36]; coordinates described in [6.41]
[f] Taken from [6.39]
[g] Taken from [6.37]
[h] Taken from [6.37, 41]
[i] Taken from [6.41]
[k] Charge of N in NO$^+$ is −0.14, and charge of O is +0.473
[l] The three oxygens in the xy-plane produce D$_{3h}$ point symmetry of the central iron
[m] The two chlorines and the two oxygens in the xy plane are situated between the x and y-axes

6.4.3. Temperature Dependent Quadrupole Splitting

In α-FeSO$_4$ [6.47] and several other ferrous high-spin compounds [6.48–52] the quadrupole splitting was found to be considerably temperature dependent. This is mainly due to the fact that many-electron terms Ψ_i (i = 1, ..., 5) are energetically close to each other, and therefore allow thermal electronic population according to Boltzmann statistics. Ingalls [6.48] and other authors [6.49, 53] derived an expression for $\Delta E_Q(T)$ on the basis of crystal field arguments, which may be written

$$\Delta E_Q(T) = \frac{2}{7} e^2 Q (1 - R) \langle r^{-3} \rangle \alpha^2 F(E_i, \alpha^2 \lambda_0, T). \tag{6.41}$$

The temperature sensitive factor F in (6.41) depends on the energy separations E_i between the Ψ_i (which are built up only from an iron 3d basis set in this approximation), on the isotropic covalency factor α^2 (which influences the expectation value of $1/r^3$ as well as the free ion spin-orbit coupling constant λ_0), and on the temperature T. The main limitation of this approach is due to the rigorous assumptions made on α^2, on the iron 4p, on the overlap contribution, and on the lattice (ligand) contribution. Thus, this level of approximation for calculating $\Delta E_Q(T)$ seems only plausible for highly symmetric and highly ionic compounds.

A more rigorous analysis of $\Delta E_Q(T)$ may be carried out using many-electron cluster wave functions Ψ_i with the inclusion of all iron and ligand valence orbitals as basis set wave functions ψ_μ. For ferrous high-spin compounds the Ψ_i are characterized by a total spin of S = 2. The fivefold spin degeneracy of each Ψ_i is lifted by spin-orbit interaction, expressed by the operator $\hat{H}_{s.o.}$, which may lead to a considerable mixing of Ψ_i terms. The resulting eigenvectors $|e_\alpha >$ of this problem are linear combinations

$$|e_\alpha > = \sum_{im} C_{\alpha im} \Psi_i, \tag{6.42}$$

with m = 2, 1, 0, −1, −2. The coefficients $C_{\alpha im}$ and energies E_α are obtained from diagonalization of the total Hamiltonian including $\hat{H}_{s.o.}$. Since $\hat{H}_{s.o.}$ can be expressed in terms of single-electron operators [6.54], the matrix elements $\langle \Psi_{im} | \hat{H}_{s.o.} | \Psi_{jm'} \rangle$ can be derived in a similar way as described by (6.7). Thus, the remaining problems for calculating $\Delta E_Q(T)$ are

(i) to calculate the relevant (V_{pq}) tensor for each state $|e_\alpha >$

$$(V_{pq})^\alpha = \langle e_\alpha | (\hat{V}_{pq}) | e_\alpha \rangle, \tag{6.43}$$

(ii) to temperature-average the tensors $(V_{pq})^\alpha$ of the individual substates $|e_\alpha >$ according to Boltzmann statistics (if we are in the fast relaxation limit)

$$\langle (V_{pq}) \rangle_T = \sum_\alpha (V_{pq})^\alpha \exp(-E_\alpha/kT) / \sum_\alpha \exp(-E_\alpha/kT), \tag{6.44}$$

and

(iii) to diagonalize $\langle (V_{pq}) \rangle_T$, which leads to $V_{zz}(T)$, $\eta(T)$, to the orientation of the principal axes system of the EFG, and finally through (3.39) to $\Delta E_Q(T)$.

For α-FeSO$_4$, which was represented by a $[Fe(SO_4)_6]^{10-}$ cluster with coordinates as derived from x-ray structure analysis [6.55], MO calculations and the subsequent computation of $\Delta E_Q(T)$ were carried out [6.56] and compared with experimental $\Delta E_Q(T)$ values (Fig. 6.7). The three solid curves in Fig. 6.7 correspond to three dif-

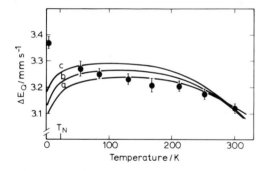

Fig. 6.7. Temperature-dependent quadrupole splitting, $\Delta E_Q(T)$, for α-FeSO$_4$. Experimental values are taken from [6.47]. Solid curves are taken from [6.56] and calculated as described in the text; they correspond to spin-orbit coupling constants λ of -100 cm^{-1} (a), -90 cm^{-1} (b), and -80 cm^{-1} (c). $T_N =$ 23.5 K (Néel temperature)

ferent spin-orbit coupling constants λ. (Similar results have been obtained for other ferrous high-spin compounds like deoxy-hemoglobin [6.11] and FeF$_2$ [6.57]). In the paramagnetic region of α-FeSO$_4$, $\Delta E_Q^{calc}(T)$ and $\Delta E_Q^{exp}(T)$ values coincided nearly quantitatively.

The disagreement (cf. Fig. 6.7) of the calculated ΔE_Q value of 3.15 mm s^{-1} at 4.2 K with the measured quadrupole splitting of 3.37 $\pm$ 0.024 mm s^{-1} (at the same temperature) finds a plausible explanation from inspection of the low-temperature properties of α-FeSO$_4$, i.e., from magnetically induced quadrupole splittings, which will be discussed in the following section.

6.4.4. Magnetically Induced Quadrupole Splitting

At temperatures $T < T_N = 23.5$ K, α-FeSO$_4$ exhibits antiferromagnetic behaviour [6.58]. From Mössbauer powder spectra at 4.2 K [6.47], information was obtained about the quadrupole splitting ΔE_Q, the asymmetry parameter η, the magnitude of the internal magnetic field $\vec{H}^{int}$ at the nuclear site of iron, and the orientation of $\vec{H}^{int}$ relative to the principal axes system of (V_{pq}) (polar angles ϑ^H and φ^H)

(i) $\vartheta^H = 25.4 \pm 0.5°$, $\varphi^H = 0°$, $\eta = 0.29 \pm 0.08$.

(ii) $\vartheta^H = 18.7 \pm 0.5°$, $\varphi^H = 90°$, $\eta = 0.77 \pm 0.08$,

and $\Delta E_Q(4.2K) = 3.37 \pm 0.02K$ mm s^{-1}, $H^{int}(4.2K) = 21.2 \pm 0.3$ Tesla.

The ambiguity of finding two possible parameter sets comes from the fact that powder spectra deliver only a limited amount of information. Taking the neutron diffraction result [6.59] that the spin direction in α-FeSO$_4$ is parallel to the b-axis, and making use of the condition that one of the principal axes of the electric field gradient tensor must be parallel to the twofold axis (a), the manyfold of possible fit-solutions of the powder spectrum could be reduced to the two cases (i) and (ii) above.

The presence of a magnetic field $\vec{H}$ reorients the spin states $|e_\alpha>$ according to the interaction of the electron spin $\vec{S}$ with $\vec{H}$. Depending on the magnitude of $\vec{H}$ we may therefore arrive at appreciable mixing among the zero field states $|e_\alpha>$. Thus the

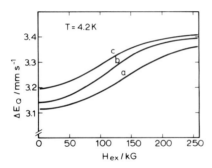

Fig. 6.8. Calculated quadrupole splittings, ΔE_Q, at 4.2 K for α-FeSO$_4$, depending on the magnitude of the exchange field, H^{exch} (taken from [6.56]). Curves a, b, and c correspond to λ-values as given in Fig. 6.7

quadrupole splitting under the presence of $\hat{H}$ might be different from that in zero magnetic field. Fig. 6.8 shows the field dependence of ΔE_Q (4.2 K) from calculations in which the electronic Hamiltonian contains also the interaction $\mu_B(\vec{L} + 2\vec{S})\vec{H}$ [6.56].

In the molecular field approximation for ferromagnetic or antiferromagnetic materials the exchange interaction is represented by an exchange field $\vec{H}^{exch}$. This field in the present type of approximation is proportional to the temperature average of S_z at temperature T [6.60–62]

$$H^{exch} = h\langle\hat{S}_z\rangle_T, \tag{6.45}$$

where h is a constant. The particular value of $\langle\hat{S}_z\rangle_T$ is obtained by the self-consistency condition of molecular field theory [6.60–62]. In the present case this requirement takes the following form: at temperature T the value of H^{exch} inserted in the electronic Hamiltonian in order to calculate the reorientation of $|e_\alpha\rangle$ must be such as to yield a value of $\langle\hat{S}_z\rangle_T$ satisfying (6.45). The solution of this problem can be found graphically (Fig. 6.9). The intersections of the straight line (care of (6.45)) and the $\langle\hat{S}_z(H^{exch})\rangle_T$ curves yield the self-consistent values $\langle\hat{S}_z\rangle_T$ for different temperatures T. The straight line is determined by the slope of $\langle\hat{S}_z(H^{exch})\rangle_T$ at T = T$_N$ and H^{exch} = 0

$$\frac{d\langle\hat{S}_z\rangle_T}{d\,H^{exch}} = \frac{1}{h}. \tag{6.46}$$

From Fig. 6.9 one finds for T = 4.2 K

$\langle\hat{S}_z\rangle_{4.2\,K}$ = 1.98 and H^{exch} = 23.0 Tesla.

Going with this value of H^{exch} into Fig. 6.8, the magnetically induced quadrupole splitting ΔE_Q at 4.2 K takes the values 3.34, 3.38, and 3.39 mm s^{-1} for the cases a, b, and c, respectively. This result now is in very good agreement with the experimental low-temperature value $\Delta E_Q^{exp}(4.2)$ = 3.37 ± 0.024 mm s^{-1}. From the calculated asymmetry parameter, η = 0.2 ± 0.05 (under the presence of H), it is fur-

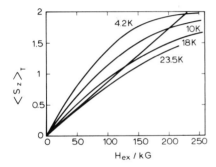

Fig. 6.9. Graphic solution for self-consistent values of $\langle S_z \rangle_T$ and $H^{exch}(T)$ in the molecular field approximation. The picture corresponds to case b of Figs. 6.7 and 6.8 (nearly the same graphs result for cases a and c) (taken from [6.56])

ther found that Wehner's [6.47] parameter set (i) describes the low-temperature properties of α-FeSO$_4$ more realistically than the set (ii).

From the present section it is concluded that high magnetic fields (> 10 Tesla), applied externally or present internally like in α-FeSO$_4$ below 23.4 K, may drastically influence the low-temperature properties of (V_{pq}), especially in cases where spin-orbit coupling is present. As a further example of magnetically induced nuclear quadrupole interaction Kamal et al. [6.63] have investigated cubic iron (II) systems.

6.5. Magnetic Hyperfine Interaction

6.5.1. General

The magnetic hyperfine interaction is represented by the operator $\hat{H}_M$ [6.30]

$$\hat{H}_M = -g_N \mu_N \sum_{p=x,y,z} \hat{H}_p^{eff} \hat{I}_p, \tag{6.47}$$

which describes the interaction between $\vec{H}^{eff}$, the effective magnetic field at the nucleus, and $\vec{I}$, the nuclear spin. g_N is the nuclear g-factor and μ_N the nuclear magneton.

The problem of how to calculate Mössbauer spectra if $\vec{H}^{eff}$ is known, or the reverse case of how to find $\vec{H}^{eff}$ from an experimental Mössbauer spectrum, has been discussed for ^{57}Fe in Section 3.2. In the present section we are concerned with the various contributions to $\vec{H}^{eff}$.

$\vec{H}^{eff}$ is the vector sum of the externally applied magnetic field $\vec{H}^{ext}$ and the internal magnetic field $\vec{H}^{int}$

$$\vec{H}^{eff} = \vec{H}^{ext} + \vec{H}^{int}. \tag{6.48}$$

The latter consists of basically three parts [6.64]

$$\vec{H}^{int} = \vec{H}^L + \vec{H}^d + \vec{H}^c. \tag{6.49}$$

(i) $\vec{H}^L$ is the contribution from the orbital motion of the electrons (orbital term)

$$\hat{H}_p^L = -2\mu_B \langle r^{-3}\rangle_L \, \hat{L}_p, \tag{6.50}$$

with L_p being the p(x, y, z)-component of the expectation value of the orbital angular momentum operator $\hat{L}$ and μ_B the Bohr magneton. The orbital momentum L may be quenched by the influence of the ligand field; in such cases $\vec{H}_L$ becomes zero.

(ii) $\vec{H}^d$ is the contribution of the magnetic moment of the spin of the electrons outside the nucleus (spin-dipolar term)

$$\hat{\vec{H}}^d = 2\mu_B \left(\frac{3\hat{\vec{r}}\,(\hat{\vec{S}}\hat{\vec{r}})}{r^5} - \frac{\hat{\vec{S}}}{r^3} \right), \tag{6.51a}$$

which is equivalent to

$$\hat{H}_p^d = 2\mu_B \sum_{q=x,y,z} \frac{3\hat{p}\hat{q} - r^2 \delta_{pq}}{r^5} \, \hat{S}_q. \tag{6.51b}$$

Using the operator equivalence [6.30], which leads us from (6.34) to (6.36), we derive for (6.51b)

$$\hat{H}_p^d = -2\mu_B \xi \langle r^{-3}\rangle_d \sum_{q=x,y,z} \left[\frac{1}{2}(\hat{l}_p\hat{l}_q + \hat{l}_q\hat{l}_p) - \frac{l(l+1)}{3}\delta_{pq} \right] \hat{S}_q, \tag{6.52}$$

with $\xi = \dfrac{2}{7}$ for d-electrons. S_q is the q(x,y,z)-component of the electron spin $\vec{S}$. The values to be used for $\langle r^{-3}\rangle_L$ and $\langle r^{-3}\rangle_d$ are effective values which should include shielding corrections and potential effects from different electronic configurations as described in (1) of Section 6.4.1. Strictly the $\langle r^{-3}\rangle$ values involved in H^L and H^d are different from each other and from that used in (6.36); however, Freeman and Watson [6.64] have calculated $\langle r^{-3}\rangle$ values for the two interactions and have shown that they are the same to within about 10%. Thus one may use the value of 56 Tesla for both, $2\mu_B\langle r^{-3}\rangle_L$ and $2\mu_B\langle r^{-3}\rangle_d$.

(iii) $\vec{H}^c$ is the contribution of the spin-density at the nucleus (Fermi contact term) [6.65]

$$H^c = \frac{8}{3}\pi \mu_B \, g(\rho^\uparrow(o) - \rho^\downarrow(o)), \tag{6.53}$$

with $\rho^\uparrow(o)$ and $\rho^\downarrow(o)$ being the spin-up and spin-down spin densities at the nucleus, respectively. $\vec{H}^c$ turns out to be parallel to the external field $\vec{H}^{ext}$ if $\rho^\uparrow(o) > \rho^\downarrow(o)$ is valid, and vice versa.

All three contributions to $\vec{H}^{int}$ can in principle be derived from (6.7) using appropriate operators $\hat{o}$ and the formalism described in Sections 6.3 and 6.4. The similarity in computing $\rho(o)$ and H^c especially can be seen from (6.53). Thus it is obvious that the effective magnetic field at the nucleus, $\vec{H}^{eff}$, and through (6.47) the magnetic hyperfine interaction depend considerably on the many-electron wave function Ψ (which describes the electronic structure and the chemical environment of the Mössbauer probe), i.e., on symmetry, covalency, valency, spin, temperature, etc.

In several publications the magnetic coupling between the electrons and the nucleus is expressed in terms of the hyperfine interaction tensor $\widetilde{A}$

$$\hat{H}_M = \hat{\vec{S}}\,\widetilde{A}\,\hat{\vec{I}} - g_N \mu_N \hat{\vec{H}}^{ext}\hat{\vec{I}}. \tag{6.54}$$

Comparing (6.47) and (6.54) we find that the internal field $\vec{H}^{int}$ is related to $\widetilde{A}$ by

$$\vec{H}^{int} = -\frac{\widetilde{A}\,\langle\hat{\vec{S}}\rangle}{g_N \mu_N}.$$

The description of $\hat{H}_M$ by $\widetilde{A}$ is useful in cases where $\widetilde{A}$ is supplied from other measurements (for example ENDOR spectroscopy). If this is not so, it seems more favourable for the understanding of the electronic and magnetic structure of the compound under study to use the formalism described by (6.7) and (6.49) through (6.53), rather than to lump everything together in $\widetilde{A}$.

The magnetic hyperfine interaction of many iron-containing chemical compounds has been studied so far; to our knowledge, however, in nearly all cases the many-electron cluster wave functions Ψ of (6.7) have been approximated drastically. In most of the relevant publications there was made use of ligand field theory, i.e., Ψ has been cut down to iron-only orbitals, and the next-nearest neighbour and nearest neighbour effects have been packed into isotropic covalency parameters. This type of approximation is only useful to derive a qualitative understanding of the electronic and magnetic structure of a compound and gives reasonable results especially for ionic materials. Examples in this field have been the investigation of the ferrous high-spin compounds $FeSiF_6 \cdot 6\,H_2O$ [6.66, 67], $[FeL_3] \cdot (ClO_4)_2$ with $L = \alpha,\alpha'$−dipyridyl [6.68], $KFeF_3$ [6.62], $RbFeF_3$ [6.62], $Mn(Fe)CO_3$ [6.69], of an $Fe^{(IV)}$ $(S = 1)$ compound [6.70] (which exhibits an interesting temperature dependence of $\vec{H}^{int}$), of several biological ferric high-spin compounds [6.71], etc. For a complete reference list we refer to the Mössbauer Data Index [1.6].

6.5.2. Example (α-FeSO$_4$)

Considering α-FeSO$_4$ as an example we discuss the various contributions of the internal field, $\vec{H}^{int}$.

1. In an approximate description of $\rho^{\uparrow}(o) - \rho^{\downarrow}(o)$ similar to that for $\rho(o)$ in Section 6.3.2, H^c is given [6.15, 72, 73] by a reduced free ion contribution, $H_{Free} \cdot \langle S \rangle/S$, depending on the actual iron 3d spin $\langle S \rangle$ of the compound, by a covalency contribution, H_{cov}, which corrects $\rho^{\uparrow}(o) - \rho^{\downarrow}(o)$ due to overlap distortion effects, and which also accounts for direct valence contributions of the iron 4s shell, and by the supertransferred hyperfine field, H_{STHF}, which takes into account the Fe-ligand-Fe spin-coupling

$$H^c = H_{Free}\frac{\langle\hat{S}\rangle}{S} + H_{cov} + H_{STHF}. \tag{6.55}$$

The numerical values for H_{Free} have been calculated by Watson et al. [6.65]

$$H_{Free}(Fe\ 3d^6, S = 2) = -55.0\ \text{Tesla, and}$$

$$H_{Free}\left(Fe\ 3d^5, S = \frac{5}{2}\right) = -63.0\ \text{Tesla.}$$

These values have been reported to be approximately proportional to the spin value S [6.65, 74], leading to the numbers for $H'_{Free} = H_{Free}/2S$

$$H'_{Free}(3d^6) = -13.75\ \text{Tesla, and}$$

$$H'_{Free}(3d^5) = -12.60\ \text{Tesla}$$

per unpaired spin in the iron 3d shell. The reason for $|H'_{Free}(3d^6)|$ being larger than $|H'_{Free}(3d^5)|$ has to do with the radial distribution of the iron 3d-electrons with respect to the iron core ns electrons. 1s- and 2s-electrons with spin being parallel to the partially filled iron 3d-shell ($\uparrow$) are attracted towards the 3d-region thus decreasing the spin-up density and hence increasing the spin-down density at the nucleus. 4s-electrons are mainly lying "outside" the 3d-region, and therefore the exchange coupling has just the reverse effect: the 4s-polarization contributes with an increased spin-up density to H'_{Free}. The situation for 3s-polarization is more complicated, since the 3s-shell lies neither "inside" nor "outside" the 3d-shell. The outside part, however, gives a positive and predominant contribution to H'_{Free}. For iron $3d^5$, the individual 1s-, 2s-, and 3s-contributions (of full ns-shells) to the resulting H'_{Free} value of -12.6 Tesla per unpaired spin are -1.0, -35.8, and $+24.2$ Tesla, respectively. Going from $3d^5$ to $3d^6$ configuration increases the mean radius of the 3d-shell as we already know from the changes in $\langle r^{-3} \rangle$ from Section 6.4.1. Therefore the positive contribution of 3s-polarization and the absolute contribution of 1s- and 2s-polarization become weaker for $3d^6$ configuration compared with $3d^5$. The iron ns-contributions to H'_{Free} $= -13.75$ Tesla of the corresponding iron $3d^6$ case are -0.85, -32.6, and $+19.7$ Tesla, respectively. This situation is also reflected by the investigation of the internal field in α-$FeSO_4$, for which the electronic configuration $3d^{6.066}\ 4s^{0.132}\ 4p^{0.322}$ was found from MO calculations [6.56, 75], and for which an effective spin value of 1.98 was derived [6.76] (compare with Sec. 6.4.4). With these numbers the contribution from the core polarization of iron ns AO's (n = 1, 2, 3) to H^c turns out to be -54.4 Tesla.

The dominant contribution to H_{cov} comes from the spin-polarization of the iron 4s-electrons [6.15, 73]; a full 4s-shell is expected to give a field of about $+12$ Tesla per unpaired 3d-spin [6.15, 65]. In iron oxides the charge transfer into the 4s-shell turns out to be about 0.15 electron charges [6.75] from configuration interaction calculations. Thus in the case of α-$FeSO_4$, where we have $q_{4s} = 0.132$ and $\langle S \rangle = 1.98$, the contribution H_{cov} to H^c is 6.2 Tesla.

Extending the cluster under study beyond the next-nearest-neighbour shell of the Mössbauer nucleus (in our example beyond the SO_4-ligands), H^c may also include the so-called supertransferred hyperfine field, H_{STHF}. (It is called supertransferred because the spin-density here is transferred from a distant ion to the central iron ion via an intermediate ion in a similar process as that proposed for superexchange). This feature has been extensively discussed by Šimanek et al. [6.73] and Sawatzky et al. [6.15], who studied clusters composed of a central ferric high-spin iron ion in an octahedron of six O^{2-} ions which in turn were surrounded by six ferric high-spin iron ions. There H_{STHF} is caused by the overlap distortions of the central iron s-orbitals by the ligand O^{2-} orbitals which have been unpaired by transfer into unoccupied 3d-orbitals of the neighboring iron ions. For the ferric iron-oxygen-iron cluster in $LaFeO_3$, H_{STHF} was estimated to be $H_{STHF} = -5.5$ Tesla [6.73]. The direction of H_{STHF} is parallel to the magnetic moment of the neighboring cations, and therefore enhances the contact field H^c if the magnetic moments in Fe-ligand-Fe are coupled antiferromagnetically. Since the nature of magnetic ordering in α-$FeSO_4$ is known to be of the antiferromagnetic type (see Sec. 6.4.4), it is clear that H_{STHF} is negative in this case; its absolute value of 0.15 Tesla is relatively small, because there are only two Fe-O-Fe chains per cluster.

2. The orbital contribution H^L to H^{int} is given by (6.50). Because of the relatively large energy splitting of orbital terms in α-$FeSO_4$ (i.e., weak spin-orbit coupling) [6.56], it is a reasonable approximation for low temperatures to derive $\langle \hat{L}_p \rangle$ in (6.50) in terms of $\langle \hat{S}_p \rangle$ by [6.77] (the same procedure was applied to other ferrous high-spin compounds [6.78])

$$\langle \hat{L}_p \rangle = (g_{pp} - 2) \langle \hat{S}_p \rangle, \tag{6.56}$$

with

$$g_{pp} = -2\lambda \sum_{k \neq 1} \frac{|\langle 1 | \hat{L}_p | k \rangle|^2}{E_k - E_1} . \tag{6.57}$$

$|1>$ is the many-electron orbital ground state, and $|k>$ stands for excited orbital states; λ is the spin-orbit coupling constant. (The approximation in deriving $\langle \hat{L}_p \rangle$ from (6.56), however, is no longer valid if g is not close to the free-electron value as for example in ferricyanides, as described in Section 6.5.3.) For α-$FeSO_4$, H_p^L takes the values $H_z^L = +17.7$ Tesla and $H_x^L = H_y^L = 0$.

3. The dipolar contribution H^D to H^{int} is described by (6.52). In α-$FeSO_4$ the dipolar term H^D is the only one which contributes with more components than the z-component only: $H_z^D = 11.3$ Tesla, $H_x^D = -4.3$ Tesla, $H_y^D = 0$.

From these results for α-$FeSO_4$ the following conclusions may be drawn:

(i) The main contribution to H^{int} comes from the Fe ns (n = 1, 2, 3) polarization; therefore $H_z^{int} < 0$.

(ii) The contributions from H_z^L and H_z^D are positive and of the same order of magnitude as in a similar compound (ferrous fluorosilicate hexahydrate [6.66]).

(iii) Even small amounts of charge transfer into the Fe 4s atomic orbital give considerable H_{cov} values owing to Fe 4s-polarization.

(iv) H_{STHF} turns out to be negative because of antiferromagnetic spin coupling in α-$FeSO_4$ [6.58]. Its absolute value is relatively small since we are concerned with only two Fe-O-Fe chains per cluster.

(v) The calculated absolute value H_{calc}^{int} = 21.3 Tesla is fairly close to the experimental value H_{exp}^{int} (4.2 K) = 21.2 ± 0.3 Tesla.

(vi) The deviation of H^{int} from the spin direction is due to the dipolar term H_x^D.

(vii) The calculated angle between the z axis of the principal axes system of the EFG and $\vec{H}^{int}$ is ϑ_{calc} = 27.5°, compared to the experimental value ϑ_{exp} = 25.4° ± 0.5° (see Sec. 6.4.4).

6.5.3. Example (Ferricyanide)

From the investigation of ferricyanide there is considerable contradiction in the literature concerning the sign of the experimentally determined value H_{exp}^{int} (either +25.5 Tesla [6.80, 81] or −25.5 Tesla [6.79]) and therefore also concerning the various calculated contributions to H_{cal}^{int}.

Hrynkiewicz et al. [6.79] interpret their experimental H_{exp}^{int} values for various ferricyanides (Table 6.3) in terms of iron core ns-polarization and supertransferred hyperfine contributions. They quote that the values between −16.0 and −27.0 Tesla, which are relatively large compared to the free ion $3d^5$ value of −12.6 Tesla, are due to two effects:

(i) Since it is known from MO calculations [6.18, 84] that the electronic configuration in ferricyanides is rather $3d^7$ than $3d^5$, the enhanced H^{int} field with respect to the value of $H_{Free} \left(3d^5, S = \frac{1}{2} \right) = -16.6$ Tesla seems to be partially due to an increased mean radius of the 3d-shell.

(ii) The contribution H_{STHF} to H^c is assumed to be positive for the various ferricyanides in Table 6.3 in order to explain the tendency that H^{int} becomes smaller if metal ions with increased net spin are substituted. This assumption, however, would imply that the spin-spin coupling is of the ferromagnetic type, in contradiction to the findings of other authors.

Hazony's [6.85] interpretation of H_{exp}^{int} is more extended compared to the work of Hrynkiewicz; however it uses approximations for the estimate of H^L of

Table 6.3. Experimental values[a] for H^{int} in $Me_3[Fe(CN)_6]_2$; taken from [6.79]

Me	Transition temp. (K)	H^{int}(Tesla)[b]
$Mn^{II}(S = 5/2)$	15.8 ± 0.5	−19.5 ± 1.5
$Co^{II}(S = 3/2)$	> 4.2	−16.0 ± 1.5
$Ni^{II}(S = 1)$	18.9 ± 0.5	−26.9 ± 1.0
$Cu^{II}(S = 1/2)$	16.2 ± 0.7	−26.6 ± 1.0
$Fe^{III}(S = 1/2)$	130	−25.5

[a]   Comparable investigations have been performed with iron oxygen compounds [6.88] in order to study the influence of H_{STHF} on H^{int} by substituting non-magnetic ions for the paramagnetic iron ions

[b]   The sign of H^{int} was reported to be positive in more recent publications [6.80, 81]

the type mentioned in connection with (6.56) in Section 6.5.2. Ferricyanides are characterized by relatively strong spin-orbit coupling [6.84, 86], thus (6.56) no longer is a fair approximation for $\langle \hat{L}_p \rangle$.

Lang et al. [6.87] gave a more rigorous interpretation of their experimental H_{exp}^{int} value, which they quote to be *positive*. [A positive sign of H_{exp}^{int} is consistent with the findings of other authors [6.80, 81], that the magnetic ordering of ferricyanide is of the antiferromagnetic type [6.82, 83] (see the argument (ii) from above)]. They do not use the approximations involved in Hazony's work; however, they neglect covalent and supertransferred hyperfine contributions to H^c; and their Fermi contact contributions seem to be somewhat too small, since they do not take into account the enhancement of H_{Free} due to an increased mean radius of the iron 3d-shell for the typical iron $3d^7$ configuration of ferricyanides.

6.6. Theoretical Methods for the Description of the Electronic and Magnetic Structure

At this stage it seems appropriate to give a brief survey of methods for the description of the electronic (and magnetic) structure of a compound.

1. Crystal field approach [6.49, 89–91]. Works with a very limited basis set (in the case of Fe containing compounds only with Fe 3d AO's), neglects valence orbitals like Fe 4s and Fe 4p AO's and Fe core orbitals, and takes into account nearest neighbor effects through an isotropic covalency parameter. Advantage: easy to include spin-orbit coupling, useful to study trends of temperature-dependent quadrupole splittings and energy term schemes of highly ionic compounds (was especially applied to ferrous high-spin compounds), not expensive.

2. Heitler-London approach [6.92]. Is an extension of (1) insofar as molecule A (for example heme) and molecule B (for example O_2) are regarded as "atoms" constituting the "bi-atomic molecule" AB (heme-O_2) in an explicit MO-procedure. Advantage: takes into account at least one covalent bond explicitly, handles a fairly large number of electronic configurations without being expensive.

3. MO-cluster approach. Takes into account the whole molecule or at least part of it (up to about 50 atoms or 200 basis AO's).

a) The semi-empirical methods (like "self-consistent charge" calculations [6.18, 32, 93, 94], are based on empirical results of isolated atoms, ions, or radicals and neglect or approximate various parts of multicenter integrals. Advantage: applicable to relatively large systems without being extremely computer time consuming.

b) First principle calculations have in common an extended AO basis set (including core orbitals) and a Hamiltonian, the various interaction parts of which are calculated numerically on the basis of theoretical methods like Hartree-Fock, Hartree-Fock-Slater, Dirac-Fock, Dirac-Slater methods, etc.; a detailed description of these methods is given in [6.95]. Advantage: self-consistent method, independent from experimental input parameters, direct calculation of $\rho(o)$ and V_{pq} without subsequent overlap distortion, potential distortion, or shielding correction is possible; however, very expensive for large clusters.

4. Energy band calculations. Take into account the whole periodic solid; periodic potentials are included, wave functions reflect the periodicity of the solid (Bloch functions). One distinguishes between empirical approaches and first principles calculations like under 3. Comparisons between first principles calculations and experimental observables are very scarce. Since it is generally too time consuming to follow the Hartree-Fock scheme in computing potentials and exchange integrals within a N-electron system of a solid, one normally assumes a set of ordinary and exchange potentials and ends up with a one-electron Schrödinger equation. This has the advantage that one stays within a reasonable time limit for an energy band structure calculation; however, of course this leads to a lack of generality. (Details of band theory and its various approaches can be found in textbooks, for example [6.96].)

6.7. Fluctuations and Transitions

From the discussion of electric and magnetic hyperfine interactions so far, one might be tempted to believe that time-dependent effects in Mössbauer spectroscopy are unusual, and are an exception to the "normal" static situation. This is not the case; the importance of fluctuations of charge density, electric field gradient and magnetic field for the analysis of Mössbauer spectra is reflected by the numerous publications (see Mössbauer Data Index) in this field. For the present purpose we only give a rough explanation (i) of the influence of relaxation upon Mössbauer spectra, and (ii) of the main relaxation mechanisms. For a more detailed study of these effects we refer to the literature [6.97–106].

6.7.1. Simple Example for a Mössbauer Relaxation Spectrum

Taking the electron charge density $\rho(o)$, the electric field gradient V_{pq}, or the magnetic field $\vec{H}^{eff}$ at the nucleus to be time dependent, then the nuclear Hamiltonian will be so, too, and, of course, the resulting transition energies and intensities will also be time dependent. A simple example to illustrate the influence of relaxation upon Mössbauer spectra has been given by Wickmann and Wertheim [6.102]. Relaxation has been visualized there as an electronic spin-flip which results in a reversal of the magnetic hyperfine field $\vec{H}^{int}$; instead of (6.47) we then have (assuming the external field $\vec{H}^{ext}$ to be zero)

$$\hat{H}_M = -g_N \mu_N \sum_{p=x,y,z} f_p(t)\, \hat{H}_p^{int}\, \hat{I}_p. \tag{6.58}$$

With the electronic relaxation occurring between the $S_z = +1/2$ and the $S_z = -1/2$ states with a characteristic relaxation time τ, (6.58) reduces to

$$\hat{H}_M = -g_N \mu_N\, f(\tau)\, \hat{H}_z^{int}\, \hat{I}_z, \tag{6.59}$$

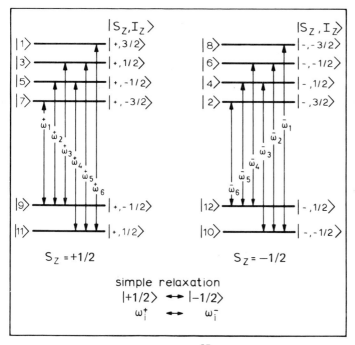

Fig. 6.10. Hyperfine levels appropriate to ^{57}Fe for the hyperfine field Hamiltonian of (6.59) (taken from [6.102])

where $f(\tau) = \pm 1$ as $S_z = \pm 1/2$. The result is a modulation of the energies of the nuclear eigenstates $|I, I_z>$ as illustrated by Fig. 6.10. Calculating Mössbauer spectra for different relaxation times τ (on the basis of NMR lineshape calculations; see Eq. 25 in [6.102c] for example) then leads for the present simple example to the relaxation spectra for ^{57}Fe as shown in Fig. 6.11. Static magnetic hyperfine splitting is observed when the Larmor precession time $\tau_{L,i}$, which is related to $\Delta\omega_i = |\omega_i^+ - \omega_i^-|$ through $\tau_{L,i} = 2\pi/\Delta\omega_i$, is smaller than $\tau (\tau \gg \tau_{L,i}$; *slow relaxation limit*, Fig. 6.11 a). The static Mössbauer spectrum consists then of a superposition of two spectra, one for each "polarity" of the internal field ($\pm H^{int}$), weighted according to the thermal population of the electronic states $S_z = +1/2$ and $S_z = -1/2$.

For shorter relaxation times, $\tau \sim \tau_{L,i}$, the overall width of the spectrum decreases and the individual lines become broader (Fig. 6.11 d, e, f). In this *intermediate relaxation range* the spectra cannot be described anymore by six Lorentzian lines with natural line width Γ (in the case of ^{57}Fe of the order of $\Gamma \sim 10^{-9}$ eV), because in this range the line shape calculations for the γ-transition under the presence of time-dependent hyperfine interactions lead to a spread in linewidths (this follows for example from the solution of relevant Bloch equations [6.102c].

In the *fast relaxation limit*, when $\tau \ll \tau_{L,i}$, the magnetic splitting collapses entirely (Fig. 6.11 h), since the time-average values of the spin and thus the magnetic field "seen" by the nucleus become zero: $\langle \hat{S}_z \rangle = \langle \hat{H}_z^{int} \rangle = 0$.

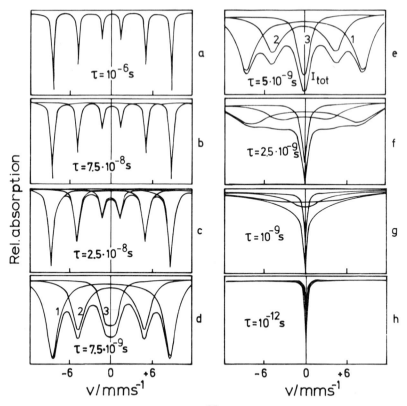

Fig. 6.11. Mössbauer relaxation spectra for ^{57}Fe for different relaxation times τ of the electronic transition $S_z = +1/2 \leftrightarrow S_z = -1/2$ (taken from [6.102]). Since we are concerned with three (i = 1, 2, 3) different frequency separations $\Delta\omega_i$ ($\Delta\omega_i = \Delta\omega_{7-i}$) in Fig. 6.10, and since the amount of motional narrowing of the spectrum depends on τ with respect to $\tau_{L,i} = 2\pi/\Delta\omega_i$ (see text), we expect for a fixed value of τ a differential collapse of the spectrum, indicated by three different lines i corresponding to $\Delta\omega_i$

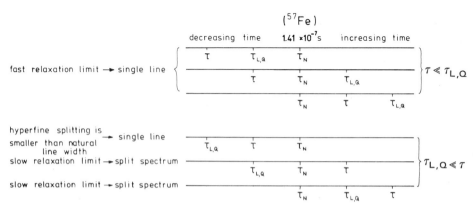

Fig. 6.12. Relation between nuclear lifetime τ_N (of the metastable nuclear state), Larmor (or quadrupole) precession time τ_L (τ_Q), and relaxation time τ for the condition of measuring a single-line spectrum or a hyperfine-split spectrum

To be more precise, we should add that $\tau \ll \tau_{L,i}$ and $\tau \gg \tau_{L,i}$ are not the only conditions for fast and slow relaxation; the consideration whether the system under investigation is in the frame of Mössbauer spectroscopy in the slow or fast relaxation limit has to include the lifetime of the excited nuclear state involved (in case of ^{57}Fe the lifetime of the first excited 14.4 keV nuclear state, $\tau_N = 1.41 \cdot 10^{-7}$ s) besides the Larmor precession time τ_L in magnetic compounds or the quadrupole precession time τ_Q in compounds with nonzero electric field gradient (see Fig. 6.12).

6.7.2. Relaxation Processes

In the following we briefly discuss the main relaxation mechanisms. Depending on the reservoir to which energy of the fluctuating spin system is transferred, one distinguishes different spin relaxation processes. Typical kinds of relaxations in solids together with their range of relaxation time and appropriate experimental tools for detection are summarized in Fig. 6.13.

6.7.2.1. Spin-Lattice Relaxation

In spin-lattice relaxation we are concerned with a transfer of energy between the spin system and the lattice, characterized by the spin-lattice relaxation time τ_1, i.e., the spin interacts via spin-orbit and orbital-phonon coupling with the phonons [6.102, 107]. And since the phonon spectrum is in principle related to fluctuations in the electrostatic crystal field potential (for example dynamic Jahn-Teller distortions or jump diffusion of vacancies, etc.) or to lattice vibrations, the spin itself becomes time dependent. Two general types of spin-lattice relaxation processes have been described in the literature [6.107]: an indirect process involving at least three spin states and two phonons, and a direct process involving a spin transition $|S_B> \rightarrow |S_A>$ with

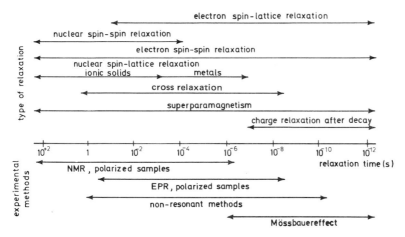

Fig. 6.13. Typical relaxation in solids (taken from [6.102a])

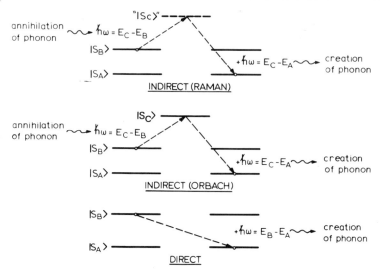

Fig. 6.14. Schematic illustration of typical spin-lattice relaxation processes. If the spin state $|S_C>$ is a well-defined eigenstate of the static electronic Hamiltonian, the relaxation is called Orbach process; if $|S_C>$ is a so-called virtual spin state, denoted by "S_C", the relaxation is called Raman process (taken from [6.102a])

the simultaneous creation of a lattice phonon of energy $E_B - E_A$ (Fig. 6.14). Since the temperature dependence of indirect and direct processes should be different according to theoretical considerations [6.108] (indirect processes dominate at high and direct processes at low temperatures), it may be possible to distinguish both mechanisms experimentally. For rare earth ions this experimental differentiation was reported to be successful [6.109]. Although spin-lattice relaxation has mainly been studied in the rare earth region, there are also examples for iron compounds.

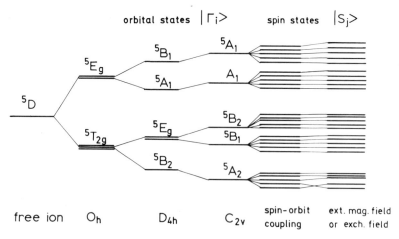

Fig. 6.15. Effect of crystal field distortions (O_h, D_{4h}, and C_{2v} symmetry), spin-orbit coupling, and magnetic field on the orbital and spin degeneracy of a ^5D state (e.g., Fe^{++} ion)

For a better understanding of the influence of spin-lattice relaxation processes upon Mössbauer spectra we discuss some examples involving ferrous high-spin iron, because the dominant relaxation mechanism for paramagnetic ferrous complexes is based on spin-lattice relaxation [6.110, 111]. In such complexes we have various possibilities to lift the orbital and spin degeneracy of the original ^{5}D state of the ferrous ion (Fig. 6.15), depending on the symmetry and the strength of the ligand field potential, on spin-orbit coupling, on random strains and on the magnitude of an externally applied magnetic field, or an internally produced exchange field. As the orbital states $|\Gamma_i>$, originating from the ^{5}D term in Fig. 6.15, are strongly coupled to phonons,[7] and since spin-orbit coupling is present, one does not observe magnetic hyperfine split (hfs) Mössbauer spectra of ferrous high-spin compounds (as far as spontaneous magnetization can be neglected like in α-FeSO$_4$ [6.56] or FeF$_2$ [6.112] above the Néel temperature, or in magnetically dilute systems like deoxyhemoglobin [6.11]) although the total spin of each spin state $|S_j>$ is 2, producing, via spin-polarization of Fe core ns AO's, an internal magnetic field at the iron nucleus of the order of -40 T (see Sec. 6.5), which, however, fluctuates with $\tau \ll 10^{-10}$ s due to direct spin-lattice relaxation.

Now the question may arise: why does one measure a quadrupole split spectrum in these compounds where the strong lattice-orbital and orbital-spin coupling cause the fluctuating spins to be in the fast relaxation range? The reason is the following: whereas a fluctuating spin averages to zero in the fast relaxation limit (Fig. 6.12), a fluctuating electric field gradient may average to a nonzero value in the fast relaxation range, because the time-averaged position of the Mössbauer atom can be such that it may be described by a noncubic static ligand field potential which then results in a nonvanishing electric field gradient. An example for this situation has been described by Kreber et al. [6.113] who investigated the bipyramidal lattice site in BaFe$_{12}$O$_{19}$ (Fig. 6.16). At low temperatures (T $<$ 60 K) the iron is located in one of the two equivalent sites (4e) 0.156 Å above or below the mirror plane, while at elevated temperatures (T $>$ 60 K) the iron atom oscillates between the two displaced positions. In the fast relaxation limit the iron now sees an "average environment" which is comparable to the environment seen by the iron located in the mirror plane (2b), and which produces a nonzero electric field gradient (as exemplified by MO

---

7   Matrix elements of the orbital-lattice interaction ($\hat{H}_{0\text{-lat}}$) are zero if L = 0. This situation is at first sight not very different for both cases, (i) a non-S-state iron ion the orbital angular momentum of which is quenched by the crystalline field, or (ii) a S-state iron ion. Nonzero matrix elements of $\hat{H}_{0\text{-lat}}$ are obtained for case (i) only after mixing in excited states via spin-orbit coupling. The excited states in case (i) may have the same total spin as the ground state; then there will be matrix elements of the orbital-lattice interaction between the ground state and the admixed excited states. Since the admixture is proportional to the inverse of the energy separation Δ of the excited state relative to the ground state, and to the spin-orbit coupling constant λ, the matrix element itself will be proportional to λ/Δ. In a S-state ion, however, the spin-orbit coupling will mix only states with different total spin values into the ground state. This implies that the matrix elements of the orbital-lattice interaction, which is independent of spin coordinates, will be of the order of $(\lambda/\Delta)^2$. Thus, relaxation times for S-state ions are longer (by about 10^2) than for corresponding non-S-state-ions at low temperatures. [Taken from M. Blume and R. Orbach Phys. Rev. *127*, 1587 (1962)].

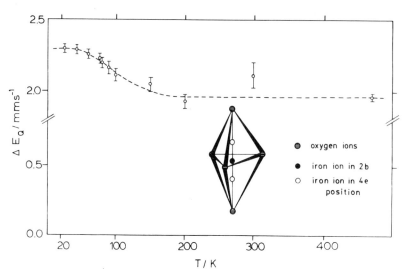

Fig. 6.16. The bipyramidal lattice site in $BaFe_{12}O_{19}$. Quadrupole splitting of the subspectrum corresponding to the bipyramidal lattice site (2b position) of $BaFe_{12}O_{19}$ as a function of temperature (taken from [6.113])

calculations [6.113]). This situation is reflected by the temperature dependence of the quadrupole splitting of the bipyramidal lattice site in $BaFe_{12}O_{19}$ (Fig. 6.16).

We now turn to the question why in many ferrous compounds (like in α-$FeSO_4$, FeF_2 and deoxyhemoglobin in the paramagnetic range) one measures only a single quadrupole doublet, although several of the spin states $|S_j >$ in Fig. 6.15 are populated at elevated temperatures, and the populated states $|S_j >$ may produce different partial electric field gradients, while in some other cases (like $Fe(phen)_2 (NCS)_2$, phen = 1,10-phenanthroline) one measures a superposition of two quadrupole doublets, though in a first approximation all iron sites are believed to be identical. The explanation of this situation is based on the possible (direct or indirect) relaxation between spin states $|S_j >$. To estimate the rate of relaxation among these states one has to calculate appropriate transition probabilities $W_{S_j S_k}$. For a direct process, which is relatively easy to handle, this would be [6.114]

$$W^{(1)}_{S_j S_k} = \frac{2\pi}{\hbar} |\langle S_j, N - 1|\hat{V}^{(1)}\epsilon|S_k, N\rangle|^2 \rho(\omega = \Delta/\hbar) \tag{6.60}$$

$$= \frac{2\pi}{\hbar} |\langle S_j|\hat{V}^{(1)}|S_k\rangle|^2 |\langle N - 1|\epsilon|N\rangle|^2 \rho(\omega = \Delta/\hbar),$$

where N denotes the occupation number of phonons, $\rho(\omega = \Delta/\hbar)$ the phonon density of final states, Δ the energy separation of states $|S_j >$ and $|S_k >$, and $\hat{V}^{(1)}\epsilon$ originates from an expansion of the ligand field potential $\hat{V}$ [6.114], which, besides the static contribution $\hat{V}^{(0)}$, contains higher terms $\hat{V}^{(1)}, \hat{V}^{(2)}$... describing orbital-lattice interactions

$$\hat{V} = \hat{V}^{(0)} + \hat{V}^{(1)}\epsilon + \hat{V}^{(2)}\epsilon^2 + \ldots . \tag{6.61}$$

The quantity ϵ represents the isotropic average of the strain tensor. Since $\hat{V}^{(1)}$ can be derived in terms of $\hat{D}^{(1)}S_jS_k$ in the spin hamiltonian approximation [6.117], the matrix elements $\langle S_j|\hat{V}^{(1)}|S_k\rangle$ in (6.60) reduce to $\langle j|\hat{D}^{(1)}|k\rangle S_jS_k$, with $|j\rangle$ and $|k\rangle$ being the orbital part of spin states $|S_j\rangle$ and $|S_k\rangle$, respectively, and with S_j and S_k standing for the effective spins of these states. (This includes the assumption that the effective spin can be approximated by S_z; $S_x = S_y \sim 0$.) The real part of the correlation function $W^{(1)}_{S_jS_k}$ yields the relaxation rate $1/\tau_1$

$$W^{(1)}_{S_jS_k} = \frac{1}{\tau_1} = \frac{3S_jS_k}{2\pi\hbar^4\rho_cv^5}\,|\langle j|\hat{D}^{(1)}|k\rangle|^2\Delta^3N(\Delta). \tag{6.62}$$

ρ_c is the density of the material under investigation, and v is the velocity of propagation of acoustic waves. The function $\Delta^3N(\Delta)$ represents the energy density of the phonons up to a constant factor; it is proportional to Δ^2 for $\Delta \ll kT$ and to Δ^3 exp $(-\Delta/kT)$ for $\Delta \gg kT$. With (6.62) we are able to understand (at least qualitatively) why one observes a superposition of two quadrupole doublets in cases where ferrous high-spin and ferrous low-spin states are energetically close together (Fig. 6.17): The relaxation rate $W_{{}^5B_2{}^1A_1}$ becomes zero, because the effective spin of the diamagnetic term $^1A_1$ is zero, thus no relaxation between $^1A_1$ and all other states takes place. However, the transition probabilities $W_{S_jS_k}$ with $S_j \sim S_k \sim 2$ among the spin-quintet states are certainly not zero; this results in a fast relaxation between the high-spin states $|S_j\rangle$ and $|S_k\rangle$ if at elevated temperature the states $|S_j\rangle$ and $|S_k\rangle$ are populated. The overall effect then is that one measures (i) a separate quadrupole doublet originating from the $^1A_1$ state and (ii) a separate quadrupole doublet originating from a thermal average of the populated high-spin states. The real situation, of course, is more complicated since besides direct processes indirect relaxation processes may also be involved; we shall come back to this later. Many examples have been reported, e.g., [6.115, 116] for the parallel existence of high-spin and low-spin quadrupole doublets in ferrous compounds, one of them is shown in Fig. 6.18.

The spin transition phenomena have been interpreted on the basis of various mechanisms; a somewhat deeper insight into this problem will be given in Section 6.7.3.

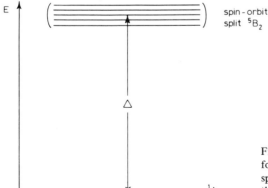

Fig. 6.17. Possible energy term scheme for ferrous compounds exhibiting high-spin low-spin transition spectra (Δ is of the order of kT)

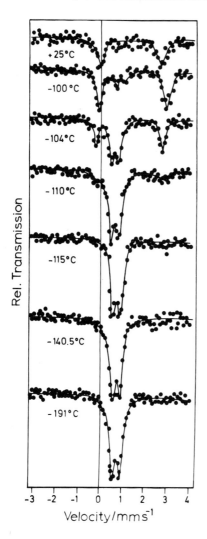

Fig. 6.18. Example for parallel existence of high-spin and low-spin quadrupole doublets: Mössbauer spectra of Fe(phen)$_2$(NCS)$_2$ as a function of temperature (taken from [6.116])

The application of an external magnetic field $\vec{H}^{ext}$ of appropriate magnitude to a ferrous high-spin compound forces the electronic spin to align itself more or less antiparallel[8] to $\vec{H}^{ext}$. Although the system may be with $|\vec{H}^{ext}|$ in the saturation range, one does not observe a superposition of five magnetic spectra if, for example, the five spin-orbit split states $|S_j >$ (j = 1, ... 5) of the $^5B_2$ orbital term are populated. Due to fast relaxation between the states $|S_j >$ one has to thermal average over the contributions to the internal magnetic field $\vec{H}^{int}$, as a result of which one measures a single Mössbauer spectrum according to $\langle \hat{H}^{int} \rangle_T$. This was the procedure of calculating $\langle \hat{S}_z \rangle_T$

---

8   The magnetic moment $\vec{\mu}$ orients itself parallel to $\vec{H}^{ext}$, but $\vec{\mu}$ and the electronic spin $\hat{S}$ are antiparallel: $\vec{\mu} = -g\beta\hat{S}$ with g being the electronic g-factor, and $\beta = e\hbar/2mc$ the Bohr magneton.

for α-FeSO$_4$ [6.56] in order to derive the exchange field $\vec{H}^{exch}$ and the magnetically induced quadrupole splitting in Section 6.4.4; this also explains the temperature dependence of $\vec{H}^{int}$ in the Fe (IV) (S = 1) compound [6.70] mentioned in Section 6.5.

We now come back to the indirect process. Zimmermann et al. [6.117] recently reported on the interpretation of experimental Mössbauer spectra obtained from a ferrous high-spin compound in an external magnetic field by taking into account direct and indirect processes at 4.2 K. Another example, where the indirect process plays an important role, has been described by Ham [6.118] and Chappert et al. [6.119] for ^{57}Fe in MgO and KMgF$_3$, respectively.

The transition probability $W^{(2)}_{S_jS_k}$ for a two-step process via real states $|S_t>$ (Orbach process) involves the product of transition probabilities $W^{(1)}_{S_jS_t}$ and $W^{(1)}_{S_tS_k}$

$$W^{(2)}_{S_kS_j} = \frac{W^{(1)}_{S_kS_t} \, W^{(1)}_{S_tS_j}}{W^{(1)}_{S_kS_t} + W^{(1)}_{S_tS_j}} \qquad (6.63)$$

In the case of ^{57}Fe in MgO and KMgF$_3$, where random strains lift the threefold degeneracy of the Γ_{5g} spin-orbit triplet, the electronic transitions among these three states may occur by way of the Orbach process involving the first excited spin-orbit level $|S_t>$ (Fig. 6.19a), leading to a relaxation rate of

$$\frac{1}{\tau_1} = \frac{6 \, G^2_{44} \, \Delta^3}{\pi \rho \hbar^4 v^5_T} \left| 1 + \frac{2}{3}\left(\frac{v_T}{v_L}\right)^5 \right| \exp(-\Delta/kT). \qquad (6.64)$$

$\Delta \sim 2|\lambda| \gg kT$ ($\lambda = 103$ cm^{-1} is the spin-orbit coupling constant of the free ferrous ion), ρ = density of the material, v_T and v_L are the velocities of propagation of transverse and longitudinal acoustic waves, G_{44} is the strain coupling coefficient for Fe^{2+} in MgO or KMgF$_3$. With the appropriate numerical values for G_{44}, Δ, v_T, v_L, and ρ (cf. [6.119]) it turns out for ^{57}Fe in KMgF$_3$ (for MgO similar results were obtained) that at 12 K the relaxation time τ_1 is equal to the nuclear quadrupole precession time $\tau_Q = \hbar/\Delta E_Q = 3 \cdot 10^{-8}$ s, with $\Delta E_Q = 0.44$ mm s^{-1}. From the temperature depen-

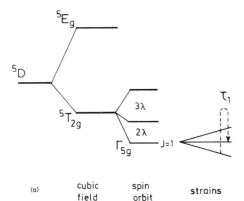

Fig. 6.19. (a) Energy scheme of the ^{5}D term of Fe^{2+} in MgO and KMgF$_3$

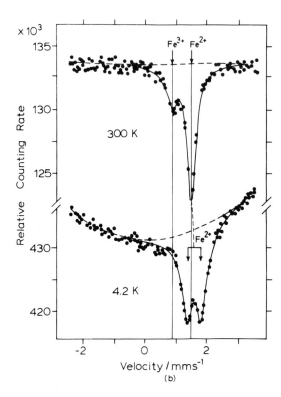

Fig. 6.19. (b) Experimental Mössbauer spectra of ^{57}Fe doped into
$KMgF_3$ at 300 K and 4.2 K showing
the onset of an Fe^{2+} quadrupole
splitting at low temperature (taken
from [6.119])

dence of τ_1 from (6.64) it follows that the system is in the slow relaxation limit below
12 K producing a quadrupole split spectrum, and in the fast relaxation limit above
12 K leading to a single line spectrum (Fig. 6.19b), because the thermal average of
the spin-orbit triplet Γ_{5g} produces a zero electric field gradient. The fact that Δ was
only about 100 cm^{-1} instead of $2|\lambda| \sim 200$ cm^{-1} was attributed to a Jahn-Teller
coupling between the various states involved [6.118, 119].

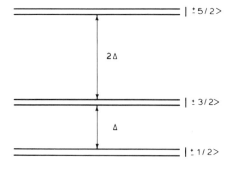

Fig. 6.20. ^6S state of high-spin Fe^{3+} split by
(axial) crystalline fields into three Kramers
doublets. The splittings are 2Δ and Δ, respectively, in case of including only the second-
order crystal field parameter B_2^0 in the crystal
field Hamiltonian, and neglecting the other
second-order term B_2^2 and fourth-other terms
[6.121]

6.7.2.2. Spin-Spin Relaxation

In spin-spin relaxation the energy transfer occurs between interacting spins (via dipole or exchange coupling) with relaxation time τ_2. Most of the relevant studies deal with compounds of high-spin iron (III) [6.102, 103, 120]. In these compounds the spin-lattice interaction is usually considered to be negligible compared to spin-spin inter-action, since the electronic ^6S state is characterized by vanishing orbital momentum, $\langle L \rangle = 0$ (see footnote 7 on page 89).

Concentration Dependence of Spin-Spin Relaxation

We first consider the case of magnetically dilute systems (like ^{57}Fe in Al_2O_3 in the concentration range of $< 0.1\%$ Fe), in which the paramagnetic Fe^{3+} ions are so far apart from each other that dipole or exchange coupling becomes negligible, and thus with spin-spin transitions we are in the slow relaxation limit. The ^6S state of a ferric high-spin ion may be split into three Kramers doublets either by crystalline fields or by second-order spin-orbit coupling with an energetically low-lying spin quartet. If the spin quartet is relatively far above the spin sextet (by some $1,000$ cm^{-1}) spin-orbit coupling may be neglected and the splitting is only due to second- (or higher) order effects of the crystal field [6.121]. In this case the zero field splitting between the $|\pm 5/2 >$, $|\pm 3/2 >$, and $|\pm 1/2 >$ Kramers doublets (Fig. 6.20) is of the order of 1 cm^{-1} and smaller (Fe^{3+} in Al_2O_3 [6.122]: $\Delta \sim 0.3$ cm^{-1}), and a magnetic hyperfine split (hfs) Mössbauer spectrum is measured [6.123] (Fig. 6.21a), which is composed of three separate spectra originating from the electronic $|\pm 5/2 >$, $|\pm 3/2 >$, and $|\pm 1/2 >$ states.[9]

In the first instant it may be surprising that degenerate Kramers doublets give rise to an effective magnetic field at the iron nucleus, and thus to a magnetic hfs spectrum. The situation, however, becomes transparent, if one keeps in mind that in solids the nuclear spin relaxes in times $\tau > 10^{-4}$ s only (cf. Fig. 6.13), and that the nuclear lifetime of the 14.4 keV state of ^{57}Fe is $0.98 \cdot 10^{-7}$ s. Therefore, the nuclear spin "sees" either the $|+S_z >$ or the $|-S_z >$ state ($S_z = 5/2, 3/2, 1/2$), if the fluctuation from $+S_z$ to $-S_z$ is slow compared to τ_L (Fig. 6.12). Since (i) we are in the slow spin-lattice relaxation range for $|\pm S_z >$ states (because of vanishing orbital momentum of the ^6S term), and since (ii) no relaxation between the three Kramers doublets can occur because the three doublets are already eigenfunctions of the crystal field potential and therefore transitions among these states, which require crystal field energy, are excluded, and finally since (iii) the paramagnetic spin system is very dilute (yielding vanishing spin-spin relaxation) the fluctuation of spins indeed is very slow. In total, one should observe six magnetic hfs spectra [6.123] in Al_2O_3 (0.08% Fe), but since the $+S_z$ spectrum is identical to the $-S_z$ spectrum only three separate spectra are observed (Fig. 6.21a).

---

[9] The spectrum corresponding to the electronic $|\pm 1/2 >$ state is a complicated 11-line-pattern due to the specific situation that the combined electronic-nuclear terms $|\pm S_z, \pm I_z >$ with $S_z = 1/2$ produce off-diagonal matrix elements of the Hamiltonian $\hat{H} = A [I_z S_z + 1/2 (\hat{S}_+ \hat{I}_- + \hat{S}_- \hat{I}_+)]$.

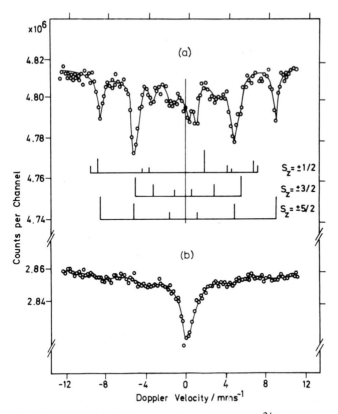

Fig. 6.21. (a) The hfs Mössbauer spectrum of 0.08% Fe^{3+} in Al_2O_3-powder at 78 K. The theoretical hyperfine structures for the three Kramers doublets are shown by stick spectra; they are obtained by assuming the centroid and the quadrupole interaction to be the same for all three spectra. (b) The effect of concentration of Fe^{3+} in Al_2O_3-powder at 78 K (0.9% Fe) on the spin-spin relaxation (see text) (taken from [6.123])

With increasing concentration of paramagnetic Fe^{3+} sites in the diamagnetic matrix Al_2O_3, one finds that dipole-dipole interaction between spins of adjacent ions more and more influences the relaxation rate of spins:[10] with increasing concentration the spin-spin relaxation time τ_2 becomes smaller, and in passing over a certain concentration threshold the system reaches the fast relaxation limit which is associated with an entire collapse of the magnetic hfs spectrum [6.123] (Fig. 6.21 b). A similar concentration dependence of spin-spin relaxation has been observed in hemin [6.123], in catalase [6.125], and in several other ferric high-spin compounds [6.126–131].

---

[10] It is interesting to note that for Fe^{3+} ions in frozen solutions a dependence on R^6 for the spin-spin relaxation time has been found; R is the distance between interacting spins. (See F. Sontheimer, H. Wegener and D. Seyboth in „Jahresbericht des Physikalischen Instituts der Universität Erlangen-Nürnberg (1975), p. 66.)

Temperature Dependence of Spin-Spin Relaxation

The system of Fe^{3+} in Al_2O_3 is an example for temperature independent spin-spin relaxation in compounds which are characterized by small zero-field splitting Δ, i.e., by negligible spin-orbit coupling between spin-sextet and spin-quartet states. In the following we describe an example where spin-spin relaxation can become temperature dependent if (second-order) spin-orbit coupling strongly dominates over crystal field effects on the splitting of the ^6S term into three Kramers doublets. In hemin [6.132] and also in $FeCl_3 \cdot 6H_2O$ [6.133] a relatively low lying excited spin-quartet term splits the ^6S term into Kramers doublets through spin-orbit interaction (Fig. 6.22a); the splitting parameter Δ is of the order of 10 cm^{-1}. At low temperatures only the $|\pm 1/2 >$ and $|\pm 3/2 >$ doublets are populated, which in this case are characterized by fast spin-lattice relaxation, because the spin-quartet $^4A_2$ (which by itself shows fast spin-lattice relaxation; $\langle L_{4A_2} \rangle \neq 0!$) is mixed into these doublets via spin-orbit inter-action. So, at low temperatures no magnetic hypferine split spectra but only a well resolved quadrupole doublet is measured [6.134] (Fig. 6.22b). With increasing tem-perature, however, the $|\pm 5/2 >$ doublet, which does not contain any admixture from the $^4A_2$ term and thus should be in the slow relaxation limit, becomes populated. Nevertheless one does not observe, at elevated temperature (300 K), a well resolved magnetic hfs spectrum due to the $|\pm 5/2 >$ state, but only a broadened line [6.134] (Fig. 6.22b). The reason is the following:

The concentration of paramagnetic Fe^{3+} sites in hemin (or $FeCl_3 \cdot 6 H_2O$) is high enough that spin-spin coupling between the spins of adjacent iron ions is pos-sible. With the assumption that the predominant mechanism for spin-spin transitions is due to dipole-dipole interaction between the spin (S) of the ion which contains the Mössbauer nucleus and the spin (S_j) of other ions in the sample, this coupling is pro-portional [6.132] to $r_j^{-5} [(\hat{S} \hat{S}_j) r_j^2 - 3 (\hat{S} \hat{r}_j) (\hat{S}_j \hat{r}_j)]$. Possible transitions will be in-duced by terms of the form $r_j^{-3} [(1 - 3 \cos^2 \theta_j) \cdot (\hat{S}_+ \hat{S}_{j-} + \hat{S}_- \hat{S}_{j+})]$. From this we see an important feature of spin-spin relaxation. In order for one ion to go from

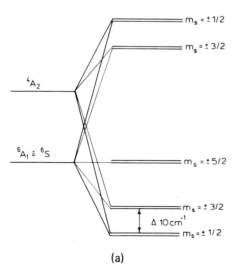

(a)

Fig. 6.22. (a) The low-lying energy terms ^6S and $^4A_2$ within the $3d^5$ configuration of hemin. The point symmetry is predominantly C_{4v}. The two terms are split into five Kramers doublets by virtue of spin-orbit interaction

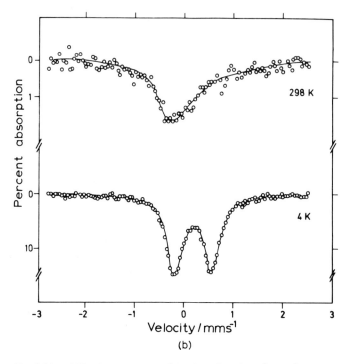

Fig. 6.22. (b) Mössbauer spectra of hemin (taken from [6.134])

the $|+5/2>$ to the $|+3/2>$ state simultaneously a neighboring ion must perform a transition from the $|+3/2>$ to the $|+5/2>$ state (due to the operator $\hat{S}_+\hat{S}_{j-}$), and accordingly, due to the operator $\hat{S}_-\hat{S}_{j+}$, the transition from the $|-5/2>$ to the $|-3/2>$ state on the one ion and simultaneously the transition from the $|-3/2>$ to the $|-5/2>$ state at a neighboring ion must take place.

6.7.2.3. Cross Relaxation

A further type of relaxation which involves mutual spin flips (which are nearly energy conserving) at different ionic sites in the lattice has been termed cross relaxation, with relaxation time τ_{CR}. In cases where the two sites are identical and the spin flip process is totally energy conserving – like in hemin, described in the preceeding Chapter – this relaxation is termed "resonant cross relaxation", other cases, where fluctuations are not entirely energy conserving, are simply called "cross relaxation". Examples of cross relaxation have been described by Wertheim et al. [6.123], by Date [6.131, 135] and by Bates [6.136]. Here we follow the work of Date and describe the situation of ^{57}Fe in Al_2O_3, which in the low concentration case shows a magnetic hfs spectrum (Fig. 6.21a). After heating in vacuo of the Al_2O_3 (^{57}Fe) absorber the original magnetic spectrum collapses and two quadrupole split spectra are measured (Fig. 6.23), one due to Fe^{3+} ions occupying substitutional (Al^{3+}) lattice sites in

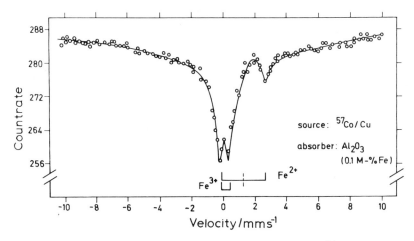

Fig. 6.23. The effect of heat treatment on the relaxation time for Fe^{3+} in Al_2O_3 at 80 K (compare with Fig. 6.21a). The present sample is vacuum heated; by this procedure interstitial Fe^{2+} sites are produced in the Al_2O_3 lattice in addition to the substitutional Fe^{3+} sites. The collapse of the original magnetic hfs spectrum of Fe^{3+} ions of Fig. 6.21a is due to cross relaxation between the spins of Fe^{3+} ions and those of Fe^{2+} ions (see text) (taken from [6.131])

Al_2O_3 and one due to Fe^{2+} ions occupying interstitial positions in the lattice. It should be mentioned again that the Fe^{3+} ions are "S" state ions with negligible spin-lattice relaxation ($\tau_1 \sim 10^{-5}$ s), and the Fe^{2+} ions are "D" state ions with considerable spin-lattice relaxation ($\tau_1 \ll 10^{-10}$ s). The absence of a magnetic hfs spectrum in this situation where Fe^{3+} and Fe^{2+} ions are simultaneously present in the lattice is ascribed to spin-spin interaction between the Fe^{3+} spin and the Fe^{2+} spin (a cross relaxation process similar to that described for hemin in the preceeding chapter), by virtue of which the relaxation time of the Fe^{3+} spin is drastically lowered.

6.7.2.4. Superparamagnetism

Mössbauer absorption spectra of ferri-, ferro-, or antiferromagnetically ordered samples usually show a static magnetic hfs spectrum, where $\overrightarrow{H}^{int}$ is caused by contributions which have already been discussed in Section 6.5. In ultrafine particles of magnetically ordered samples, however, the magnetization and therefore $\overrightarrow{H}^{int}$ may fluctuate very fast by thermal excitation. Due to anisotropies each magnetic one-domain particle may have one or more easy directions for the magnetization. (In an isotropic particle all directions would be equally probable for the magnetization vector.) Therefore the thermal fluctuations of the magnetization vector can be hindered, giving rise to an increase of the relaxation time; thus the magnitude of the relaxation time τ depends [6.137] on the balance of energies KV (anisotropy energy) and kT (thermal energy)

$$\tau = \frac{1}{f_0} \exp\left(-KV/kT\right). \tag{6.65}$$

K is a constant describing the lattice anisotropy, V is the particle volume, and **T** the temperature. For $KV/kT < 25$ the thermal fluctuations of the magnetization vectors take place without showing hysteresis effects [6.1.38]; this phenomenon is called "superparamagnetism", f_0 is a temperature-independent constant in the order of $10^8 - 10^{12}$ s^{-1}. More specifically, Krop et al. [6.139] conclude that (i) for particles with uniaxial anisotropy and exponential factor $KV/kT > 0.7$ the preexponential factor f_0 takes a value of the order of 10^9 s^{-1}, and (ii) for particles with cubic anisotropy (6.65) should only be used in the range $6 < KV/kT < 25$ with f_0 in the order of 10^8 s^{-1}. Again, as described before, the order of magnitude of the relaxation time τ with respect to the Larmor precession time τ_L and the nuclear lifetime τ_N (see Fig. 6.12) determines whether the system under investigation is in the fast, intermediate or slow relaxation limit. Measuring in the intermediate range has the advantage to determine for each temperature T the relaxation time $\tau(T)$ using Wickman's explicit formula for a relaxation dependent Mössbauer spectrum (see Eq. (25) in [6.102]). Thus by performing measurements at different temperatures T in the intermediate range it should be possible to determine f_0, K, and V.

Since τ is volume dependent, the Mössbauer spectra reflect the magnitude of the particle size V in the relevant absorbers. Takada et al. [6.140] have studied the Mössbauer absorption of α-FeOOH as a function of V. Fig. 6.24 shows their result obtained at room temperature, indicating that the critical volume, for which the motional narrowing of the magnetic hyperfine spectrum begins, is about $3 \cdot 10^{-17}$ cm^3 at 294 K for α-FeOOH.

In many practical situations one oversimplifies in assuming that all particles have nearly identical volume V. It appears to be a better approximation to work with a distribution function F(V) which enters (6.65). On the basis of a logarithmic Gauss distribution,

$$F(V) = \exp-(\ln V/V_0)^2/b^2, \tag{6.66}$$

a satisfactory explanation of the superparamagnetic behavior of α-FeSO$_4$ particles has been reported [6.141].

6.7.3. Spin Transitions in Iron Complex Compounds

It is well known in the coordination chemistry of the first-row transition elements that metal ions with $3d^4$ up to $3d^7$ electron configuration in complexes with (pseudo) octahedral symmetry, depending on the strength of the crystal field splitting Δ as compared to the mean spin pairing energy P, form either high-spin compounds ($\Delta < P$, obeying Hund's first rule of maximum spin multiplicity) or low-spin compounds ($\Delta > P$, not following Hund's first rule). In relatively few cases, however, the energy separation between the low-spin state and the high-spin state competes in magnitude with the mean spin pairing energy, crudely speaking $\Delta \simeq P$ (in a more accurate treatment one has, of course, to consider all kinds of perturbations like axial and rhombic field distortions, spin-orbit coupling, and configuration interaction). Such complexes exhibit temperature-dependent spin transition: the central ion shows low-

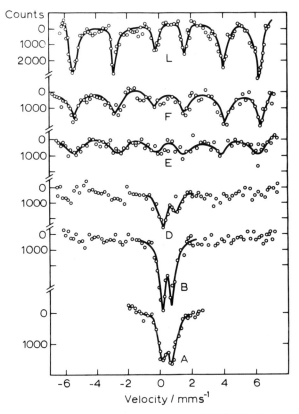

Fig. 6.24. Mössbauer absorption spectra of α-FeOOH of various particle sizes at 294 K (taken from [6.140]). The estimated volumes according to (6.65) of the text are $3 \cdot 10^{-18}$, $7 \cdot 10^{-18}$, $2 \cdot 10^{-17}$, $5 \cdot 10^{-17}$, $2 \cdot 10^{-16}$, and $> 10^{-15}$ cm^3 for sample A, B, D, E, F, and L, respectively

spin (LS) behaviour at sufficiently low temperatures and high-spin (HS) behaviour at higher temperatures. Such complexes are also known as magnetic cross-over systems [6.142].

There are several physical methods which can be and have been used to follow the spin transition LS $\rightleftharpoons$ HS as a function of temperature, e.g., magnetic susceptibility measurements, calorimetric methods, UV/visible and far-infrared spectroscopy, ESCA, and last but not least, Mössbauer spectroscopy. Mössbauer spectroscopy is a microscopic tool probing individual complex molecules of a given Mössbauer atom.

Among the early applications of Mössbauer spectroscopy to spin transition in iron complexes is the work of Dézsi et al. [6.116] and that of Jesson et al. [6.143, 144]. The former authors recorded ^{57}Fe Mössbauer spectra of Fe^{II}(phen)$_2$(NCS)$_2$ as a function of temperature. Their spectra, which are reproduced in Fig. 6.18, demonstrate well the transition from the high-spin ($^5T_2$) state (quadrupole doublet of outer two lines with $\Delta E_Q \approx 3$ mm s^{-1}) to the low-spin ($^1A_1$) state (quadrupole doublet of inner two lines with $\Delta E_Q \approx 0.5$ mm s^{-1}) with decreasing temperature. Jesson and coworkers [6.143, 144] studied the $^5T_2 \rightleftharpoons {}^1A_1$ transition in a ferrous complex with hydro-tris(1-pyrazolyl)borate as ligands and found also well resolved [57]Fe Mössbauer spectra with increasing intensity of the low spin ($^1A_1$) quadrupole doublet and decreasing intensity of the high-spin ($^5T_2$) doublet upon lowering the temperature (cf. Fig. 5 of [6.144]).

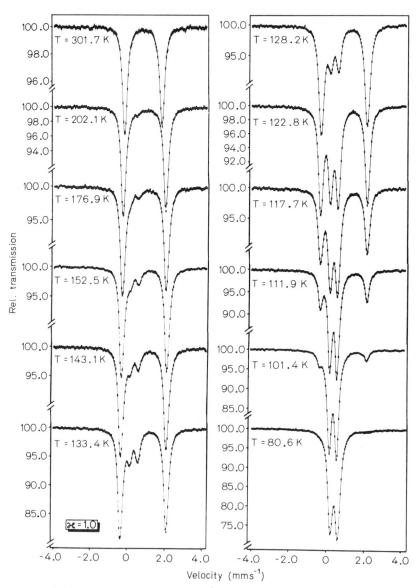

Fig. 6.25. [57]Fe Mössbauer spectra of [Fe(2-pic)$_3$]Cl$_2$ · C$_2$H$_5$OH as a function of temperature (source: [57]Co/Cu at room temperature). The outer two lines represent the quadrupole doublet of the $^5T_2$ state, the inner two lines that of the $^1A_1$ state. The solid lines are theoretical spectra obtained by a least-squares computer fit of Lorentzian lines to the experimental spectra (from [6.157])

Examples for the ^{57}Fe Mössbauer effect evidence of a $^6A_1 \rightleftharpoons {}^2T_2$ transition in ferric complexes have been published by Rickards, Johnson and Hill [6.145], who recorded variable-temperature Mössbauer spectra of some iron(III) tris(dithiocarbamate) complexes, for which the existence of a temperature dependent magnetic cross-over has been known for long [6.146]. The spin equilibrium in ferric tris(dithio-

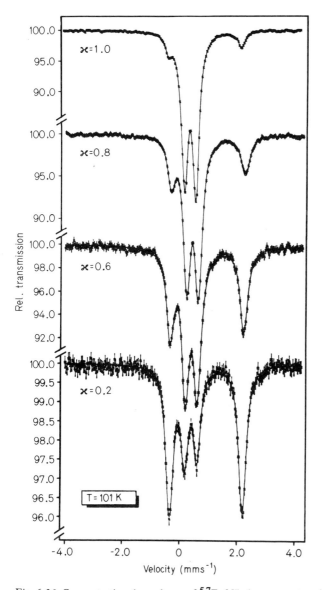

Fig. 6.26. Concentration dependence of ^{57}Fe Mössbauer spectra of $[Fe_xZn_{1-x}(2\text{-pic})_3]Cl_2 \cdot C_2H_5OH$ at 101 K. The spectra clearly reveal that the $^5T_2$ state is favoured upon substitution of iron for zinc (from [6.157])

carbamate) complexes has recently been revisited by Hall and Hendrickson [6.147] using various physical methods.

A fairly large number of spin transition systems have been reported on during the past ten years, mostly diimine complexes of iron(II) [6.148]. Much has been contributed in this area by Goodwin and co-workers [6.148, 149] from the preparative point of view, and by König, Ritter and other collaborators [6.150–153], by Hoselton, Wilson, and Drago [6.154], by Sorai and Seki [6.155] and recently from Gütlich's group [6.156–161] through physical, particularly Mössbauer, measurements and also preparative work. This list is by no means complete.

Considerable progress towards understanding the driving force and the mechanism of the spin transition in ferrous diimine complexes was achieved recently by a systematic ^{57}Fe Mössbauer study of the influence of substituting iron for nonmagnetic zinc ions in the system $[Fe_x^{II}Zn_{1-x}(2\text{-pic})_3]Cl_2 \cdot C_2H_5OH$ in the range $0.0009 \leqslant x \leqslant 1$ [6.157, 160] and the influence of the crystal solvent effect in the system $[Fe(2\text{-pic})_3]Cl_2 \cdot Sol$ (Sol $= C_2H_5OH$, CH_3OH, H_2O, $2H_2O$; CD_3OD) [6.158, 6.161] on the $^5T_2 \rightleftharpoons {}^1A_1$ transition (2-pic = α-picolylamine). Some of the most important results of these investigations are shown in Figs. 6.25–29. From these studies and the earlier work of Sorai and Seki [6.155] it is strongly suggested that the spin transition is cooperative in nature, i.e., that the spin transition takes place through a significant coupling between the electronic state and the lattice phonon system, and that the conversion of the electronic state occurs simultaneously in a group of molecules which form a "cooperative region" (domain). The phonon system is considered to include both intramolecular and lattice (intermolecular) vibrations. At sufficiently low temperatures, all the iron complex molecules are in the low-spin state and there exist normal mode vibrations characteristic of this spin state. With increasing temperature the spin state of a certain amount of molecules will be changed by thermal

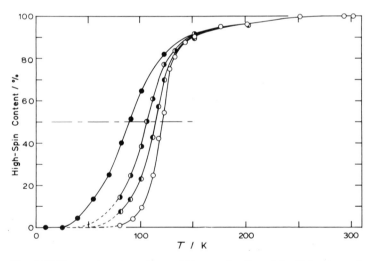

Fig. 6.27. Temperature dependence of the area fraction of the high-spin quadrupole double for $[Fe_xZn_{1-x}(2\text{-pic})_3]Cl_2 \cdot C_2H_5OH$. With increasing x the transition curves are shifted towards lower temperatures and become less steep (from [6.157])

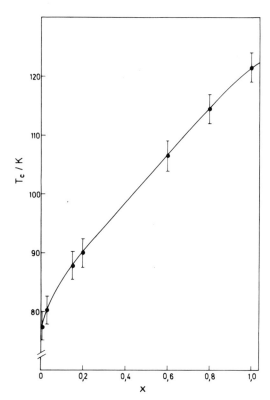

Fig. 6.28. Concentration dependence of the phase transition temperature T_c (temperature of equal area fractions for the high-spin and the low-spin state) for $[Fe_xZn_{1-x}(2\text{-pic})_3]\,Cl_2 \cdot C_2H_5OH$ (from [6.157, 160]). The solid line is the result of a model calculation by Steinhäuser [6.160] based on a phenomenological thermodynamic treatment considering intra- and inter-molecular enthalpy changes and entropy changes as parameters

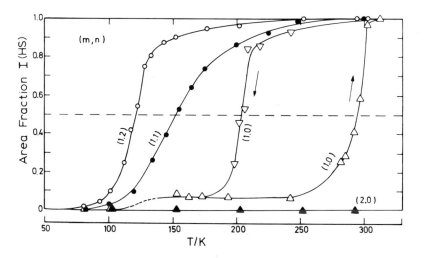

Fig. 6.29. Temperature dependence of the high-spin area fraction I(HS) for $[Fe(2\text{-pic})_3]Cl_2 \cdot mC_nH_{2n+1}OH$: (○) C_2H_5OH, (○) CH_3OH, (△) H_2O on heating, (▽) H_2O on cooling and (△) $2H_2O$. The influence of the various crystal solvent molecules on the high-spin ⇌ low-spin transition is evident from this diagram (from [6.158])

excitation from low-spin to high-spin. As could be found by Sorai and Seki by variable-temperature vibrational spectroscopy (f.i.r.) and heat capacity measurements on the spin cross-over system $Fe(phen)_2(NCS)_2$ [6.155], the spin transition from low-spin to high-spin upon rising temperature is essentially entropy driven, the entropy of the high-spin state both in the electronic part and the vibrational part being higher than in the low-spin state. Some normal modes of lattice vibration are expected to be modulated to some extent upon the spin transition. Through the modulated normal modes the information concerning the change of the spin state will be communicated to other molecules, whereby these molecules tend to change their spin state, too. The number of complex molecules of like spin state forming a domain depends strongly on the nature of the system. It has been estimated by Sorai and Seki [6.155] to be in the order of 80–100 in the strongly coupled system $Fe(phen)_2(NCS)_2$, the transition curve of which changes very abruptly (within a few degrees Kelvin) around 190 K. For the weakly coupled system $[Fe(2\text{-pic})_3]Cl_2 \cdot C_2H_5OH$ this number has been calculated to be in the order of 5 [6.160].

Various model calculations have been performed recently, e.g. by Zimmermann and König [6.162] and in one of the present author's (P. G.) laboratory [6.160], which demonstrate that intermolecular interactions are of utmost importance for spin transition processes, in full agreement with the "cooperative domain" model suggested by Sorai and Seki [6.155].

We wish to express our thanks to Dr. S. K. Date and Dr. V. Marathe for helpful discussions.

References

[6.1] Trautwein, A., Harris, F. E., Freeman, A. J., Desclaux, J. P.: Phys. Rev. *B 11*, 4101 (1975)

[6.2] Regnard, J. R., Pelzl, J.: Phys. Stat. Sol. *56*, 281 (1973)

[6.3] Post, D., van Duinen, P. Th., Nieuwport, W. C.: to be published

[6.4] Erickson, N. E.: in: *The Mössbauer effect and its Application in Chemistry*, R. F. Gould (ed.). (Advances in Chemistry, Series 68, American Chem. Soc., Washington DC 1967), p. 86

[6.5] Harris, F. E.: J. Chem. Phys. *46*, 2769 (1967); Advances in Quantum Chemistry *3*, 61 (1967)

[6.6] Ballhausen, C. J.: in *Introduction to Ligand Field Theory*. New York: Mc Graw-Hill 1962

[6.7] Sugano, S., Tanabe, Y., Kamimura, H.: In *Multiplets of Transition-Metal Ions in Crystals*. New York: Academic Press, 1970

[6.8] Griffith, J. G.: in *The Theory of Transition-Metal Ions*. Cambridge: University Press 1964

[6.9] Shirley, D. A.: Rev. Mod. Phys. *36*, 339 (1964)

[6.10] Zerner, M., Gouterman, M., Kobayashi, H.: Theoret. Chim. Acta *6*, 363 (1966)

[6.11] Trautwein, A., Maeda, Y., Harris, F. E., Formanek, H.: Theoret. Chim. Acta *36*, 67 (1974)
Trautwein, A., Zimmermann, R., Harris, F. E.: Theoret. Chim. Acta *37*, 89 (1975)

[6.12] Flygare, W. H., Hafemeister, D. W.: J. Chem. Phys. *43*, 789 (1965)

[6.13] Simanek, E., Sroubeck, Z.: Phys. Rev. *163*, 275 (1967)

[6.14] Trautwein, A., Regnard, J. R., Harris, F. E., Maeda, Y.: Phys. Rev. *B7*, 947 (1973)

[6.15] Sawatzky, G. A., van der Woude, F.: J. de Physique *12*, C6–47 (1974)

[6.16] Pople, J. A., Beveridge, D. L.: in *Approximate Molecular Orbital Theory*. New York: Mc Graw-Hill 1970

[6.17] Chappert, J., Frankel, R. B., Misetich, A., Blum, N. A.: Phys. Rev. *179*, 578 (1969)

[6.18] Trautwein, A., Harris, F. E.: Theoret. Chim. Acta *30*, 45 (1973)

[6.19] Trautwein, A., Harris, F. E.: Theoret. Chim. Acta *38*, 65 (1975)

[6.20] Walch, P. F., Ellis, D. E.: Phys. Rev. *B7*, 903 (1973)

[6.21] Duff, K. J.: Phys. Rev. *B9*, 66 (1974)

[6.22] Reschke, R., Trautwein, A.: J. de Physique *12*, C6–459 (1976)

[6.23] Blomquist, J., Roos, B., Sundban, M.: University of Stockholm, Inst. of Physics, Report *71–07*, Aug. 1971

[6.24] Gupta, G. P., Lal, K. C.: Phys. Stat. Sol. *51*, 233 (1972)

[6.25] Einstein, A.: Ann. Phys. (4. Folge) *35*, 898 (1911)

[6.26] Lustig, H.: Amer. J. Phys. *29*, 1 (1961)

[6.27] Taylor, R. D., Craig, P. P.: Phys. Rev. *175*, 782 (1968)

[6.28] Sternheimer, R. M.: Phys. Rev. *92*, 1460 (1953)

[6.29] Ingalls, R.: Phys. Rev. *128*, 1155 (1962)

[6.30] Bleaney, B.: *Hyperfine Structure and Electron Paramagnetic Resonance*, A. J. Freeman, R. B. Frankel (eds.). Hyperfine Interactions. New York: Academic Press 1967

[6.31] Weissbluth, M., Maling, J. E.: J. Chem. Phys. *47*, 4166 (1967)

[6.32] de Vries, J. L. K. F., Kreijzers, C. P., de Boer, F.: Inorg. Chem. *11*, 1343 (1972)

[6.33] Trautwein, A., Harris, F. E.: Phys. Rev. *B7*, 4755 (1973)

[6.34] Rein, R., Clarke, G. A., Harris, F. E.: in *Quantum Aspects of Heterocyclic Compounds in Chemistry and Biochemistry II*. (Israel Academy of Sciences and Humanities 1970)

[6.35] Trautwein, A., Reschke, R., Dezsi, I., Harris, F. E.: J. de Physique *12*, C6–463 (1976)

[6.36] Foley, H. M., Sternheimer, R. M., Tycko, D.: Phys. Rev. *93*, 734 (1954)

[6.37] Clark, M. G., Bancroft, G. M., Stone, A. J.: J. Chem. Phys. *47*, 4250 (1967)

[6.38] Osterhuis, W. T., Lang, G.: J. Chem. Phys. *50*, 4381 (1969)

[6.39] Kreber, E., Gonser, U., Trautwein, A., Harris, F. E.: J. Chem. Phys. Solids *36*, 263 (1975); Trautwein, A., Kreber, E., Gonser, U., Harris, F. E.: J. Chem. Phys. Solids *36*, 325 (1975)

[6.40] Dézsi, I., Molnar, B., Srolay, T., Iaszberesyi, I.: Chem. Phys. Lett. *18*, 598 (1973)

[6.41] Grant, R. W., Wiedersich, H., Housley, R. M., Espinosa, G. P.: Phys. Rev. *B3*, 678 (1971)

[6.42] Trautwein, A., Harris, F. E., Dézsi, I.: Theoret. Chim. Acta *35*, 231 (1974)

[6.43] Dorn, W.: Dissertation, Universität Hamburg, 1974

[6.44] Zory, P.: Phys. Rev. *140*, A1401 (1965)

[6.45] Trautwein, A., Maeda, Y., Gonser, U., Parak, F., Formanek, H.: in *Proceedings of the 5th International Conference in Mössbauer Spectroscopy*. Bratislava (CSSR), Sept. 1973; Gonser, U., Maeda, Y., Trautwein, A., Parak, F., Formanek, H.: Z. Naturforsch. *296*, 241 (1974); Zimmermann, R.: Nucl. Instr. Meth. *128*, 537 (1975)

[6.46] Trautwein, A., Reschke, R., Zimmermann, R., Dézsi, I., Harris, F. E.: J. de Physique *12*, C6–235 (1974)

[6.47] Wehner, H.: Dissertation, Universität Erlangen-Nürnberg, 1973

[6.48] Ingalls, R.: Phys. Rev. *133*, A787 (1964)

[6.49] Eicher, H., Trautwein, A.: J. Chem. Phys. *50*, 2540 (1969)

[6.50] Champion, P. M., Münck, E., Debrunner, P. G., Hollenberg, P. F., Hager, L. P.: Biochemistry, *12*, 426 (1973)

[6.51] Edwards, P. R., Johnson, C. E., Williams, R. J. P.: J. Chem. Phys. *47*, 2074 (1967)

[6.52] For a more detailed list of ferrous high-spin compounds which exhibit a temperature-dependent quadrupole splitting see chapter 6.A in: Greenwood, N. N., Gibb, T. C.: *Mössbauer Spectroscopy*. London: Chapman and Hall Ltd. 1971

[6.53] Huynh, B. H., Papaefthymiou, G. C., Yen, C. S., Groves, J. L., Wu, C. S.: J. Chem. Phys. *61*, 3750 (1974)

[6.54] Misetich, A. A., Buch, T.: J. Chem. Phys. *41*, 2524 (1964)

[6.55] Samaras, D., Coing-Boyat, J.: Bull. Soc. Fr. Min. Crist. *93*, 190 (1970)

[6.56] Zimmermann, R., Trautwein, A., Harris, F. E.: Phys. Rev. B *12*, 3902 (1975)

[6.57] Reschke, R., Trautwein, A., Harris, F. E.: Phys. Rev. B *15*, 2708 (1977)

[6.58] Borovik-Romanov, A. S., Karasik, V. R., Kreines, N. M.: JEPT *4*, 109 (1957)

[6.59] Frazer, B. C., Brown, P. J.: Phys. Rev. *125*, 1283 (1962)

[6.60] Weiss, P.: J. Phys. 6, 667 (1907)

[6.61] Kittel, C.: *Introduction to Solid State Physics*, 3rd edition. New York: Wiley & Sons 1966

[6.62] Davidson, G. R., Eibschütz, M., Guggenheim, H. J.: Phys. Rev. *B8*, 1864 (1973)

[6.63] Kamal, R., Mendriatta, R. G.: J. Phys. Chem. Solids *31*, 872 (1969)

[6.64] Freeman, A. J., Watson, R. E.: Phys. Rev. *131*, 2566 (1963)

[6.65] Watson, R. E., Freeman, A. J.: Phys. Rev. *123*, 2027 (1961)

[6.66] Johnson, C. E.: Proc. Phys. Soc. *92*, 748 (1967)

[6.67] Spiering, H., Zimmermann, R., Ritter, G.: Phys. Stat. Sol. (b) *62*, 123 (1974)

[6.68] Zimmermann, R., Spiering, H., Ritter, G.: Chem. Phys. *4*, 133 (1974)

[6.69] Price, D. C., Maartense, I., Morrish, A. H.: Phys. Rev. B *9*, 281 (1974)

[6.70] Paez, E. A., Waever, D. L., Osterhuis, W. T.: J. Chem. Phys. *57*, 3709 (1972)

[6.71] See section *Biological Studies* in *Applications of Mössbauer Spectroscopy*, Vol. 1, R. L. Cohen (ed.). New York: Academic Press 1976

[6.72] Boekma, C., van der Woude, F., Sawatzky, G. A.: Int. J. of Magnetism *3*, 341 (1972)

[6.73] Šimanek, E., Huang, N. L., Orbach, R.: J. Appl. Phys. *38*, 1072 (1967)

[6.74] Abragam, A., Horowitz, I., Pryce, M. H. L.: Proc. Roy. Soc. (London) *A230*, 169 (1955)

[6.75] Reschke, R., Trautwein, A., Desclaux, J. P.: J. Phys. Chem. Solids *38*, 837 (1977)

[6.76] Trautwein, A., Zimmermann, R.: Phys. Rev. *B13*, 2238 (1976)

[6.77] Abragam, A., Pryce, M. H. L.: Proc. Roy. Soc. (London) *A205*, 135 (1951)

[6.78] Johnson, C. E.: Symp. Faraday Soc. *1*, 7 (1967)

[6.79] Hrynkiewicz, A., Sawicka, B., Sawicki, J.: Report No. 685/PL of the Institute of Nuclear Physics, Cracow (Poland) 1969; Proceedings of the *Conference on Applications of the Mössbauer Effect* Tihany, I. Dézsi (ed.). Budapest (Hungary) 1971

[6.80] Osterhuis, W. T., Lang, G.: Phys. Rev. *178*, 439 (1969)

[6.81] Ono, K.: Phys. Rev. Lett. *24*, 770 (1970); Chappert, J., Sawicka, B., Sawicki, J.: Phys. Stat. Sol. (b) *72*, K139 (1975)

[6.82] Figgis, B. N., Gerloch, M., Mason, R.: Proc. Roy. Soc. *A309*, 91 (1969)

[6.83] Baker, I. M., Bleaney, B., Bowers, K. B.: Proc. Phys. Soc. *B69*, 1205 (1956)

[6.84] Osterhuis, W. T., Lang, G.: J. Chem. Phys. *50*, 4381 (1969)

[6.85] Hazony, Y.: J. Phys. C: Solids, *5*, 2267 (1972)

[6.86] Reschke, R.: Dissertation, Universität des Saarlandes (1976)

[6.87] Lang, G., Dale, B. W.: J. Phys. C: Solids, *6*, L80 (1973)

[6.88] Sawatzky, G. A., van der Woude, F., Morrish, A. H.: Phys. Rev. *187*, 747 (1969); Phys. Let. *25A*, 147 (1967); J. Appl. Phys. *39*, 1204 (1968)

[6.89] Ingalls, R.: Phys. Rev. *133*, A787 (1964)

[6.90] Eicher, H., Trautwein, A.: J. Chem. Phys. *52*, 932 (1970)

[6.91] Fleisch, J., Gütlich, P., Hasselbach, K. M.: Inorg. Chem. *16*, 1979 (1977); Fleisch, J., Gütlich, P., Hasselbach, K. M., Müller, W.: Inorg. Chem., *15*, 958 (1976); Fleisch, J., Gütlich, P., Hasselbach, K. M.: Inorg. Chim. Acta, *17*, 51 (1976)

[6.92] Seno, Y., Otsuka, J., Matsuoka, O., Fuchikami, N.: J. Phys. Soc. of Japan *33*, 1645 (1972); Otsuka, J., Matsuoka, O., Fuchikami, N., Seno, Y.: J. Phys. Soc. of Japan *35*, 854 (1973)

[6.93] Zerner, M., Gouterman, M., Kobayashi, H.: Theor. Chim. Acta *6*, 363 (1966)

[6.94] Loew, G. H., Lo, D. Y.: Theoret. Chim. Acta *32*, 217 (1974)

[6.95] Freeman, A. J., Ellis, D. E.: private communication

[6.96] Callaway, J.: *Energy Band Theory*, New York: Academic Press 1964

[6.97] Afanas'ev, A. M., Kagan, Y.: Sov. Phys. JEPT *18*, 1139 (1964)

[6.98] Bradford, E., Marshall, W.: Proc. Phys. Soc. (London) *87*, 731 (1966)

[6.99] Gabriel, H.: Phys. Stat. Sol. *23*, 195 (1967); Bosse, I., Gabriel, H., Vollmann, W.: Phys. Stat. Sol. (b) *68*, 81 (1975)

[6.100] Wegener, H.: Z. Phys. *186*, 498 (1965)

[6.101] Blume, M.: in *Hyperfine Structure and Nuclear Radiation*, E. Matthias, D. A. Shirley (eds.). Amsterdam: North-Holland 1968; and references therein

[6.102] Wickmann, H. H.: in *Hyperfine Structure and Nuclear Radiation*, E. Matthias, D. A. Shirley (eds.). Amsterdam: North-Holland 1968;
Wickmann, H. H., Wertheim, G. K.: in *Chemical Applications of Mössbauer Spectroscopy*, V. I. Goldanskii, R. H. Herber (eds.). New York: Academic Press 1968;
Wickmann, H. H.: in *Mössbauer Effect Methodology*, Vol. 2, I. J. Gruverman (ed.). New York: Plenum Press 1966, p. 39

[6.103] Mørup, S., Thrane, N.: Phys. Rev. *B4*, 2087 (1971);
Phys. Rev. *B8*, 1020 (1973);
Mørup, S.: in *Mössbauer Effect Studies of Electronic Relaxation in Ferric Compounds*, preprint (1974)

[6.104] Schwegler, H.: Fortschr. Phys. *20*, 251 (1972)

[6.105] Poole, C. R., Farach, H. A.: in *Relaxation in Magnetic Resonances*. New York: Academic Press 1971

[6.106] Belakhovsky, M.: manuscript *Relaxation et Effet Mössbauer, Approche Stochastique*. Laboratoire de diffraction neutronique, Centre d'études nucléaires de Grenoble/France, 1969

[6.107] Manenkov, A. A., Orbach, R.: in *Spin Lattice Relaxation in Ionic Solids*. Harper and Row 1966

[6.108] Orbach, R.: Proc. Phys. Soc. (London) *A 77*, 821 (1961)

[6.109] Scott, P. L., Jeffries, C. D.: Phys. Rev. *127*, 32 (1962)

[6.110] Al'tshuler, S. A., Kochelaev, B. J., Lenshin, A. M.: Usp. Fiz. Nauk *75*, 459 (1961) (engl. transl. Sov. Phys. Usp. *4*, 880 (1962))

[6.111] Abragam, A., Bleaney, B.: in *Electron paramagnetic resonance of transition ions*, W. Marshall, D. H. Wilkinson (eds.). Oxford: Clarendon Press 1970, Chap. 10

[6.112] Reschke, R., Trautwein, A., Harris, F. E.: Phys. Rev. B *15*, 2708 (1977)

[6.113] Kreber, E., Gonser, U., Trautwein, A., Harris, F. E.: J. Phys. Chem. Solids *36*, 263 (1975)

[6.114] Orbach, R.: Proc. Roy. Soc. (London) *A264*, 458 (1961)

[6.115] König, E., Ritter, G.: in *Mössbauer Effect Methodology*, Vol. 9, I. J. Gruverman (ed.). New York: Plenum Press 1974, p. 1;
Sorai, M., Ensling, J., Gütlich, P.: Chem. Phys. *18*, 199 (1976);
Sorai, M., Ensling, J., Hasselbach, K. M., Gütlich, P.: Chem. Phys. *20*, 197 (1977)

[6.116] Dézsi, I., Molnar, B., Tarnozci, T., Tompy, K.: J. Inorg. Nuclear Chem. *29*, 2486 (1967)

[6.117] Zimmermann, R., Spiering, H., Ritter, G.: Chem. Phys. *4*, 133 (1974)

[6.118] Ham, F. S.: Phys. Rev. *138A*, 1727 (1965);
Ham, F. S.: Phys. Rev. *160*, 328 (1967);
Ham, F. S., Schwarz, W. M., O'Brian, M. C. M.: Phys. Rev. *185*, 548 (1969)

[6.119] Chappert, J., Frankel, R. B., Misetich, A., Blum, N. A.: Phys. Rev. *179*, 578 (1969);
Chappert, J., Frankel, R. B., Misetich, A., Blum, N. A.: Intern. Conference on Magnetism, Grenoble/France (1970)

[6.120] Simopoulos, A., Petridis, D., Kostikas, A., Wickman, H. H.: Chem. Phys. *2*, 452 (1973)

[6.121] Wallace, W. E.: in *Rare Earth Intermetallics*. New York: Academic Press 1973

[6.122] Kornienko, L. S., Prokhorov, A. M.: Zh. Eksper. i. Teor. Fiz. (USSR) *40*, 1594 (1961); (Engl. trans. Sov. Phys. JEPT *13*, 1120 (1961))

[6.123] Wertheim, G. K., Remeika, J. P.: Proc. XIII Coloque Ampère, Univ. of Leuven (1964), North-Holland, Amsterdam (1965)

[6.124] Lang, G., Asakura, T., Yonetani, T.: Phys. Rev. Lett. *24*, 981 (1970)

[6.125] Maeda, Y., Trautwein, A., Gonser, U., Yoshida, K., Kikuchi-Torii, K., Homma, T., Ogura, Y.: Biochim. Biophys. Acta *303*, 230 (1973)

[6.126] Wertheim, G. K., Remeika, J. P.: Phys. Lett. *10*, 14 (1964)

[6.127] Wignall, J. W. G.: J. Chem. Phys. *44*, 2462 (1966)

[6.128] Alan, A., Chandra, S., Hoy, G. R.: Phys. Lett. *22*, 26 (1966)

[6.129] Champell, L., de Benedetti, S.: Phys. Lett. *20*, 102 (1966)

[6.130] Wickman, H. H., Trozollo, A. M.: Phys. Rev. Lett. *16*, 162 (1966)

[6.131] Date, S. K.: PhD-Thesis, University of Bombay, India (1969)

[6.132] Blume, M.: Phys. Rev. Lett. *18*, 305 (1967)

[6.133] Thrane, N., Trumpy, G.: Phys. Rev. *B1*, 153 (1970)
[6.134] Shulman, R. G., Wertheim, G. K.: Rev. Mod. Phys. *36*, 459 (1964)
[6.135] Date, S. K.: J. Phys. Soc. Japan *30*, 1203 (1971)
[6.136] Bates, C. A., Gavaix, G., Steggles, P., Vasson, A., Vasson, A. M.: J. Phys. C: Solid State Phys. *8*, 2300 (1975)
[6.137] Néel, L.: Ann. Geophys. *5*, 99 (1949)
[6.138] Müller, K., Thurley, F.: Int. J. Magnetism *5*, 203 (1973)
[6.139] Krop, K., Korecki, J., Zukrowski, J., Karas, W.: Int. J. Magnetism *6*, 19 (1974)
[6.140] Takada, T., Kiyama, M., Bando, Y., Shinjo, T.: Bull. of the Institute for Chem. Res. (Kyoto Univ., Japan) *47*, 298 (1969)
[6.141] Irler, W., Wegener, H., Wehner, H. L.: in *Jahresbericht des Physikalischen Institutes der Universität Erlangen-Nürnberg* (1975), p. 64
[6.142] Martin, R. L., White, A. H.: in *Transition Metal Chemistry*, R. L. Carlin (ed.), Vol. 4. New York: Marcel Dekker Inc. 1968, p. 113
[6.143] Jesson, J. P., Weiher, J. F.: J. Chem. Phys. *46*, 1995 (1967)
[6.144] Jesson, J. P., Weiher, J. F., Trofimenko, S.: J. Chem. Phys. *48*, 2058 (1968)
[6.145] Rickards, R., Johnson, C. E., Hill, H. A. O.: J. Chem. Phys. *48*, 5231 (1968)
[6.146] Cambi, L., Cagnasso, A.: Atti. Accad. Naz. Lincei *13*, 809 (1931); Cambi, L., Szego, L.: Ber. Dtsch. Chem. Ges. *64*, 2591 (1931)
[6.147] Hall, G., Hendrickson, D. N.: Inorg. Chem. *15*, 607 (1976)
[6.148] Goodwin, H. A.: Coord. Chem. Rev. *18*, 293 (1976)
[6.149] Goodwin, H. A., Mather, D. W., Smith, F. E.: Aust. J. Chem. *28*, 33 (1975)
[6.150] König, E., Ritter, G., Spiering, H., Kremer, S., Madeja, K., Rosenkranz, A.: J. Chem. Phys. *56*, 3139 (1972)
[6.151] König, E.: Ber. Bunsenges. Physik. Chem. *76*, 975 (1972)
[6.152] König, E., Ritter, G., Madeja, K., Böhmer, W. H.: Ber. Bunsenges. Physik. Chem. *77*, 390 (1973)
[6.153] König, E., Ritter, G., Kanellakopulos, B.: J. Phys. C: Solid State Phys. *7*, 2681 (1974)
[6.154] Hoselton, M. A., Wilson, L. J., Drago, R. S.: J. Amer. Chem. Soc. *97*, 1722 (1975)
[6.155] Sorai, M., Seki, S.: J. Phys. Chem. Solids *35*, 555 (1974)
[6.156] Fleisch, J., Gütlich, P., Hasselbach, K. M., Müller, W.: Inorg. Chem. *15*, 958 (1976)
[6.157] Sorai, M., Ensling, J., Gütlich, P.: Chem. Phys. *18*, 199 (1976)
[6.158] Sorai, M., Ensling, J., Hasselbach, K. M., Gütlich, P.: Chem. Phys. *20*, 197 (1977)
[6.159] Fleisch, J., Gütlich, P.: Inorg. Chem. to be published
[6.160] Steinhäuser, H. G.: Diplomarbeit, Department of Chemistry, University of Mainz, 1977 Gütlich, P., Köppen, H., Link, R., and Steinhäuser, H. G., submitted to J. Chem. Phys.
[6.161] Steinhäuser, H. G., Sudheimer, G., Link, R., Gütlich, P.: the work on the effect of deuterated methanol CD3OD to be published; Sudheimer, G.: Diplomarbeit, Department of Chemistry, Technische Hochschule Darmstadt, 1976
[6.162] Zimmermann, R., König, E.: J. Phys. Chem. Solids *38*, 779 (1977)

7. Mössbauer-Active Transition-Metals Other than Iron

Chapter 6 was exclusively devoted to the measurements and interpretation of ^{57}Fe spectra of various iron containing systems. The chemistry of iron is by far the most extensively explored field compared with all other Mössbauer-active elements, because the Mössbauer effect of ^{57}Fe is very easy to observe and the spectra are in general well resolved and reflect important information about bonding and structural properties. Beside iron there are a good number of other transition metals suitable for Mössbauer spectroscopy, which are, however, less extensively studied because of technical and/or spectral resolution problems. In recent years many of these difficulties have been overcome, and we shall see in the following that a good deal of successful work has been done in Mössbauer spectroscopy on compounds of nickel (^{61}Ni), ruthenium (^{99}Ru, ^{101}Ru), tantalum (^{181}Ta), tungsten (^{180}W, ^{182}W, ^{183}W, ^{184}W, ^{186}W), osmium (^{186}Os, ^{188}Os, ^{189}Os, ^{190}Os), iridium (^{191}Ir, ^{193}Ir), platinum (^{195}Pt), and gold (^{197}Au). The nuclear gamma resonance effect on the rest of the Mössbauer-active transition metals – zinc (^{67}Zn), technetium (^{99}Tc), silver (^{107}Ag), hafnium (^{176}Hf, ^{177}Hf, ^{178}Hf, ^{180}Hf), rhenium (^{187}Re), mercury (^{199}Hg, ^{201}Hg) – has been of relatively little use to the chemist so far. There are various reasons responsible for this, viz., (a) extraordinary difficulties in measuring the resonance effect because of the long lifetime of the excited Mössbauer level and hence the extreme smallness of the transition line width (e.g., in ^{67}Zn), (b) very poor resolution of the resonance lines due to either very small nuclear moments or very short lifetime of the excited Mössbauer level resulting into very broad resonance lines, (c) insufficient resonance effect due to unusually high transition energy between the excited and the ground nuclear levels, which in turn increases the recoil energy and thus reduces the recoilless fraction of emitted and observed γ-rays.

In Table 7.1 (inside cover at the end of the book) nuclear data are collected for those Mössbauer transitions of transition metal nuclides which are used in Mössbauer spectroscopy. The symbols used in this table have the following meanings:

a	natural abundance of resonant nuclide (in %);
E_γ	energy of γ-ray transition (in keV);
$t_{1/2}$	half-life of excited nuclear state (in ns);
Γ	natural line width (in mm s^{-1});
I_g, I_e	nuclear spin quantum number of ground (g) and excited (e) state (the sign refers to the parity);
α_T	total internal conversion coefficient;
E_R	recoil energy (in 10^{-3} eV);
σ_0	resonant absorption cross section (in 10^{-20} cm^2);

MP   multipolarity of γ-radiation;

μ_g, μ_e nuclear magnetic dipole moment of ground (g) and excited (e) state (in nuclear magnetons, n. m.);

Q_g, Q_e nuclear electric quadrupole moment of ground (g) and excited (e) state (in barn = 10^{-28} m^2);

$\dfrac{\Delta\langle r^2 \rangle}{\langle r^2 \rangle}$ = $(\langle r^2 \rangle_e - \langle r^2 \rangle_g)/\langle r^2 \rangle$, relative change of mean-square nuclear radius in going from excited (e) to ground (g) state (in 10^{-4}).

In the last column of Table 7.1 the most popular radioactive precursor nuclide is given together with the nuclear decay process (EC = electron capture, β^- = beta particle emission) feeding the Mössbauer excited nuclear level.

In the following sections we shall present and discuss the decay schemes for all Mössbauer-active transition metal nuclides other than iron. For the sake of completeness, the decay scheme for ^{57}Fe (see Fig. 7.1) is inserted here. The relevant nuclear data, which are of interest in ^{57}Fe Mössbauer spectroscopy may be taken from Table 7.1 (end of the book).

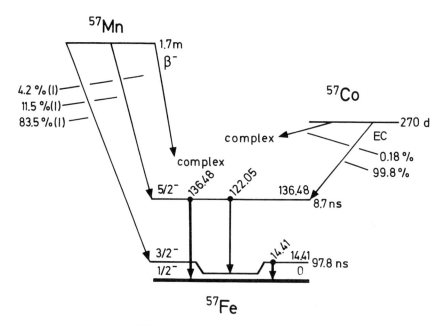

Fig. 7.1. Decay scheme of ^{57}Fe and parent nuclei, from [7.2.8]

7.1. Nickel (^{61}Ni)

The nuclear gamma resonance effect in ^{61}Ni was first observed in 1960 by Obenshain and Wegener [7.1.1]. The practical application to the study of nickel compounds, however, was hampered for several years by (1) the lack of a suitable single-line source, (2) the poor resolution of the overlapping broad hyperfine lines due to the short excited state lifetime, and (3) the difficulties in producing and handling the short-lived Mössbauer sources containing the ^{61}Co and ^{61}Cu parent nuclides, respectively.

Single-line sources are now available which cut down the number of resonance lines in a spectrum and thereby reduce the resolution problems considerably. Nowadays many laboratories have access to electron and ion accelerators to produce the parent nuclides ^{61}Co and ^{61}Cu. So the major obstacles of the experimental part of ^{61}Ni spectroscopy have been overcome and a good deal of successful work has been performed in recent years.

7.1.1. Some Practical Aspects

The sources used in ^{61}Ni Mössbauer work mostly contain ^{61}Co as the parent nuclide of ^{61}Ni; in a few cases ^{61}Cu sources have been used. Although the half-life of ^{61}Co is relatively short (99 m) this nuclide is much superior to ^{61}Cu, because it decays via β^- emission directly to the 67.4 keV Mössbauer level (cf. Fig. 7.1.1), whereas ^{61}Cu ($t_{1/2}$ = 3.32 h) decays in a complex way with only about 2.4% populating the 67.4 keV level. There are a number of nuclear reactions leading to ^{61}Co [7.1.2]; the most popular ones are ^{62}Ni(γ, p)^{61}Co with the bremsstrahlung (ca. 100 MeV) from an electron accelerator, or ^{64}Ni(p, α)^{61}Co via proton irradiation of ^{64}Ni in a cyclotron.

In nickel metal and many metallic systems the nuclear spin of ^{61}Ni undergoes magnetic hyperfine interaction with relatively strong magnetic fields causing a splitting of the 67.4 keV emission line into 12 partially resolved lines (cf. Fig. 7.1.2). The use of such a source for the study of magnetically ordered nickel compounds as absorbers would give rise to a large number of resonance lines, which mostly overlap and make the evaluation of the spectra rather cumbersome [7.1.1]. Much effort has therefore been invested into the development of nonmagnetic single-line sources. The alloy $Ni_{0.85}Cr_{0.15}$ has been found to provide a single-line source at temperatures down to 4 K [7.1.3–6], yielding an experimental line width of 0.97 mm s^{-1} and a resonance effect of about 10% at 77 K [7.1.6]. Obenshain et al. prefer a $Ni_{0.86}V_{0.14}$ alloy, which is also nonmagnetic at 4.2 K, and activate it by proton irradiation to give ^{64}Ni(p, α)^{61}Co [7.1.7, 8]; the line width is very near the natural width. A Japanese group has worked out a method of preparing a single-line source containing ^{61}Cu [7.1.9].

The relatively low Debye-Waller factors (f) encountered in ^{61}Ni Mössbauer spectroscopy (a diagram showing the Debye-Waller factor as a function of Debye temperature is given in [7.1.6]) require the cooling of both source and absorber, preferentially to temperatures $\lesssim$ 80 K. Cryogenic systems have been described, e.g., in [7.1, 2].

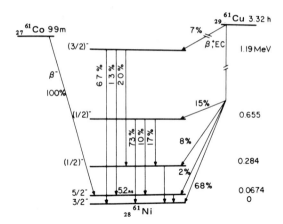

Fig. 7.1.1. Decay schemes of the parent nuclides of ^{61}Ni (from [7.1.6])

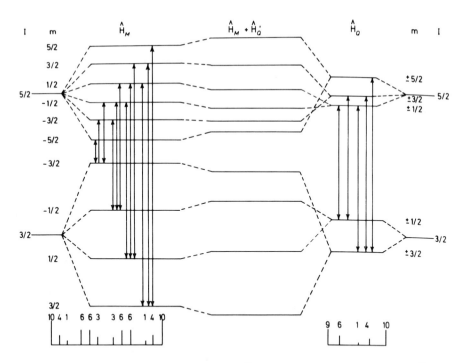

Fig. 7.1.2. Effect of magnetic dipole interaction ($\hat{H}_M$), electric quadrupole interaction ($\hat{H}_Q$), and combined interaction, $\hat{H} = \hat{H}_M + \hat{H}_Q$, $E_M > E_Q$ on the Mössbauer nuclear levels of ^{61}Ni. The larger spacings between the sublevels of the ground state are due to the somewhat larger magnetic dipole moment of the nuclear ground state as compared to the excited state. The relative transition probabilities for a powder sample as well as the relative positions of the transition lines are indicated by the stick spectra below

7.1.2. Hyperfine Interactions in ^{61}Ni

a) Isomer Shifts

The essential nuclear data of ^{61}Ni may be taken from Table 7.1 (end of the book).
More about nuclear properties has been summarized in [7.1.2].

Isomer shifts have been measured in a variety of nickel compounds. In most cases,
however, the information concerning chemical bond properties has not been very im-
pressive, because the second-order Doppler (SOD) shift is in many systems of a comparable magnitude to the real chemical isomer shift, which causes serious difficulties
in correlating the electron density at the nickel nucleus with chemically important
concepts like electronegativity of the coordinated ligand and ionicity of the nickel-
ligand bond. If, however, the lattice dynamical properties are sufficiently known from
infrared and optical spectroscopy one may calculate the second-order Doppler shift
and substract it from the experimentally observed energy shift to find the real chemical isomer shift δ_{IS} according to [7.1.7, 8]

$$\delta_{exp} = \delta_{IS}{}^{(A)} - \delta_{IS}{}^{(S)} + \delta_{SOD}{}^{(A)} - \delta_{SOD}{}^{(S)}, \tag{7.1.1}$$

where the superscripts stand for the absorber (A) and the source (S). For cases where
information about the lattice dynamics is not available, Love et al. [7.1.7] have sug-
gested estimating δ_{SOD} according to $\delta_{SOD} = 1.23\, \delta_{SOD}^{(Debye)}$, where $\delta_{SOD}^{(Debye)} =$
$+178/\ln f$ (in $\mu m\, s^{-1}$) may be calculated in the framework of a Debye model. Following
this approach Obenshain et al. [7.1.7, 8] have determined the isomer shifts for a number of inorganic compounds as listed in Table 7.1, and have attempted to correlate
the corrected isomer shift δ_{IS} with the electronegativity (Pauling) of the ligand as
well as with the ionicity of the nickel-ligand bond (cf. Fig. 7.1.3 taken from [7.1.8]).
As $\Delta\langle r^2\rangle/\langle r^2\rangle$ is negative for ^{61}Ni, as is the case for ^{57}Fe (cf. Table 7.1), the isomer
shifts observed in nickel compounds follow the same trends as in ^{57}Fe spectroscopy.
It is obvious from Fig. 7.1.3 that the nickel fluorides with the highest values for the
ligand electronegativity and the bond ionicity show the most positive isomer shifts
and thus the smallest electron densities at the ^{61}Ni nucleus, due to minimal population in the 4 s valence shell. Following the plots A and B to the left we find nickel
compounds of more complex bond properties with electron densities resulting from
the combined σ- and π-interaction between nickel and the ligands. At any rate, a quantitative discussion of the isomer shift in relation to bond properties should be performed in connection with molecular orbital calculations of the type described in
Section 6.3.

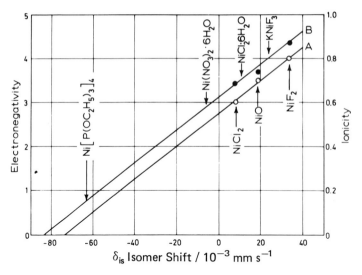

Fig. 7.1.3. Isomer shift (relative to Ni metal and corrected for SOD) as a function of ligand electro-negativity (plot A) and as a function of nickel-ligand ionicity (plot B) (from [7.1.8])

^{61}Ni isomer shifts have also been measured of other nickel compounds like K$_2$NiF$_4$ [7.1.5], Ni(PCl$_3$)$_4$ [7.1.4], (NH$_4$)$_6$[NiMo$_9$O$_{32}$] [7.1.5], K$_2$[Ni(CN)$_4$] [7.1.4] K$_4$Ni$_2$(CN)$_6$ [7.1.7], NiSO$_4$ · 7H$_2$O [7.1.7], (Et$_4$N)[Ph$_3$PNiBr$_3$] [7.1.10], (Ph$_3$MeAs)$_2$ [NiCl$_4$] [7.1.10], (diap)$_2$Ni (diap = N, N′-diphenyl-1-amino-3-iminopropene) [7.1.10], Ni(en)$_2$Cl$_2$ [7.1.4], [Ni(en)$_3$]Cl$_2$ · 6H$_2$O [7.1.10], Ni(CO)$_4$ [7.1.4], (Et$_4$N)$_2$[NiBr$_4$] [7.1.10], and of a number of alloys like CuNi(2%) [7.1.5], CuNi(20%) [7.1.5], FeNi(1.5%) [7.1.5], Ni$_2$R (R = Gd, Ho, Er, Tm, Yb) [7.1.4]. Unfortunately, the iso-mer shift data reported for most of these compounds have not been corrected for SOD (except those given in [7.1.7, 8]) and therefore cannot be included in the corre-lation diagram of Fig. 7.1.3. For meaningful discussions ^{61}Ni isomer shift studies should always include SOD corrections. SOD corrected δ-values should also be used in setting up an isomer shift vs electron configuration diagram for ^{61}Ni as has been attempted by Travis and Spijkerman [7.1.6] in analogy to the Walker-Wertheim-Jacca-rino diagram for ^{57}Fe. Such diagrams, however, are much inferior to a quantitative MO-evaluation of the contributions to the electron density from nickel 4s and all other atomic orbitals of nickel and the ligands taking part in the molecular orbitals.

b) Magnetic Interactions

Most nickel containing substances are paramagnetic, ferromagnetic, or antiferromagnetic because of a relatively large spin due to unpaired Ni 3d electrons. For instance, S = 1 is common in Ni^{2+} compounds with octahedral, tetrahedral, or square planar local symmetry around the nickel atom. The relatively large electronic spin gives rise to a correspondingly large internal magnetic field at the nickel nucleus *via* corepolarization. Other contributions to the local field may come from the orbital motion of Ni 3d electrons and from conduction electron polarization, which is in most nickel alloys the dominating contribution.

The local magnetic field interacts with the relatively large magnetic dipole moments μ of the ^{61}Ni nucleus in both the 67.4 keV excited (μ_e) and the ground state (μ_g) (μ_e is about 4 times larger than μ_e of ^{57}Fe, μ_g is about 8 times larger than μ_g of ^{57}Fe; the signs are opposite, cf. Table 7.1). However, as we have already pointed out in Section 6.7, the magnetic interaction is only observable in the Mössbauer spectrum if the internal field fluctuations going along with the electronic spin fluctuations (in the absence of an external field) are sufficiently slow compared with the lifetime τ_N of the 67.4 keV nuclear state of ^{61}Ni and the time τ_L corresponding to the Larmor precession frequency ω_L of the nuclear spin.

^{61}Ni Mössbauer spectra with pure magnetic dipole interaction (in addition to electric monopole interaction) generally show a more or less resolved 4-line pattern; each of the four broad and unsymmetric lines represents an unresolved triplet arising from the transitions shown in Fig. 7.1.2. Despite the relatively large nuclear magnetic moments of the ^{61}Ni Mössbauer states, the resolution of the magnetically split spectra is often quite poor. The reason is twofold: (a) relatively weak magnetic fields, and (b) the broad line width, which is about 14 times broader than in case of ^{57}Fe. Two examples for magnetically split ^{61}Ni spectra are shown in Fig. 7.1.4 a, b. The spectrum of the $Fe_{0.95}Ni_{0.05}$ alloy (Fig. 7.1.4 a, reproduced from [7.1.2]) is reasonably well resolved due to a relatively large internal magnetic field of 23.4 Tesla at the ^{61}Ni nuclei. The theoretical spectrum [7.1.7] matches the experimental data reasonably well; it consists of twelve individual Lorentzian lines (also shown in the figure) of different intensity referring to the allowed transitions between substates of the 67.4 keV level and of the ground state (cf. Fig. 7.1.2, case of H_M). The magnitude of the internal magnetic field has been found to vary with the composition of the Fe_xNi_{1-x} alloys [7.1.11, 7]. In NiO the internal magnetic field is much smaller (1.4 Tesla) leading to a less pronounced magnetic splitting which in turn results in a poorly resolved spectrum (cf. Fig. 7.1.4 b, produced from [7.1.6]).

c) Electric Quadrupole Interaction

Both the ground and the 67.4 keV nuclear states of ^{61}Ni possess a nonzero electric quadrupole moment. If placed in an inhomogeneous electric field (electric field gradient, EFG $\neq$ 0) the ^{61}Ni nucleus undergoes electric quadrupole interaction with the EFG at the nucleus, as a result of which the 67.4 keV level will split into three substates $|I, \pm m_I\rangle = |5/2, \pm 5/2\rangle$, $|5/2, \pm 3/2\rangle$, and $|5/2, \pm 1/2\rangle$, and the ground level will

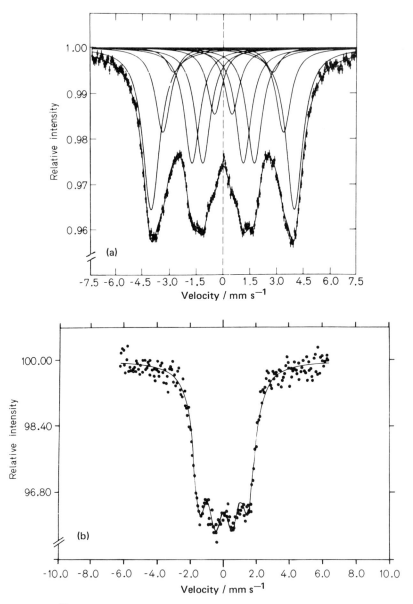

Fig. 7.1.4. ^{61}Ni Mössbauer spectra of (a) $Fe_{0.95}Ni_{0.05}$ obtained at 4.2 K with a $Ni_{0.86}V_{0.14}$ source (reproduced from [7.1.2]), (b) NiO obtained at 80 K with a $Ni_{0.85}Cr_{0.15}$ source (reproduced from [7.1.6])

split into two substates $|3/2, \pm3/2\rangle$ and $|3/2, \pm1/2\rangle$. In practical ^{61}Ni spectroscopy, one has observed electric quadrupole interaction in a few examples only. The reason is that the contributions to the EFG from noncubic valence electron distributions vanish in most cases (we shall consider some examples below), and the contributions from noncubic lattice surroundings are relatively small because of the $1/R^3$ effect (see Sec.

6.4). Furthermore, the quadrupole moment of the ground state, known from other spectroscopic measurements (+0.16 b [7.1.12]), turns out to be rather small. In the few examples of quadrupole interaction observed so far, e.g. in $(NH_4)_6NiMo_9O_{32}$ studied by Travis et al. [7.1.6], and the spectrum of $NiCr_2O_4$ (Fig. 7.1.5a) measured at 77 and 4.2 K by Göring [7.1.13], the interaction manifested itself only in the asymmetry of the Mössbauer spectrum and additional broadening of the resonance lines. Göring [7.1.13] determined the quadrupole moment of the 67.4 keV excited nuclear state to be Q = – 0.20 b. The theoretical spectra were evaluated by considering 5 transition lines between the quadrupole split ground and excited states (see case $\hat{H}_Q$ on the far right side of Fig. 7.1.2).

An instructive description of the first-order perturbation treatment of the quadrupole interaction in ^{61}Ni has been given by Travis and Spijkerman [7.1.6]. These

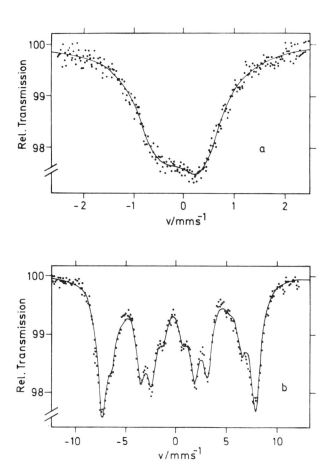

Fig. 7.1.5. ^{61}Ni Mössbauer spectra of $NiCr_2O_4$ using a $Ni_{0.85}Cr_{0.15}$ source. (a) Source and absorber at 77 K (pure quadrupole interaction). (b) Source and absorber at 4 K (combined magnetic and quadrupole interaction) (from [7.1.13])

authors also show in graphical form the quadrupole-spectrum line positions and the quadrupole-spectrum as a function of the asymmetry parameter η; they give eigenvector coefficients and show the orientation dependence of the quadrupole-spectrum line intensities for a single crystal of a ^{61}Ni compound. The reader should also consult the very useful article by Dunlap [7.1.14] about electric quadrupole interaction in general.

A quantitative consideration of the origin of the EFG should be based on reliable results from molecular orbital calculations, e.g., of the SCC-MO-LCAO type including spin-orbit coupling, as we have pointed out in detail in Section 6.4 for ^{57}Fe Mössbauer spectroscopy. For a qualitative discussion, however, it will suffice to use the easy-to-handle one-electron approximation of the crystal field model. In this framework it is easy to realize that in nickel(II) complexes of O_h and T_d symmetry and in tetragonally distorted octahedral nickel (II) complexes, no valence electron contribution to the EFG should be expected (cf. Fig. 7.1.6 and use Table 3.2 of Sec. 3.1.3). A temperature-dependent valence electron contribution is to be expected in distorted tetrahedral nickel(II) complexes; for tetragonal distortion, e.g., $V_{zz} = -(4/7)e\langle r^{-3}\rangle_{3d}$ for compression along the z-axis, and $V_{zz} = +(2/7)e\langle r^{-3}\rangle_{3d}$ for elongation along the z-axis; e is the positive elementary charge. Valence electron contributions to the EFG are also expected in square-planar nickel(II) complexes; $V_{zz} = -(8/7)e\langle r^{-3}\rangle_{3d}$ both for high-spin (S = 1) and low-spin (S = 0) configurations. The lattice contributions to the EFG, of course, have to be treated separately, e.g., using the simple Townes-Dailey

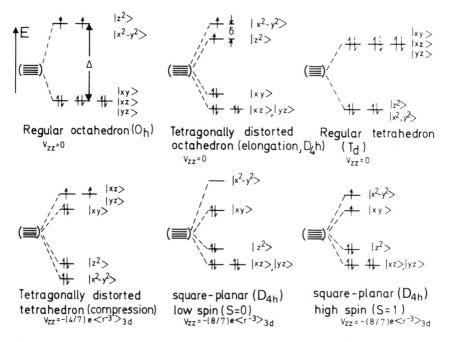

Fig. 7.1.6. Schematic diagrams for common electron configurations of Ni^{2+} complexes in the one-electron approximation. The resulting valence electron contributions V_{zz} are obtained by use of Table 3.2

method [7.1.15, 16]. However, it should be emphasized again that much more reliable results are arrived at by finding the wave functions of the molecular orbitals involved and taking the Boltzmann weighted sum over all orbitals involved for each EFG tensor element.

With the known quadrupole moments of the excited and the ground nuclear state of ^{61}Ni it is possible to extract the sign of the EFG (V_{zz}) directly from the quadrupole-perturbed Mössbauer spectrum of a polycrystalline sample, because the spacings in the quadrupole spectrum of ^{61}Ni are unequal (cf. Fig. 7.1.2).

d) Combined Magnetic and Quadrupole Interaction

The low temperature Mössbauer spectra of the spinel type oxides NiCr$_2$O$_4$ [7.1.13, 17] (Fig. 7.1.5 b) and NiFe$_2$O$_4$ [7.1.6, 17] have been found to exhibit combined magnetic dipole and electric quadrupole interaction (Fig. 7.1.6). For the evaluation of these spectra the authors have assumed a small quadrupolar perturbation and a large magnetic interaction, as depicted in Fig. 7.1.2 and represented by the Hamiltonian [7.1.6]

$$\hat{H}_1 = \hat{H}_M + \hat{H}_Q', \quad \text{with} \quad \hat{H}_M = -g\beta_N \hat{H}_z \hat{I}_z, \tag{7.1.2}$$

$$\hat{H}_Q' = [eQV_{zz}/4I(2I-1)][3(-\sin\theta\,\hat{I}_x + \cos\theta\,\hat{I}_z)^2 - \hat{I}^2].$$

First-order perturbation theory yields the eigenvalues (diagonal elements of the perturbation matrix added to the eigenvalues of the major interaction)

$$E_m^1 = -g\beta_N H_z m_i + [eQV_{zz}/4I(2I-1)][3m_I^2 - I(I+1)](3\cos^2\theta - 1)/2,$$

$$m_I = -I, -I+1, ..., I. \tag{7.1.3}$$

θ is the angle between the quadrupolar quantization axis and that of the magnetic interaction in systems of axial symmetry (asymmetry parameter $\eta = 0$).

A rigorous quantum mechanical treatment of the combined interaction is laborious and computer-time consuming, because each interaction tends to have its own set of quantization axes, whereas the total perturbation does not quantize along either of the two sets of axes.

The alternative case of approximation analogous to the one mentioned above in (7.1.2) assumes a small magnetic perturbation and a large quadrupole interaction. This case, which is very rare and has not yet been observed in nickel systems, is expressed by the Hamiltonian [7.1.6]

$$\hat{H}_2 = \hat{H}_Q + \hat{H}_M', \quad \text{with} \quad \hat{H}_Q = [eQV_{zz}/4I(2I-1)](3\hat{I}_z^2 - \hat{I}^2), \tag{7.1.4}$$

$$\hat{H}_M' = g\beta_N H_z (-\sin\theta\,\hat{I}_x + \cos\theta\,\hat{I}_z).$$

The first-order eigenvalues for axial systems ($\eta = 0$) are

$$E_m^2 = [eQV_{zz}/4I(2I-1)]\,[3m_I^2 - I(I+1)] - g\beta_N H_z m_I \cos\theta. \qquad (7.1.5)$$

Here the magnetic perturbation is projected onto the quantization axes of $\hat{H}_Q$ as the major interaction.

7.1.3. Applications

From the applications of ^{61}Ni Mössbauer spectroscopy in solid-state research one has learned (i) that informations from isomer shift studies generally are not very reliable because of the smallness of the observed isomer shifts and the necessity of second-order Doppler shift corrections which turn out to be difficult, and (ii) that useful informations about magnetic properties and site symmetry is obtained from spectra reflecting magnetic and/or quadrupolar interactions.

In addition to the examples mentioned in Section 7.1.2 we add in the following a list of reports on ^{61}Ni Mössbauer studies of solid-state problems published recently. We deliberately omit details like parameter data and experimental conditions; the reader who is interested in those is referred to the Mössbauer Effect Data Index [7.1.2] and the original reports. Under "Remarks" some essential features and solid-state problems of interest are quoted briefly.

Reference	System	Remarks
[7.1.11]	$ErNi_2$ $GdNi_2$ Ni_xFe_{1-x}	$\langle H^{int}\rangle$ at Ni sites, dependence of $\langle H^{int}\rangle$ on x
[7.1.18]	Ni_xPd_{1-x}	Dependence of $\langle H^{int}\rangle$ on x
[7.1.19]	Ni–Pt Ni–Pd	
[7.1.20]	$Ni_{1-x}Pd_x(x:0-0.995)$	Absorber recoilless fraction, energy shift, and $|H^{int}|$ as function of x; δ values from temperature dependence of SOD; calculation of $\langle H^{int}\rangle$ and comparison with measurements in applied fields
[7.1.21]	$Ni_{1-x}Mn_x(x:0.05 - 0.25)$	$\langle H^{int}\rangle$ at Ni sites as function of x
[7.1.22]	$Ni_{1-x}Pt(x:0.1 - 0.5)$ $Ni_{1-x}Pd_x(x0.05 - 0.9)$	$\langle H^{int}\rangle$ at Ni sites as function of x
[7.1.17]	$NiFe_2O_4$ (a) $NiCo_2O_4$ (b) $NiMn_2O_4$ (c) $GeNi_2O_4$ (d) $NiCr_2O_4$ (e) $NiRh_2O_4$ (f)	$|H^{int}| = 7-10$ Tesla for Ni in octahedral sites (a–d), ca. 45 Tesla for Ni in tetrahedral sites (e); quadrupolar interaction in (e) and (f)

Reference	System	Remarks
[7.1.7]	$K_4Ni(CN)_6$ NiF_2 $KNiF_3$ NiO $NiSO_4 \cdot 7H_2O$ $(NH_4)_{12}[NiMo_9O_{32}] \cdot 13H_2O$ $NiAl$ $Fe_xNi_{1-x}(x:0.2-0.99)$ $Co_xNi_{1-x}(x:0.1-0.5)$ $Cu_xNi_{1-x}(x:0.15-0.7)$	δ from measured energy shifts after SOD correction, comparison with calculated (relativistic) values; H^{int} (^{61}Ni) in antiferromagnetic d^8 compounds; $\langle H^{int}\rangle$ (^{61}Ni) as function of x
[7.1.23]	NiS	Vibrational, magnetic, and electronic properties in semimetallic antiferromagnetic and in metallic phase; phase transition study
[7.1.24]	NiS_2	Study of magnetic dipole and electric quadrupole interactions, two Ni sites differing in angle between H and EFG axis; phase transitions in $NiS_{2.00}$ and $NiS_{1.96}$
[7.1.25]	NiS_2 $NiS_x(1.91 \leqslant x \leqslant 2.1)$ $Ni_{1-y}Cu_yS_{1.93}$ $(0.03 \leqslant y \leqslant 0.1)$	Investigation of structural, electronic, and magnetic properties by means of x-ray diffraction, densitometry, resistivity, susceptibility and ^{61}Ni Mössbauer spectroscopy as function of x; temperature of phase transition from semimetallic to metallic state as function of x; different Ni sites with different $\langle|H^{int}|\rangle$ and different angle between H and EFG axis; effect of Cu impurities
[7.1.26]	$NiCr_2O_4$	Study of magnetic hyperfine and quadrupole splitting, $H(T)/H(T_0) = f(T/T_N)$, fit to Brillouin function with $S = 1$

References

[7.1.1] Obenshain, F. E., Wegener, H. H. F.: Phys. Rev. *121*, 1344 (1961)

[7.1.2] Obenshain, F. E.: Nuclear Gamma Resonance with ^{61}Ni, in *Mössbauer Effect Data Index*, J. G. Stevens, V. E. Stevens (eds.). London: Adam Hilger 1972, p. 13

[7.1.3] Erich, U., Quitmann, D.: in *Hyperfine Struct. Nucl. Radiat.*, Proceedings, E. Matthias, D. A. Shirley (eds.). Amsterdam: North-Holland 1968, p. 130

[7.1.4] Erich, U.: Z. Phys. *227*, 25 (1969)

[7.1.5] Spijkerman, J. J.: in *The Mössbauer Effect*, Symp. Faraday Soc., No. 1, 1967. London: Butterworth 1968, p. 134

[7.1.6] Travis, J. C., Spijkerman, J. J.: in *Mössbauer Effect Methodology*, Vol. 4, I. J. Gruverman (ed.). New York: Plenum Press 1970, p. 237

[7.1.7] Love, J. C., Obenshain, F. E., Czjzek, G.: Phys. Rev. B. *3*, 2827 (1971)

[7.1.8] Obenshain, F. E., Williams, J. C., Houk, L. W.: J. Inorg. Nucl. Chem. *38*, 19 (1976)

[7.1.9] Ambe, F., Ambe, S., Takeda, M., Wei, H. H., Ohki, K., Saito, N.: Raidochem. Radioanal. Letters *1* (5), 341 (1969)

[7.1.10] Erich, U., Fröhlich, K., Gütlich, P., Webb, G. A.: Inorg. Nucl. Chem. Lett. *5*, 855 (1969)

[7.1.11] Erich, U., Kankeleit, E., Prange, H., Hüfner, S.: J. Appl. Phys. *40*, 1491 (1969)

[7.1.12] Childs, W. J., Goodman, L. S.: Phys. Rev. *170*, 136 (1968)

[7.1.13] Göring, J.: Z. Naturforsch. *26a*, 1931 (1971)

[7.1.14] Dunlap, B. D.: An Introduction to Electric Quadrupole Interactions in Mössbauer Spectroscopy, in *Mössbauer Effect Data Index*, 1970, J. G. Stevens, V. E. Stevens (eds.). London: Adam Hilger 1972

[7.1.15] Townes, C. H., Dailey, B. P.: J. Chem. Phys. *17*, 782 (1949)

[7.1.16] Dailey, B. P., Townes, C. H.: J. Chem. Phys. *23*, 118 (1955)

[7.1.17] Sekizawa, H., Okada, T., Okamoto, S., Ambe, F.: J. Phys. (Paris), Colloq. 32 Suppl., C1−326 (1971)

[7.1.18] Erich, U., Göring, J., Hüfner, S., Kankeleit, E.: Phys. Lett. *31A*, 492 (1970)

[7.1.19] Ferrando, W. A., Segnan, R., Schindler, A. I.: Phys. Rev. *B5*, 4657 (1972)

[7.1.20] Tansil, J. E., Obenshain, F. E., Czjzek, G.: Phys. Rev. *B 6*, 2796 (1972)

[7.1.21] Göring, J.: Phys. Stat. Solidi *B 57*, K 11 (1973)

[7.1.22] Göring, J.: Phys. Stat. Solidi *B 57*, K 7 (1973)

[7.1.23] Fink, J., Czjzek, G., Schmidt, H., Rübenbauer, K., Coey, J. M. D., Brusetti, R.: J. Phys. (Paris), Colloq. C 6, suppl. Vol. 35, C 6−657 (1974)

[7.1.24] Czjzek, G., Fink, J., Schmidt, H., Krill, G., Gautier, F., Lapierre, M. F., Robert, C.: J. Phys. (Paris), Colloq. C 6, suppl. Vol. 35, C 6−621 (1974)

[7.1.25] Krill, G., Lapierre, M. F., Gautier, F., Robert, C., Czjzek, G., Fink, J., Schmidt, H.: J. Phys. C: Solid State Phys. *9*, 761 (1976)

[7.1.26] Göring, J., Wurtinger, W., Link, R.: submitted to J. Appl. Phys.

7.2. Zinc (^{67}Zn)

Craig, Nagle and co-workers [7.2.1, 2] were the first to demonstrate the existence of recoilless nuclear resonance absorption for the 99.31 keV transition in ^{67}Zn. In an on-off type of experiment they merely clamped the source of ^{67}Ga/ZnO rigidly to a ZnO absorber, both at 2 K. The absorber was placed in a magnetic field varying between zero and 0.07 T (700 G). As the magnetic field was increased, a variation in the transmission of the 99.3 keV γ-rays (maximum relative absorption was 0.3%) due to increasing magnetic hyperfine splitting in the absorber was detected. In a similar experiment a Russian group also succeeded in observing nuclear resonance absorption in ^{67}Zn [7.2.3]. Attempts to record conventional velocity spectra were first undertaken by Alfimenkov et al. [7.2.4] using a piezoelectric quartz drive system, but unambiguous results could not yet be obtained. Considerable progress has been made by de Waard and Perlow [7.2.5], who succeeded in recording well-resolved ^{67}Zn Mössbauer hyperfine spectra, also using a piezoelectric quartz drive system.

The 93 keV resonance in ^{67}Zn has the highest relative energy resolution of the established Mössbauer isotopes. The comparatively long half-life of $t_{1/2} = 9.1$ μs of the 93 keV level yields an extremely small natural line width of $2\Gamma = 0.32$ μm s^{-1}. The relative natural width $\Gamma/E_\gamma = 5.2 \cdot 10^{-16}$ is about 600 times smaller than that of the 14.4 keV level of ^{57}Fe. The noise vibration velocities in most spectrometers are generally two orders of magnitude larger than the natural line width of the 93 keV transition in ^{67}Zn and render Mössbauer spectroscopy of zinc compounds a difficult task to perform. Good spectra can only be obtained with a specially designed spectrometer with an unconventional drive system, e.g., the piezoelectric quartz drive system as described in [7.2.5, 6], or a transducer containing commercially available cylinder of PZT-4 (lead zirconate-titanate) fed with a sinusoidal voltage to produce periodical elongation and contraction of the cylinder [7.2.7]. Both source and absorber are rigidly clamped to the piezoelectric crystal and the transducer is kept at 4.2 K. Calibration is based on the known piezomodulus at that temperature.

The 93 keV Mössbauer level is populated either by β^- decay of ^{67}Cu or by EC decay of ^{67}Ga. The decay scheme, reproduced from [7.2.8], is shown in Fig. 7.2.1. The nuclear data of interest for ^{67}Zn Mössbauer spectroscopy may be taken from Table 7.1 (end of the book).

The sources most commonly used so far consisted of sintered disks containing about 100 mg ZnO enriched to 90% in ^{66}Zn. The disks were irradiated with 12 MeV deuterons or 30 MeV ^{3}He particles, to yield the 78 h activity of ^{67}Ga, and annealed after irradiation by heating them in oxygen to 700–1,000 K for about 12 hours and cooling down slowly (about 50 K/h) to room temperature. A NaI scintillation counter 2–3 mm thick is suitable for the detection of the 93 keV γ-rays. Because of the relatively high transition energy, both source and absorber are generally kept at liquid helium temperature.

Most ^{67}Zn Mössbauer experiments so far have been carried out with ZnO as absorber. De Waard and Perlow [7.2.5] used polycrystalline ZnO enriched to 90% in ^{67}Zn with various pretreatments. They intended to determine (1) the quadrupole splitting in ZnO, (2) the influence of source and absorber preparation on the width and depth of a resonance, (3) the second-order Doppler shift, and (4) the influence of pressure on the source.

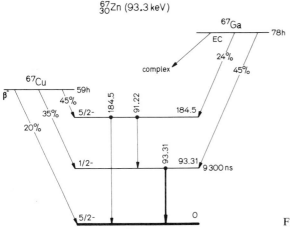

Fig. 7.2.1. Decay scheme of ^{67}Zn, from [7.2.8]

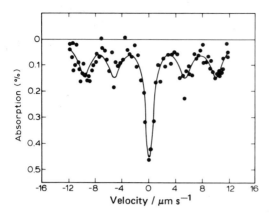

Fig. 7.2.2. ^{67}Zn Mössbauer spectrum of ^{67}ZnO at 4.2 K. Source: ^{67}Ga in ^{66}ZnO at 4.2 K. Absorber: 2.11 g/cm^2 ^{67}ZnO enriched to 90% in ^{67}Zn (from [7.2.5])

Figure 7.2.2 shows the ^{67}Zn Mössbauer spectrum of hexagonal ^{67}ZnO obtained with a ^{67}Ga/^{66}ZnO source by de Waard and Perlow [7.2.5, 6]. The full spectrum expected for electric quadrupole splitting in both source and absorber should contain seven resonance lines with intensity ratios of 1:1:1:3:1:1:1, according to the nine equally probable quadrupole resonance components for the pure E2 transition between the 5/2$^-$ ground state and the 1/2$^-$ excited state. Three of the nine components are degenerate and form the intense central line. The ^{67}ZnO spectrum of de Waard and Perlow does not show the outer two resonance lines because of the limited velocity range from -12 to $+12$ μm s^{-1} they were able to scan. From the positions of the lines these authors derived a quadrupole coupling constant $e^2qQ = 2.47 \pm 0.03$ MHz corresponding to 32.8 μm s^{-1}. From a small inequality of the spacings they determined the asymmetry parameter η to be 0.23, which they later corrected to 0.09 [7.2.9]. They explained the nonzero asymmetry parameter by an exceedingly small ($\sim$0.001 Å) deviation of oxygen atoms from the ideal tetrahedron. With the known ground state quadrupole moment of ^{67}Zn, Q = 0.17 b, they found for the z component of the electric field gradient at the ^{67}Zn nucleus an absolute value of eq $= V_{zz} = 6.0 \cdot 10^{16}$ V/cm^2, which agrees well with their calculated value (lattice sum of point charges) of $5.4 \cdot 10^{16}$ V/cm^2.

De Waard and Perlow [7.2.5] observed a marked influence of the method of source and absorber preparation on both the line width and the depth of resonance. For instance, the line width was found to be 2.7 μm s^{-1} in a powdered ZnO absorber as compared to 0.8 μm s^{-1} in a sintered one.

Because of the exceedingly small hyperfine splittings and shifts of resonance lines in ^{67}Zn Mössbauer spectroscopy, one should, in principle, consider the second-order Doppler shift as arising from different isotopic compositions or even more so from different chemical compositions of source and absorber. Lipkin [7.2.9] has derived a general expression for this shift, which the authors of [7.2.5] used to estimate a second-order Doppler shift for their experiments of $\Delta E_D \approx -0.006$ nm s^{-1}.

The influence of pressure on the ^{67}Ga/^{66}ZnO source was found to be remarkable [7.2.5]. Applying a pressure of about 40 kbar, the authors observed (a) a shift of about -0.11 μm s^{-1} for the intense central line of Fig. 7.2.2, (b) a (4 $\pm$ 2)% reduction of the splitting between the two outer lines, (c) about 25% broadening of

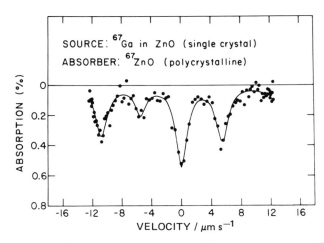

Fig. 7.2.3. ^{67}Zn Mössbauer spectrum of ^{67}ZnO using a source of ^{67}Ga in ZnO single crystal
(from [7.2.10])

the central line, and (d) a reduction of the ratio of center-line to outer-line intensity
from 3.6 ± 0.5 at zero pressure to 2.4 ± 0.3 at high pressure. All these changes were
compatible with an 8% reduction of the source quadrupole splitting as a result of
compression.

In later experiments Perlow et al. [7.2.10] used sources of ^{67}Ga in ZnO single
crystals (natural isotopic abundance, cyclotron bombardment) and a compressed
and sintered pellet of ZnO (2.1 g/cm^2, enriched to 90% in ^{67}Zn) as absorber. The
source crystals were disks of 1 cm in diameter and 0.5 mm thick, with the hexagonal
symmetry axis (c axis) perpendicular to the faces. The velocity spectrum obtained
in [7.2.10] at 4.2 K in the absence of a magnetic field is shown in Fig. 7.2.3. With
the symmetry axis of the single crystal oriented along the E2 γ-emission, the ratio
of line intensities of the full quadrupole split spectrum should be $1:2:1:3:2:0:0$
(no transitions for $\Delta m = \pm 2$). The reduced spectrum of Fig. 7.2.3 ranging from -12
to $+12$ μm s^{-1} shows four lines with relative intensities of approximately $2:1:3:2$
and is thus in very good agreement with prediction. The observed sequence of intensities shows that the quadrupole coupling is positive. Since the quadrupole moment
is known to be positive, the EFG turned out to be also positive.

In experiments with magnetic fields between 13 and 55 mT (130 to 550 G) applied to the single crystal source the authors of [7.2.10] observed magnetic dipole in
addition to electric quadrupole splitting in a reduced spectrum (cf. Fig. 7.2.4). They
determined the magnetic moment of the excited $1/2^-$ state to be $\mu(1/2^-) = + (0.58$
$\pm 0.03)\, \mu_N$.

For the purpose of the precise determination of the electric quadrupole interaction of ^{67}ZnO, Perlow et al. [7.2.11] applied the method of frequency modulation
Mössbauer spectroscopy to a ^{67}Ga/ZnO single crystal source versus polycrystalline
^{67}ZnO absorber at 4.2 K. They determined the electric quadrupole coupling constant
to be $e^2qQ = 2.408 \pm 0.006$ MHz, and the asymmetry parameter to be $\eta = 0.00$
$\left(^{+0.065}_{-0}\right)$.

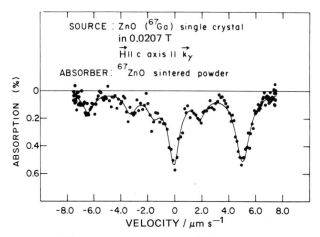

Fig. 7.2.4. ^{67}Zn Mössbauer spectrum of ^{67}ZnO as absorber with a single crystal ^{67}Ga/ZnO source in applied magnetic field (from [7.2.10])

Potzel et al. [7.2.12] used a ^{67}Ga/ZnO single crystal source in combination with a single crystal absorber of natural ZnO and observed a resonance line width of (0.36 ± 0.04) μm s^{-1} for the 93.3 keV transition in ^{67}Zn (at 4.2 K), which, after correction for finite absorber thickness, equals within the limit of error the minimum observable line width as deduced from the lifetime of 13.4 μs of the 93.3 keV state. The spectra observed by these authors are shown in Fig. 7.2.5.

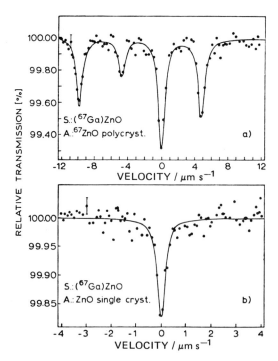

Fig. 7.2.5. ^{67}Zn Mössbauer spectra obtained with a single crystal source and (a) an enriched polycrystalline ZnO absorber (82.5% ^{67}Zn, 963 mg ^{67}Zn/cm^2, sintered in oxygen atmosphere at 1,000 K for 24 hrs), (b) a single crystal absorber of natural ZnO. Both source and absorber were at 4.2 K (from [7.2.12])

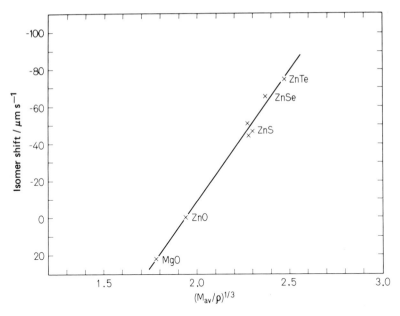

Fig. 7.2.6. Experimental isomer shifts of ^{67}Ga/XY sources (XY = ZnO, ZnS, ZnSe, ZnTe, MgO) relative to ZnO as absorber at 4.2 K is plotted versus the lattice spacing parameter $(M_{av}/\rho)^{1/3}$ (from [7.2.6]). The isomer shift for MgO was taken from [7.2.13]

Griesinger at al. [7.2.7] recorded ^{67}Zn Mössbauer spectra with sources of ^{67}Zn diffused into ZnO, ZnS (both wurtzite and sphalerite), ZnSe, ZnTe, and Cu, and an enriched ^{67}ZnO absorber. The isomer shifts extracted from their spectra cover a velocity range of 112 μm s^{-1} and are found to follow linearly the lattice spacing parameter $(M_{av}/\rho)^{1/3}$, where ρ and M_{av} are the host density and average atomic weight, respectively (cf. Fig. 7.2.6). The experimental data of Fig. 7.2.6 have been corrected for zero-point motion using Lipkin's formula [7.2.9]. The zero-point vibration calculations yielded much smaller shifts than are observed experimentally, from which the authors concluded that the shifts observed for the zinc chalcogenides were largely chemical in nature. From the observed sign of the variation in the isomer shift with $(M_{av}/\rho)^{1/3}$ and from the assumption that larger lattice spacings go along with less ionic bonds and thus with higher 4s electron density at the zinc nucleus the authors tentatively deduced the charge radius of the excited state as being greater than that of the ground state, $\Delta \langle r^2 \rangle > 0$. In agreement with this suggestion are results recently obtained in Munich, where ^{67}Zn Mössbauer experiments with a ^{67}Ga/ZnO source and single crystals of zinc chalcogenides as absorbers are currently going on [7.2.14]. For instance for ZnS as absorber, an isomer shift of 53.9 ± 0.5 μm s^{-1} with respect to a ^{67}Ga/ZnO source (both source and absorber at 4.2 K) has been measured.

References

[7.2.1] Nagle, D. E., Craig, P. P., Keller, W. E.: Nature *186*, 707 (1960)

[7.2.2] Craig, P. P., Nagle, D. E., Cochran, D. R. F.: Phys. Rev. Lett. *4*, 561 (1960)

[7.2.3] Aksenov, S. I., Alfimenkov, V. P., Lushchikov, V. I., Ostanevich, Yu. M., Shapiro, F. L., Yen, W.: Zh. Eksperim. i Teor. Fiz. *40*, 88 (1961) [Soviet Phys. JETP *13*, 63 (1961)]

[7.2.4] Alfimenkov, V. P., Ostanevich, Yu. M., Ruskov, T., Strelkov, A. V., Shapiro, F. L., Yen, W. K.: Zh. Eksperim. i Teor. Fiz. *42*, 1029 (1962) [Soviet Phys. JETP *15*, 713 (1962)]

[7.2.5] de Waard, H., Perlow, G. J.: Phys. Rev. Lett. *24*, 566 (1970)

]7.2.6] Perlow, G. J.: in *Perspectives in Mössbauer Spectroscopy*, S. G. Cohen, M. Pasternak (eds.). New York: Plenum Press 1972, pp. 221–237

[7.2.7] Griesinger, D., Pound, R. V., Vetterling, W.: Phys. Rev. B *15*, 3291 (1977)

[7.2.8] Stevens, J. G., Stevens, V. E.: *Mössbauer Effect Data Index*, Covering the 1975 Literature. New York: IFI Plenum 1976

[7.2.9] Lipkin, H. J.: Ann. Phys. (Leipz.) *23*, 28 (1963)

[7.2.10] Perlow, G. J., Campbell, L. E., Conroy, L. E., Potzel, W.: Phys. Rev. B *7*, 4044 (1973)

[7.2.11] Perlow, G. J., Potzel, W., Kash, R. M., de Waard, H.: J. Physique Collq. C6 *35*, C6–197 (1974)

[7.2.12] Potzel, W., Forster, A., Kalvius, G. M.: J. Physique Collq. C6 *37*, C6–691 (1976)

[7.2.13] Breskrovny, A. I., Lebedev, N. A., Ostanevich, Y. M.: Joint Inst. for Nuclear Research Report, Dubna, 1971 (unpublished)

[7.2.14] Forster, A. A.: Diplomarbeit, Physik-Department E15, Technische Universität München, April 1976

We thank Dr. W. Potzel for communicating to us some of their results prior to publication.

7.3. Ruthenium (^{99}Ru, ^{101}Ru)

Kistner, Monaro, and Segnan [7.3.1] were the first to observe the Mössbauer effect in a ruthenium nucleus (^{99}Ru). Kistner also reported the first example [7.3.2] to demonstrate the usefulness of ^{99}Ru Mössbauer spectroscopy in studying ruthenium compounds and alloys.

Potzel et al. [7.3.3] have established the recoil-free nuclear resonance effect in another ruthenium nuclide, ^{101}Ru. This isotope, however, is much less profitable than ^{99}Ru in the elucidation of problems of ruthenium chemistry because of the very small resonance effect as a consequence of the large transition energy (127.2 keV) and the much broader line width (about 30 times broader than the ^{99}Ru line). The relevant nuclear properties of both ruthenium isotopes are listed in Table 7.1 (end of the book). The decay schemes of Fig. 7.3.1 show the way of feeding the Mössbauer nuclear levels of ^{99}Ru (89.36 keV) and ^{101}Ru (127.2 keV), respectively.

7.3.1. Experimental Aspects

The source preparation imposes no particular difficulties except that cyclotron radiation must be accessible. The precursor of ^{99}Ru is ^{99}Rh ($t_{1/2}$ = 16 d), which is prepared by bombarding natural ruthenium metal with 10 MeV protons, ^{99}Ru (p, n)^{99}Rh, or 20 MeV deuterons, ^{99}Ru (d, 2n)^{99}Rh. Ru metal generally serves as the host lattice

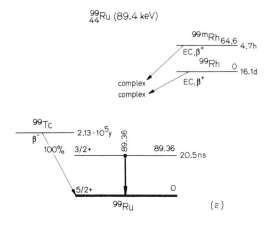

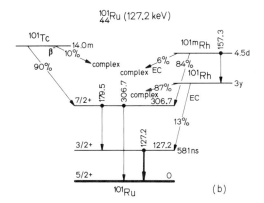

Fig. 7.3.1. Simplified decay schemes of ^{99}Ru (a) and ^{101}Ru (b) (reproduced from [7.3.4])

for ruthenium Mössbauer sources because it produces a single transition line close to the natural line width despite its hexagonal crystal structure. Spectra may be taken at liquid nitrogen temperature [7.3.1]; significantly larger effects, however, are obtained at lower temperatures using conventional helium cryostats. All results reported so far were obtained using transmission geometry and counting the 90 keV gamma rays with a NaI(Tl) scintillation detector (preferably 3 mm NaI(Tl) crystal). The absorber thickness should vary between 50 and 100 mg of natural ruthenium per cm^2 to yield sizeable absorption effects. Isomer shifts are generally given with respect to metallic ruthenium.

The precursor of ^{101}Ru is ^{101}Rh ($t_{1/2}$ = 3 y). It is prepared by irradiating natural ruthenium metal with 20 MeV deuterons, ^{100}Ru (d, n)^{101}Rh. The target is afterwards allowed to decay for several months to diminish the accompanying ^{99}Rh activity. In the only report on the ^{101}Ru Mössbauer effect published so far [7.3.3], the authors have recorded spectra of Ru metal, RuO$_2$, and [Ru(NH$_3$)$_4$ (HSO$_3$)$_2$] at liquid helium temperature in standard transmission geometry using a Ge(Li) diode to detect the 127 keV gamma rays. The absorber samples contained approximately 1 g of ruthenium per cm^2.

7.3.2. Chemical Information from ^{99}Ru Mössbauer Parameters

Kistner's early measurements of the ^{99}Ru Mössbauer effect in Ru metal, RuO_2, $Ru(C_5H_5)_2$, and $Ru_{0.023}Fe_{0.977}$ [7.3.2] have demonstrated that

(i) isomer shift changes are sufficiently large for Ru atoms in different systems to distinguish between different oxidation states and different bond properties;

(ii) quadrupolar perturbation may be revealed to be of use in studying aspherical charge distributions about Ru atoms;

(iii) magnetic hyperfine splitting may be detected and used for investigations of local magnetic fields (sign and magnitude of $\vec{H}_{int}$) in systems exhibiting cooperative magnetism.

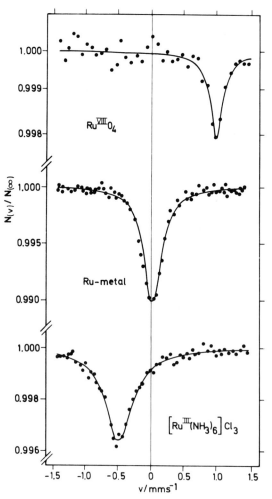

Fig. 7.3.2. ^{99}Ru Mössbauer spectra of RuO_4, Ru metal, and $[Ru^{III}(NH_3)_6]\,Cl_3$. The source (^{99}Rh in Ru metal) and the absorbers were kept at 4.2 K (from [7.3.5])

(a) Isomer Shift

The relative change of the mean-square nuclear radius in going from the excited to the ground state, $\Delta\langle r^2\rangle/\langle r^2\rangle$, is positive for ^{99}Ru. An increase in observed isomer shifts δ therefore reflects an increase of the s-electron density at the Ru nucleus caused by either an increase in the number of s-valence electrons or a decrease in the number of shielding electrons, preferentially of d-character.

One of the interesting features in ruthenium chemistry is the large variety of oxidation states (0 to +8) of ruthenium, which may well be distinguished by isomer shift measurements. As an example, Fig. 7.3.2 shows a few representative single-line spectra [7.3.5] reflecting the significant isomer shift changes for different oxidation states of ruthenium. In an early study of a variety of ruthenium compounds, Kaindl et al. [7.3.5] have correlated the observed isomer shifts with the oxidation state in a graphical representation as shown in Fig. 7.3.3. The diagram shows clearly that the isomer shift values increase monotonically with the oxidation state of ruthenium, and that each formal oxidation state embraces a certain range of isomer shifts, indicating that different bond properties in terms of σ-donation ($\sigma_L \rightarrow M$; L = ligand, M = metal) and d_π–p_π back donation ($\pi_M \rightarrow L$) are effective. It is particularly note-

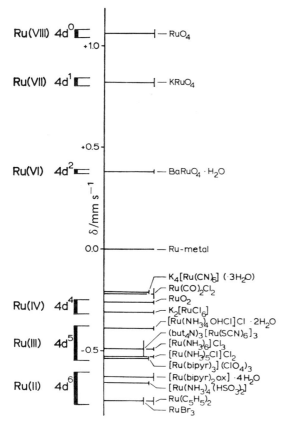

Fig. 7.3.3. Measured ^{99}Ru isomer shifts in ruthenium compounds with different oxidation states of Ru (from [7.3.5])

worthy that $K_4[Ru(CN)_6] \cdot 3H_2O$ shows an isomer shift close to the shifts of Ru(IV) compounds, from which it is inferred that practically two t_{2g} electrons have delocalized from the Ru(II) cluster to the strong π-bonding CN$^-$ ligands.

In another investigation of bond properties in ruthenium compounds Potzel et al. [7.3.6] have found that the isomer shift increases in the series

$$[Ru^{II}(CN)_5NO_2]^{4-} < [Ru^{II}(CN)_6]^{4-} < [Ru^{II}(CN)_5NO]^{2-} \text{ and}$$

$$[Ru^{II}(NH_3)_6]^{2+} < [Ru^{II}(NH_3)_5N_2]^{2-} < [Ru^{II}(NH_3)_5NO]^{3+}$$

from which they derived the following order for the back-donation power of the ligands:

$$NO_2^- < CN^- < NO^+ \text{ and}$$

$$NH_3 < N_2 \ll NO^+,$$

i.e., the nitrosyl group exhibits the strongest backbonding capability, which parallels earlier observations made in analogous iron complexes. This has also been supported by the observed isomer shift tendency in the series [7.3.6]

$$[Ru^{II}(NH_3)_6]^{2+} < [Ru^{II}(NH_3)_4pyr_2]^{2+} < [Ru^{II}(NH_3)_4(SO_3H)_2] <$$

$$< [Ru^{II}(NH_3)_4Cl(SO_2)]^+ \ll [Ru(NH_3)_4(OH)(NO)]^{2+}.$$

The isomer shift changes in the series [7.3.6]

$$[Ru^{II}Cl_5NO]^{2-} < [Ru^{II}(NH_3)_5NO]^{3+} < [Ru^{II}(CN)_5NO]^{2-} \text{ and}$$

$$[Ru^{II}(NH_3)_6]^{2+} \ll [Ru^{II}(CN)_6]^{4-}$$

reflect strong back-donation properties of the CN$^-$ ligand as compared to NH$_3$. These δ-changes also reflect that the presence of a NO$^+$ ligand provokes less change in electron density in going from the pentaammine to the pentacyano complex as compared to the transition from $[Ru^{II}(NH_3)_6]^{2+}$ to $]Ru^{II}(CN)_6]^{4-}$. This suggests that some kind of a saturation value is reached when an isomer shift change formally corresponding to about 2.5 oxidation states has been induced by π-backbonding effects. In analogy to $[Fe^{II}(CN)_5NO]^{2-}$ the $[Ru(CN)_5NO]^{2-}$ complex shows the most positive isomer shift among divalent ruthenium compounds.

From the isomer shift tendencies in the series

$$[Ru^{II}Cl_5NO]^{2-} < [Ru^{II}(NH_3)_5NO]^{3+} < [Ru^{II}(CN)_5NO]^{2-},$$

$$[Ru^{II}(NH_3)_6]^{2+} \ll [Ru^{II}(CN)_6]^{4-} \text{ and}$$

$$[Ru^{III}F_6]^{3-} < [Ru^{III}(NCS)_6]^{3-} \approx [Ru^{III}(NH_3)_6]^{3+}$$

it is obvious that the ordering of the ligands parallels the spectrochemical series with the NCS$^-$ ligand being N bonded to ruthenium.

Greatrex et al. [7.3.7] have also studied bond properties in nitrosyl ruthenium(II) compounds, [RuL$_5$(NO)]$^{n+}$, by ^{99}Ru spectroscopy and found that the isomer shift δ decreases steadily as the ligand-field strength of L decreases in the order L = CN$^-$ > > NH$_3$ > NCS$^-$ > Cl$^-$ > Br$^-$, as represented graphically in Fig. 7.3.4. From these findings the authors have derived a positive sign for $\Delta\langle r^2\rangle/\langle r^2\rangle$ in ^{99}Ru.

Up to now a large number of ruthenium complexes have been investigated by ^{99}Ru Mössbauer spectroscopy, mainly by the Munich group and the groups of M. L. Good (New Orleans, USA), N. N. Greenwood (England). It is now possible to set up much more complete isomer shift diagrams of the type shown in Fig. 7.3.3, as has been done, e.g., for ammine and halide complexes of Ru(II) and Ru(III) in an instructive review article by Good [7.3.8].

A correlation of isomer shift, electronic configuration, and calculated s-electron densities for a number of ruthenium complexes in analogy to the Walker-Wertheim-Jaccarino diagram for iron compounds has been reported by Clausen, Prados and Good [7.3.9]. Also useful is the correlation between isomer shift and electronegativity as communicated by Clausen et al. [7.3.10] for ruthenium trihalides, where the isomer shift appears to increase with increasing Mulliken electronegativity.

^{99}Ru isomer shift studies have been demonstrated to be particularly powerful in characterizing the oxidation state of ruthenium in mixed valence polynuclear ruthenium complexes. Clausen, Prados and Good [7.3.11] were able this way to distinguish

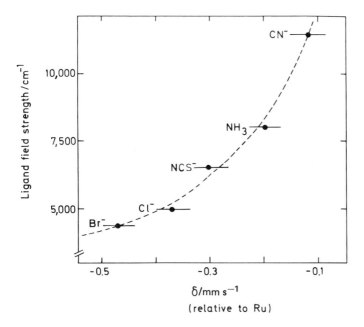

Fig. 7.3.4. Partial ligand-field strength for the ligand L in correlation with the isomer shift δ (relative to Ru metal at 4.2 K) of the nitrosylruthenium (II) compounds [RuL$_5$(NO)]$^{n\pm}$ (L = Br$^-$, Cl$^-$, NCS$^-$, NH$_3$, CN$^-$) (from [7.3.7])

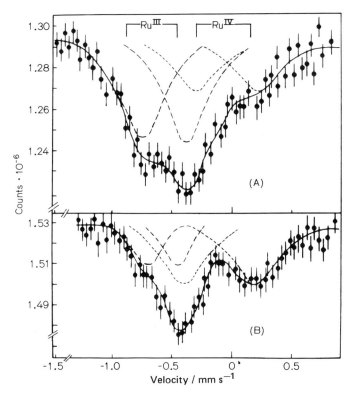

Fig. 7.3.5. ^{99}Ru Mössbauer spectra of (A) "ruthenium red" and (B) "ruthenium brown" at 4.2 K; source (^{99}Rh in Ru metal) at 4.2 K (from [7.3.11])

clearly between the +3 and +4 oxidation states in "ruthenium red", $[(NH_3)_5Ru^{III}-O-Ru^{IV}(NH_3)_4-O-Ru^{III}(NH_3)_5]^{6+}$ as well as in "ruthenium brown", $[(NH_3)_5Ru^{IV}-O-Ru^{III}(NH_3)_4-O-Ru^{IV}(NH_3)_5]^{7+}$. Their ^{99}Ru Mössbauer spectra are reproduced in Fig. 7.3.5. There are two pairs of quadrupole split lines, whose intensity ratios Ru^{III}/Ru^{IV} are in accordance with the numerical ratio of the ruthenium atoms with oxidation levels +3 and +4, viz., 2:1 in case of the ruthenium red cation, and 1:2 in case of the ruthenium brown cation. In a similar study Creutz, Good, and Chandra [7.3.12] have established the coexistence of the formal oxidation states +2 and +3 with equal intensities in the mixed valence complex

$$[(NH_3)_5Ru-N\langle\bigcirc\rangle N-Ru(NH_3)_5]^{5+}.$$

The ^{99}Ru Mössbauer spectra of the ternary oxides Na_3RuO_4, $Na_4Ru_2O_7$ and $NaRu_2O_4$ measured at 4.2 K by Greenwood, da Costa and Greatrex [7.3.13] confirm through the isomer shift data that in the first two compounds ruthenium is in the +5 oxidation state, whereas the +3 and +4 oxidation levels coexist in $NaRu_2O_4$.

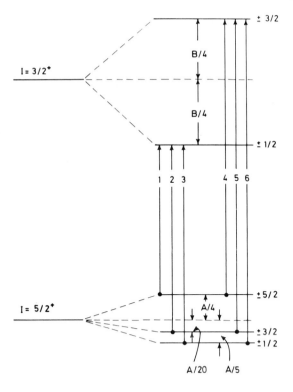

Fig. 7.3.6. Effect of a positive quadrupole interaction on the ground
and first excited states of ^{99}Ru.
The asymmetry parameter η is assumed to be zero. The ratio of the
quadrupole moments is taken to be
$Q_{3/2}/Q_{5/2} = 3$. $A = e^2qQ_{5/2}$ and
$B = e^2qQ_{3/2}$ (from [7.3.7])

(b) Quadrupole Splitting

^{99}Ru has a nuclear ground state with spin $I_g = 5/2^+$ and a first excited state with $I_e = 3/2^+$. Electric quadrupole perturbation of ^{99}Ru was first reported by Kistner [7.3.2] for RuO_2 and ruthenocene; this author has also evaluated the ratio of the nuclear quadrupole moments to be $|Q_e/Q_g| \geqslant 3$. The sign and magnitude of the individual quadrupole moments are now known and given in Table 7.1 (end of the book).

The effect of a positive quadrupole interaction on the ground and excited nuclear states of ^{99}Ru is shown schematically in Fig. 7.3.6 as adapted from a publication by Greatrex, Greenwood, and Kaspi [7.3.7]. As the transition between the $I_g = 5/2^+$ ground state and the $I_e = 3/2^+$ excited state of ^{99}Ru occurs via combined M1 and E2 radiation with a mixing ratio of $\delta^2 = 2.7 \pm 0.6$ [7.3.2], the allowed transitions are those with $\Delta m_I = 0, \pm 1, \pm 2$. Following these selection rules we see that six transition lines altogether should appear in a quadrupole-split ^{99}Ru spectrum using a single-line source; their relative intensities may be determined from the squares of the Clebsch-Gordan coefficients as given in [7.3.7]. An actual quadrupole-split ^{99}Ru spectrum, however, generally shows only two resonance lines, each of which consists of three unresolved lines due to the relatively small quadrupole moment of the ground state. Thus the measurable quadrupole splitting reflects in most cases that of the $I_e = 3/2^+$ excited state of ^{99}Ru. As an example the ^{99}Ru spectrum of Co_2RuO_4, measured at 4.2 K by Gibb et al. [7.3.14] is reproduced in Fig. 7.3.7.

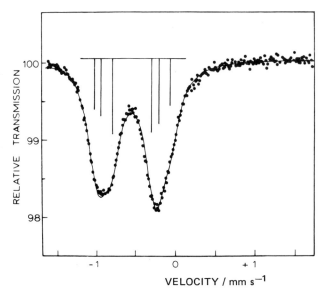

Fig. 7.3.7. Mössbauer spectrum of Co_2RuO_4 at 4.2 K with respect to the Ru metal source (from [7.3.14])

As in the case of iron chemistry, most valuable information concerning bond properties (anisotropic electron population of molecular orbitals) and local structure may be extracted from quadrupole-split ^{99}Ru spectra. This has been demonstrated in a number of communications from various groups. Clausen, Prados and Good [7.3.15], for example, have studied the cyano complexes of ruthenium(II),

I. $K_2[Ru(CN)_5NO] \cdot 2H_2O$
II. $K_4[Ru(CN)_6]$
III. $K_4[Ru(CN)_5NO_2] \cdot 2H_2O$

and interpreted the observed quadrupole splitting in I and III as arising predominately from an asymmetric expansion of the ruthenium t_{2g} electrons towards the ligands. The sign of the EFG is expected to be positive for complex I because NO^+ is known [7.3.16] to form very strong π-bonds (stronger than π-bonds of CN^-), and negative for complex III because NO_2^- is incapable of forming π-bonds. No quadrupole splitting has been observed in $K_4[Ru(CN)_6]$ because of O_h point symmetry of Ru which yields zero EFG.

The observed quadrupole splittings in the mixed-valence complexes "ruthenium red" and "ruthenium brown" [7.3.11] have been ascribed to a large contribution to the EFG from the partial filling of the t_{2g} orbitals in low-spin Ru(III) and Ru(IV).

Greatrex, Greenwood, and Kaspi [7.3.7] have observed quadrupole splittings in the octahedral nitrosylruthenium(II) complexes $[RuL_5(NO)]^{n\pm}$ with decreasing magnitudes in the order of the ligands $L = CN^- > NO^+ > NCS^- \approx Cl^- > Br^-$ and discussed these qualitatively in terms of the relative σ-donor and π-acceptor abilities of the ligands. They have found that, whereas σ-donation and π-back-donation reinforce one another in their effect on the isomer shift, they oppose one another in their effect on the quadrupole splitting.

Bancroft, Butler, and Libbey [7.3.17] have used partial quadrupole-splitting values from low-spin iron(II) compounds to obtain the sign of the quadrupole coupling constant $e^2 qQ_e$ for a number of ruthenium(II) compounds. Their correlation between $1/2 \, e^2 qQ$ of iron(II) compounds and $|1/2 \, e^2 qQ_e|$ of analogous ruthenium(II) compounds strongly suggests that the bonding in corresponding Fe and Ru complexes is reasonably similar. In the $M(NH_3)_5 L$ series ($L = N_2$, MeCN, CO) they found that $(\sigma-\pi)$ (σ-donor ability, π-acceptor ability) increases in the order $N_2 < MeCN < CO$. CO is most likely a better σ-donor and a better π-acceptor than NH_3; N_2 is probably a better π-acceptor, but poorer σ-donor than NH_3. They also established that $e^2 qQ_e$ will become more negative as $(\sigma-\pi)$ of L increases, and that the isomer shift will become more positive as $(\sigma + \pi)$ of L increases.

Foyt, Siddall, Alexander, and Good [7.3.18] interpreted the quadrupole-splitting parameters of low-spin ruthenium(II) complexes in terms of a crystal field model in the strong-field approximation with the t_{2g}^5 configuration treated as an equivalent one-electron problem. They have shown that, starting from pure octahedral symmetry with zero quadrupole splitting, ΔE_Q increases as the ratio of the axial distortion to the spin-orbit coupling increases.

The largest quadrupole splittings ever found in ^{99}Ru Mössbauer spectra have been reported by Gibb, Greatrex, Greenwood, and Meinhold [7.3.19] for some nitrido complexes of Ru(IV), $(Bu_4^n N)[RuNCl_4]$ and $(Ph_4 As)[RuNBr_4]$. From the almost resolved six-component pattern they derived new estimates for the ratio of the quadrupole moments of $Q_e/Q_g = +2.82 \pm 0.09$, with both Q_e and Q_g being positive, and for the E2/M1 mixing ratio of $\delta^2 = 2.64 \pm 0.17$.

In a recently published account on the reduction of Mössbauer data Foyt, Good, Cosgrove, and Collins [7.3.20] have described a general analysis of quadrupole-split ^{99}Ru spectra.

(c) Magnetic Splitting

Magnetic hyperfine splitting in ^{99}Ru Mössbauer spectra has already been observed by Kistner [7.3.2, 21] for an absorber of 2.3 at.-% ruthenium dissolved in metallic iron. The spectra obtained with an unpolarized absorber (a) and with polarized absorbers (b) (magnetization parallel to incident γ-rays) and (c) (magnetization perpendicular to incident γ-rays) are reproduced from Kistner's work and shown in Fig. 7.3.8. The stick spectra on top of Fig. 7.3.8a indicate the calculated [7.3.2] line positions and the relative intensities for the 18 allowed electric quadrupole (E2) transitions with the change in magnetic spin quantum number of $\Delta m = 0, \pm 1, \pm 2$ as well as for the 12 allowed magnetic dipole (M1) transitions with $\Delta m = 0, \pm 1$. The best-fit value for the mixing ratio E2/M1 in this case was found by Kistner to be $\delta^2 = 2.7$.

Extensive use of the magnetic hyperfine splitting of the ^{99}Ru Mössbauer resonance in ruthenium chemistry has been initiated by Greenwood's group in England with their study on $SrRuO_3$ [7.3.22], which orders ferromagnetically below 160 K

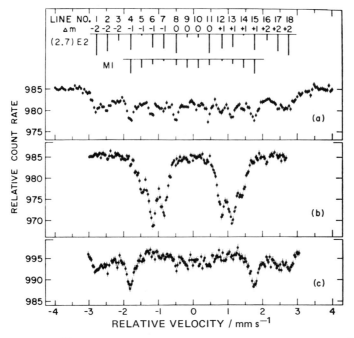

Fig. 7.3.8. ^{99}Ru Mössbauer spectra of $Ru_{0.023}Fe_{0.977}$ at 4.2 K (source: ^{99}Rh in Ru metal). (a) unpolarized absorber; (b) absorber magnetized parallel and (c) perpendicular to incident γ-rays (from [7.3.2])

and shows metallic properties. The magnetic field strength was found to be 35.2 ± 1.5 T at 4.2 K and 31.5 ± 1.5 T at 77 K. The sign of the field has not been determined, but is assumed to be negative.

In a subsequent publication the same research group reported on ^{99}Ru Mössbauer studies of the magnetic properties of the ternary oxides $CaRuO_3$, $SrRuO_3$, $BaRuO_3$ and $Y_2Ru_2O_7$, and the quaternary oxides $Sr(Ru_{1-x}Ir_x)O_3$ (x = 0.1 and 0.2) and $Sr(Ru_{0.7}Mn_{0.3})O_3$ [7.3.23]. The internal field of 35.2 T (4.2 K) in $SrRuO_3$ has been found to be compatible with the ferromagnetic moment derived from neutron diffraction and susceptibility data and confirms the collective electron magnetism model for this perovskite. The E2/M1 mixing ratio for the ^{99}Ru Mössbauer γ-transition has been reevaluated to be $\delta^2 = 2.72 \pm 0.17$. The antiferromagnetic ordering in $CaRuO_3$ reported earlier by other authors could not be confirmed. Both oxides $CaRuO_3$ and $BaRuO_3$ gave single-line spectra even at 4.2 K. $Y_2Ru_2O_7$, however, has revealed magnetic ordering at 4.2 K with a field of 12.6 T. No quadrupole splitting has been observed in any of these oxides. Substitution of ruthenium in $SrRuO_3$ has been found to reduce the magnetic field at Ru atoms in case of iridium substitution by approximately 6 T for each Ir nearest neighbour, and in case of manganese substitution by only 2.2 T for each Mn neighbour.

In a further study along this line Gibb, Greatrex, Greenwood et al. [7.3.24] have examined the magnetic superexchange interactions in the solid solution series

$Ca_xSr_{1-x}RuO_3$ ($0.1 \leqslant x \leqslant 0.5$) with distorted perovskite structure and found that the proportion of ruthenium atoms experiencing magnetic fields close to 35 T decreases as x increases. In samples with $x \geqslant 0.3$ the central "paramagnetic" component, imposed on the magnetic hyperfine split spectra, increases rapidly with increasing x, which has been interpreted as being indicative of a rapid relaxation of the total electron spin and hence the magnetic field due to weakening of the coupling between Ru atoms by Ca substitution. This result has led to the conclusion that the greater electron-pair acceptor strength (Lewis acidity) of Ca^{2+} compared to Sr^{2+} causes a more effective competition with Ru atoms for the oxygen anion orbitals involved in the superexchange mechanism.

The magnetic properties of the new solid solution series $SrFe_xRu_{1-x}O_{3-y}$ ($0 \leqslant x \leqslant 0.5$) with distorted perovskite structure, where iron substitutes exclusively as Fe^{3+}, thereby causing oxygen deficiency, has also been studied in Greenwood's group [7.3.25] using both ^{99}Ru and ^{57}Fe Mössbauer spectroscopy. Iron substitution has been found to have litte effect on the magnetic behaviour of the Ru^{4+} until $x > 0.2$.

Recently Greenwood, Costa and Greatrex [7.3.13] published a study of the ternary sodium ruthenium oxides Na_3RuO_4, $Na_4Ru_2O_7$ and $NaRu_2O_4$ by ^{99}Ru Mössbauer spectroscopy. Na_3RuO_4 shows a magnetic hyperfine pattern, from which a magnetic field at the nucleus of 58.7 T has been derived, the largest yet observed for any ruthenium system. The field arises from antiferromagnetic coupling of the Ru^{5+} ions. $Na_4Ru_2O_7$ shows a more complicated magnetic hyperfine spectrum, but with similar field strength as in Na_3RuO_4.

7.3.3. Further ^{99}Ru Studies

Other studies making use of the information content of the isomer shift and the quadrupole splitting from ^{99}Ru Mössbauer resonances deserve attention. Potzel, Wagner and Wäppling [7.3.26] have investigated ruthenium borides RuB_x (x = 1.1, 1.5, 2.1) and phosphides Ru_2P, RuP and RuP_2. The metal ion in the ruthenium borides turned out to be rather insensitive to the boron surrounding. The bonding in the iron and ruthenium borides, however, seems to be similar. The results for the phosphides also suggest similar electronic structures for the iron and ruthenium compounds.

Clausen and Good [7.3.27] have applied the ^{99}Ru Mössbauer effect to examine supported (alumina, silica) ruthenium catalysts. The Mössbauer data show that $RuCl_3 \cdot 1-3H_2O$ reacts chemically when supported onto alumina, but does not when impregnated on a silica support. Furthermore, the study has shown that a supported ruthenium catalyst converts quantitatively into RuO_2 upon calcination, and that the reduction of a supported ruthenium catalyst converts all of the ruthenium into the metallic state. One of the shortcomings in this study was that no resonance effect could be observed for very small metal particles.

References

[7.3.1] Kistner, O. C., Monaro, S., Segnan, R.: Phys. Lett. *5*, 299 (1963)

[7.3.2] Kistner, O. C.: Phys. Rev. *144*, 1022 (1966)

[7.3.3] Potzel, W., Wagner, F. E., Mössbauer, R. L., Kaindl, G., Seltzer, H. E.: Z. Phys. *241*, 179 (1971)

[7.3.4] J. G. Stevens, V. E. Stevens (eds.): Mössbauer Effect Data Index, Vols. 1973, 1975. New York: Plenum Press 1975, 1976

[7.3.5] Kaindl, G., Potzel, W., Wagner, F., Zahn, U., Mössbauer, R. L.: Z. Phys. *226*, 103 (1969)

[7.3.6] Potzel, W., Wagner, F. E., Zahn, U., Mössbauer, R. L., Danon, J.: Z. Phys. *240*, 308 (1970)

[7.3.7] Greatrex, R., Greenwood, N. N., Kaspi, P.: J. Chem. Soc. (A) *1971*, 1873

[7.3.8] Good, M. L.: in *Mössbauer Effect Data Index*, Vol. 1972, J. G. Stevens, V. E. Stevens (eds.). New York: Plenum Press 1973, p. 61

[7.3.9] Clausen, C. A., Prados, R. A., Good, M. L.: in *Mössbauer Effect Methodology*, Vol. 6, I. J. Gruverman (ed.). New York: Plenum Press 1971, p. 41

[7.3.10] Clausen, C. A., Prados, R. A., Good, M. L.: Chem. Phys. Lett. *8*, 565 (1971)

[7.3.11] Clausen, C. A., Prados, R. A., Good, M. L.: Inorg. Nucl. Chem. Lett. *7*, 485 (1971)

[7.3.12] Creutz, C., Good, M. L., Chandra, S.: Inorg. Nucl. Chem. Lett. *9*, 171 (1973)

[7.3.13] Greenwood, N. N., de A. da Costa, F. M., Greatrex, R.: Rev. Chim. Miner. *13*, 133 (1976)

[7.3.14] Gibb, T. C., Greatrex, R., Greenwood, N. N., Puxley, D. C., Snowdon, K. G.: Chem. Phys. Lett. *20*, 130 (1973)

[7.3.15] Clausen, C. A., Prados, R. A., Good, M. L.: J. Amer. Chem. Soc. *92*, 7482 (1970)

[7.3.16] Manoharan, P. T., Gray, H. B.: Inorg. Chem. *5*, 823 (1966)

[7.3.17] Bancroft, G. M., Butler, K. D., Libbey, E. T.: J. Chem. Soc. Dalton *1972*, 2643

[7.3.18] Foyt, D. C., Siddall, T. H., Alexander, C. J., Good, M. L.: Inorg. Chem. *13*, 1793 (1974)

[7.3.19] Gibb, T. C., Greatrex, R., Greenwood, N. N., Meinhold, R. H.: Chem. Phys. Lett. *29*, 379 (1974)

[7.3.20] Foyt, D. C., Good, M. L., Cosgrove, J. G., Collins, R. L.: J. Inorg. Nucl. Chem. *37*, 1913 (1975)

[7.3.21] Kistner, O. C.: in *Mössbauer Effect Methodology* Vol. 3, I. J. Gruverman (ed.). New York: Plenum Press 1967, p. 217

[7.3.22] Gibb, T. C., Greatrex, R., Greenwood, N. N., Kaspi, P.: Chem. Comm. *1971*, 319

[7.3.23] Gibb, T. C., Greatrex, R., Greenwood, N. N., Kaspi, P.: J. Chem. Soc. Dalton *1973*, 1253

[7.3.24] Gibb, T. C., Greatrex, R., Greenwood, N. N., Puxley, D. C., Snowdon, K. G.: J. Solid State Chem. *11*, 17 (1974)

[7.3.25] Gibb, T. C., Greatrex, R., Greenwood, N. N., Snowdon, K. G.: J. Solid State Chem. *14*, 193 (1975)

[7.3.26] Potzel, W., Wagner, F. E., Wäppling, R.: J. Solid State Chem. *9*, 380 (1974)

[7.3.27] Clausen, C. A., Good, M. L.: J. Catal. *38*, 92 (1975)

7.4. Hafnium (^{176,177,178,180}Hf)

The nuclear transition of the hafnium isotopes, for which the Mössbauer effect has been observed, are the 88.36, 112.97, 93.2, and 93.33 keV transitions in ^{176}Hf, ^{177}Hf, ^{178}Hf, and ^{180}Hf, respectively. Apart from the ^{177}Hf resonance, which involves $9/2^-$ and $7/2^-$ spin states, we are dealing here with $2^+ \longleftrightarrow 0^+$ electric quadrupole (E2) transitions. The decay schemes of the hafnium isotopes are given in Fig. 7.4.1. The relevant nuclear parameters are collected in Table 7.1 (end of the book). It is obvious that the high energetic transitions as well as the relatively short half-lifes

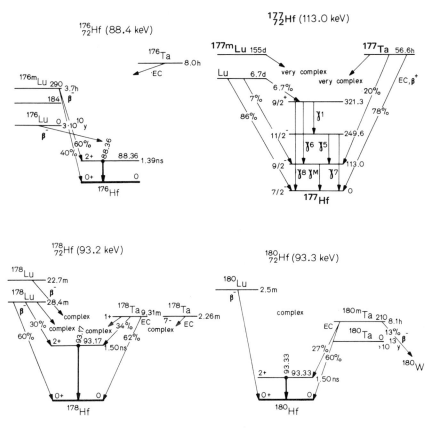

Fig. 7.4.1. Nuclear decay schemes for the hafnium isotopes ^{176,177,178,180}Hf (from [7.4.1])

of the parent nuclei prevent hafnium from being a "good" Mössbauer element. Consequently, only some ten papers have been published on this subject so far.

The first report on hafnium Mössbauer measurements by Wiedemann et al. in 1963 [7.4.2] deals with the 88.36 keV transition in ^{177}Hf. Observations of the ^{176}Hf and ^{180}Hf nuclear resonances were described in 1966, and that of the ^{178}Hf resonance in 1968 by Gerdau and co-workers [7.4.3, 4]. In the majoritiy of experiments the 93.2 keV transition in ^{178}Hf has been used because of the reasonable half-life (21.5 d) of its parent nucleus ^{178}W.

7.4.1. Practical Aspects of Hafnium Mössbauer Spectroscopy

The parent nuclei of the hafnium Mössbauer isotopes can be produced by the following reactions:

$$^{175}\text{Lu}/\text{Lu}_2\text{Rh}_2(\text{n}, \gamma) \,^{176\text{m}}\text{Lu} \xrightarrow[\beta^-]{3.7\text{h}} \,^{176}\text{Hf} \quad [7.4.5]$$

$$^{175}Lu/Lu\text{-met.}(2n, \gamma) \; ^{177}Lu \xrightarrow[\beta^-]{6.7d} \; ^{177}Hf \quad [7.4.3]$$

$$^{181}Ta/Ta\text{-met.}(^{p,\;4n}_{d,\;5n}) \; ^{178}W \xrightarrow[EC]{21.5d} \; ^{178}Ta \xrightarrow[EC,\;\beta^+]{9.4^m} \; ^{178}Hf \quad [7.4.5, 6]$$

$$^{179}Hf/HfO_2(n, \gamma) \; ^{180m}Hf \xrightarrow{5.5h} \; ^{180}Hf \quad [7.4.3, 5].$$

The experimental line width using these sources is only slightly larger than the natural line width (e.g., the thickness corrected line width of ^{178}Hf in a tantalum foil is in good agreement with the natural line width: $\Gamma_{exp} = 1.90 \pm 0.07$ mm s^{-1}, $\Gamma_{nat} = 1.99 \pm 0.04$ mm s^{-1} [7.4.12]). In addition Coulomb excitation can be used to populate the Mössbauer levels of $^{177, 178, 180}Hf$ [7.4.7–9].

The ratio of the quadrupole moments of $^{176, 178, 180}Hf$ has been determined by Snyder et al. [7.4.5] to be $Q_{2^+}(176)/Q_{2^+}(178)/Q_{2^+}(180) = (1.055 \pm 0.008)/(1.014 \pm 0.013)/1$, using the quadrupole splitting of HfB_2, HfO_2, and Hf-metal. From the known half-life of the 93 keV level in ^{178}Hf Boolchand et al. [7.4.10] estimated the quadrupole moment to be $Q_{2^+}(180) = -1.93 \pm 0.05$ b in the framework of the theory of deformed nuclei.

The isomer shifts in hafnium Mössbauer isotopes usually are of the order of some percent of the line width. Boolchand et al. [7.4.12] observed a relatively large isomer shift of $+0.19 \pm 0.06$ mm s^{-1} between cyclopentadienyl hafnium dichloride ($Hf(Cp)_2Cl_2$) and Hf metal. From a comparison with $Os(Cp)_2$ and Os-metal a value of $\delta\langle r^2\rangle(^{178}Hf) = -0.37 \cdot 10^{-3}$ fm^2 has been derived, which implies a shrinking of the nuclear radius in the excited 2^+ state. Fig. 7.4.2 shows some typical spectra for ^{178}Hf in various hafnium compounds (from [7.4.12]).

7.4.2. Magnetic Dipole and Electric Quadrupole Hyperfine Interaction

Steiner et al. have measured the magnetic hyperfine field of ^{178}Hf and ^{180}Hf in iron [7.4.6] using sources of 1 at.-% W in Fe and 1 at.-% Hf in Fe, respectively. Taking the known magnetic moments (from γ-γ angular correlation measurements) of $\mu_{2^+}(^{178}Hf) = 0.704(70)$ and $\mu_{2^+}(^{180}Hf) = 0.741(64)$ [7.4.11] a magnetic hyperfine field of $H_{eff}(4.2$ K$) = 60.6(7,0)$ Tesla and $H_{eff}(4.2$ K$) = 33.4(4.0)$ Tesla could be deduced for ^{178}Hf in a W-Fe alloy and for ^{180}Hf in a Hf-Fe alloy, respectively. The difference in the hyperfine fields is attributed to the chemical constitution of the sources used. In the case of the ^{178}W source with 1 at.-% W in Fe, a homogenous solution of ^{178}Hf in Fe is expected, whereas the field in the ^{180}Hf experiment may be due to clustering and formation of the intermetallic compound $HfFe_2$. In addition, the temperature dependence of the hyperfine field in ^{178}Hf, $H_{eff}(77$ K$)/H_{eff}(4.2$ K$) = 0.90(3)$, is much stronger than the temperature dependence of the magnetization of the iron host lattice $M(77$ K$)/M(4.2$ K$) = 0.997$. This discrepancy has not been explained.

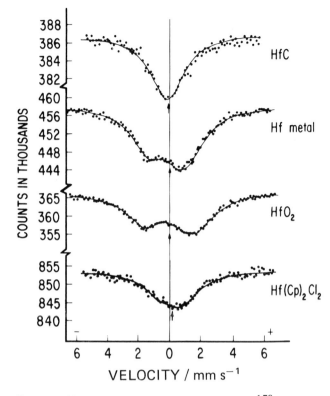

Fig. 7.4.2. Mössbauer spectra taken with a source of ^{178}W in tantalum metal using absorbers of: HfC, Hf metal, HfO$_2$, and Hf(Cp)$_2$Cl$_2$ (from [7.4.12]). The center of gravity is indicated by the arrow

Chemical information from hafnium Mössbauer spectroscopy can primarily be deduced from the quadrupole-splitting parameter. In Table 7.4.1 we have listed the quadrupole coupling constants eQV$_{zz}$ for some hafnium compounds. Schäfer et al. [7.4.13] have investigated the electric quadrupole interaction of ^{178}Hf in PbTiO$_3$ and compared their results with calculations in the framework of the ligand field model including a contribution to the electric field gradient from the permanent electric dipoles of the perovskite structure. Covalency effects have been found to play a considerable role, even in these relatively ionic compounds.

Jacobs and Hershkowitz [7.4.8] observed an increase in the $^{178, 180}$Hf quadrupole coupling constants due to recoil radiation damage following Coulomb excitation in Hf-metal, HfB$_2$, HfC, HfN, and HfO$_2$. The anomalous interactions are discussed in terms of lattice distortions resulting from vacancies in the vicinity of the recoiling Mössbauer nucleus caused by Coulomb excitation. The extent to which local vacancies can affect the EFG at the Hf nuclei could be related to the bonding properties of the above compounds.

Table 7.4.1. Quadrupole coupling constant eQV_{zz} and asymmetry parameter η for ^{178}Hf in some hafnium compounds

Compound	$eQV_{zz}/mm\ s^{-1}$	η	Reference
HfB_2	−7.18 (3)	0.42 (5)	[7.4.5]
HfO_2	−8.00 (3)	0.48 (4)	[7.4.5]
Hf-metal	−5.94 (4)	0	[7.4.5]
$Hf(NO_3)_4$	8.18 (31)	0.57 (15)	[7.4.5]
$HfOCl_2 \cdot 8H_2O$	5.96 (20)	0.86 (12)	[7.4.5]
$HfCl_4$	6.26 (57)	0.71 (36)	[7.4.5]
$(NH_4)_2HfF_6$	8.50 (15)	0.90 (5)	[7.4.4]
$Hf(Cp)_2Cl_2$	−3.54 (32)	0	[7.4.12]
$^{178}Hf/Pb(Ti_{0.9}Hf_{0.1})O_3$	−6.74 (7)	0	[7.4.13]

References

[7.4.1] J. G. Stevens, V. E. Stevens (eds.): *Mössbauer Effect Data Index* covering the 1975 Literature. New York: IFI Plenum Press 1976

[7.4.2] Wiedemann, W., Kienle, P., Stanek, F.: Z. angew. Phys. *15*, 7 (1963)

[7.4.3] Gerdau, E., Körner, H. J., Lerch, J., Steiner, P.: Z. Naturforsch. *21a*, 941 (1966)

[7.4.4] Gerdau, E., Steiner, P., Steenken, D.: in *Hyperfine Structure and Nuclear Radiations*, E. Matthias, D. A. Shirley (eds.). Amsterdam: North Holland 1968, p. 261

[7.4.5] Snyder, R. E., Ross, J. W., Bunbury, D. P.: Proc. Phys. Soc. *10*, 1662 (1968)

[7.4.6] Steiner, P., Gerdau, E., Steenken, D.: Proc. Roy. Soc. *311 A*, 177 (1969)

[7.4.7] Jacobs, C. G., Hershkowitz, N., Jeffries, J. B.: Phys. Lett. *29A*, 498 (1969)

[7.4.8] Jacobs, C. G., Hershkowitz, N.: Phys. Rev. *B 1*, 839 (1970)

[7.4.9] Hershkowitz, N., Jacobs, C. G.: in *Mössbauer Effect Methodology*, I. J. Gruverman (ed.), Vol. *6*, 143 (1971)

[7.4.10] Boolchand, P., Robinson, B. L., Jha, S.: Phys. Rev. *187*, 475 (1969)

[7.4.11] Bodenstedt, E., Körner, H. J., Gerdau, E., Radeloff, J., Auersbach, K., Mayer, L., Roggenbuck, A.: Z. Phys. *168*, 103 (1962); Bodenstedt, E., Körner, H. J., Gerdau, E., Radeloff, J., Gunther, C., Strube, G.: Z. Phys. *165*, 57 (1961)

[7.4.12] Boolchand, P., Langhammer, D., Ching-Lu Lin, Jha, S., Peek, N. F.: Phys. Rev. *C6*, 1093 (1972)

[7.4.13] Schäfer, G., Herzog, P., Wolbeck, B.: Z. Phys. *257*, 336 (1972)

7.5. Tantalum (^{181}Ta)

^{181}Ta provides two γ-transitions, the 136.25 keV transition (E2/M1 = 0.19) from the $9/2^+$ ($t_{1/2}$ = 40 ps) state to the $7/2^+$ ground state and the 6.23 keV transition (E 1) from the $9/2^-$ first excited state ($t_{1/2}$ = 6.8 ns) to the ground state, both of which are suitable for recoilless nuclear resonance; cf. Fig. 7.5.1 and Table 7.1 (end of the book).

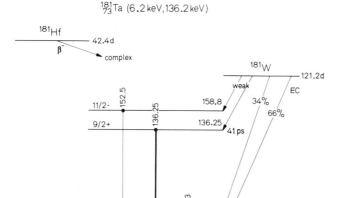

Fig. 7.5.1. Decay scheme of ^{181}Ta (from [7.5.1])

Nuclear resonance absorption for the 136 keV transition has been established by Steiner et al. [7.5.2]. The authors used a ^{181}W metal source and an absorber of metallic tantalum to determine the mean lifetime of the 136 keV level from the experimental line width ($\approx$ 52.5 mm s^{-1} for zero effective absorber thickness) and found a value of 55 ps. This has been the only report so far on the use of the 136 keV excited state of ^{181}Ta for Mössbauer experiments.

The 6.2 keV excited state of ^{181}Ta appears to be most favourable for Mössbauer spectroscopy from the point of view of its nuclear properties:

1. The small natural width of $2\Gamma = 0.0065$ mm s^{-1} implies an extremely narrow relative line width of $\Gamma/E\gamma = 1.1 \cdot 10^{-14}$, which is about 30 times narrower than that for the 14.4 keV state of ^{57}Fe.

2. The high atomic mass yields small recoil energies and thus a large recoil-free fraction even at room temperature ($f_{300} \approx 0.95$ for Ta metal [7.5.3]); this allows measurements of the resonance effect to be made over a wide temperature range up to about 2,300 K.

3. The large magnitude of the change of the mean-squared nuclear charge radius, $\Delta\langle r^2 \rangle = \langle r^2 \rangle_e - \langle r^2 \rangle_g$, between the excited state and the ground state of ^{181}Ta, and the large electric quadrupole moments in both the ground state and the excited state result in an unusually high sensitivity of the isomer shift and the quadrupole interaction to small changes in the local environment of a ^{181}Ta lattice site. On the other hand, the extremely narrow line width and the high spin of the transition renders the resonance effect very susceptible even to minor perturbations. In fact, only strongly broadened resonance lines with experimental line widths of 30–40 times the natural width could be found by the early investigators [7.5.4–6], presumably due to the unusually high sensitivity of the 6.2 keV transition to electric quadrupole interaction.

7.5.1. Experimental Aspects

The EC decay of ^{181}W ($T_{1/2}$ = 140 d), which is produced from ^{180}W by thermal neutron activation in a nuclear reactor, populates the two Mössbauer levels of ^{181}Ta at excitation energies of 6.2 and 136.2 keV.

The high sensitivity against hyperfine interactions has severe consequences with respect to source and absorber preparations. Extreme care must be taken for the purity of the samples because of the large distortions caused by small impurity concentrations. Even crystal defects have been shown to influence markedly the line width. Annealing at $\approx$ 2,000 °C under ultrahigh vacuum ($\approx 10^{-8} - 10^{-10}$ Torr) is therefore necessary in source preparation after diffusion of the ^{181}W activity into the host matrix under H_2 atmosphere at $\approx$ 1,000 °C. For further details about source and absorber preparation the reader is referred to the work of Sauer [7.5.7]. The best results concerning narrowest possible line width are obtained with single crystals avoiding crystal defects and preferred diffusion along grain boundaries. The experimental line widths observed up to now are still more than one order of magnitude larger than the minimum observable width of 2Γ. The smallest value reported up to now of Γ_{exp} = 0.069 mm s^{-1} was observed by Wortmann [7.5.8] using a source of ^{181}W in a W single crystal and a 4.6 mg/cm^2 thick tantalum foil as absorber, which was treated as described by Sauer [7.5.7]. Thin foils of tantalum are required because of the large photoelectric absorption of the 6.2 keV γ-rays; for example, a tantalum metal foil 2–3 μm thick (3–5 mg/cm^2) absorbs 65–80% of the γ-ray intensity.

The layout of the spectrometer is determined by the narrow resonance width of the ^{181}Ta (6.2 keV) transition and the wide range of line positions. A high stability drive system employing either a piezoelectric crystal transducer [7.5.4] or a conventional high quality electromechanical transducer [7.5.9] in connection with a large number of channels ($\approx$ 2,000) for recording the spectra is generally used. Special attention must be paid to rigid connections of source and absorber to avoid uncontrolled vibrations. The relatively large recoil-free fraction for the ^{181}Ta (6.2 keV) transition requires no cooling of source and absorber unless temperature effects of the hyperfine interaction are of particular interest.

A commonly used detector for the 6.2 keV γ-rays is an Ar-filled proportional counter, where the Mössbauer quanta are detected in the slope of the 8.15 keV L_α x-rays. The resolution becomes worse when working with higher counting rates, and a relatively large background of 2–3 times the counts in the 6.2 keV line has to be expected. Pfeiffer [7.5.10] reported recently on the use of Si(Li) and intrinsic germanium (IG) detectors. The 6.2 keV γ-rays and the L_α x-rays can be resolved with both types of counters, but the IG is the more favourable detector because of the lower Compton background arising from the strong K_α and K_β x-rays (56.3 and 67.0 keV) of tantalum.

As a general feature in evaluating the Mössbauer spectra of ^{181}Ta, one has to take into account in the computer fitting procedure a modification of the Lorentzian lines by the interference effect between photoelectric absorption and Mössbauer absorption followed by internal conversion [7.5.11, 12]. This effect was first observed by Sauer et al. and has been found to be particularly large for the 6.2 keV γ-rays of ^{181}Ta due to their E1 multipolarity, their low transition energy, and their high inter-

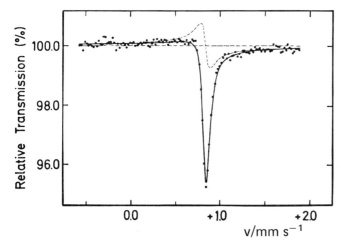

Fig. 7.5.2. ^{181}Ta (62 keV) Mössbauer spectrum of ^{181}W in W metal versus Ta metal absorber at room temperature. The solid line represents the fit of a dispersion-modified Lorentzian line to the experimental data; the dashed line shows the dispersion contribution (from [7.5.7, 13])

nal conversion coefficient. It gives rise to an additional dispersion term besides the normal Lorentzian line shape leading to an asymmetry of the observed resonance line; cf. Fig. 7.5.2. After Trammel and Hannon [7.5.11] the Mössbauer spectra are fitted with dispersion-modified Lorentzian lines of the form

$$N(v) = N(\infty)[1 - \epsilon(1 - 2\xi X)/(1 + X)^2],$$

with $X = 2(v - S)/W$. $N(v)$ is the intensity transmitted at relative velocity v, S is the position of the line, W is the full line width at half-maximum, ϵ is the magnitude of the resonance effect, and ξ is a parameter determining the relative magnitude of the dispersion term ($2\xi = -0.31 \pm 0.01$ has been found to be a good value for ^{181}Ta (6.2 keV) Mössbauer spectra).

7.5.2. Isomer Shift Studies

The first isomer shifts in ^{181}Ta (6.2 keV) Mössbauer spectroscopy were reported in [7.5.5] for the combination of a ^{181}W in W metal source and Ta metal absorber and ^{181}W in Ta source and Ta metal absorber, respectively. Sauer et al. [7.5.13] found large positive isomer shifts between 0.83 and 0.94 mm s^{-1} for various combinations of W sources and Ta absorbers at room temperature and suggested therefore a considerable change of $\Delta\langle r^2 \rangle$ in the nuclear charge radii associated with the γ-transition. Sauer et al. [7.5.7] studied the influence of plastic deformation and oxygen doping of a Ta metal absorber and observed remarkable influences on both isomer shift and line width. Kaindl, Salomon, and Wortmann [7.5.15] have summarized the results from various isomer shift studies [7.5.8, 13, 15–19]; the data are collected in Table 7.5.1. Some typical room temperature spectra showing the enormously large changes

Table 7.5.1. ^{181}Ta isomer shift, experimental line width and resonance effect, observed for sources of ^{181}W diffused into various transition metal hosts against tantalum metal as absorber, and for various tantalum compounds as absorber versus a source of ^{181}W in single-crystal W. Source and absorber in all cases at room temperature (from [7.5.14])

Source lattice	δ mm s^{-1}		Γ_{exp}(FWHM) mm s^{-1}		Effect (%)
V	−33.2	(5)	5.0	(10)	0.1
Ni	−39.5	(2)	0.50	(8)	1.6
Nb	−15.26	(10)	0.19	(6)	1.5
Mo	−22.60	(10)	0.13	(4)	3.0
Ru	−27.50	(30)	1.3	(2)	0.7
Rh	−28.80	(25)	3.4	(5)	0.3
Pd	−27.80	(25)	1.3	(3)	0.3
Hf	−0.60	(30)	1.6	(4)	0.2
Ta	−0.075	(4)	0.184	(6)	2.4
W	−0.860	(8)	0.069	(1)	20
Re	−14.00	(10)	0.60	(4)	1.3
Os	−2.35	(4)	1.8	(2)	0.8
Ir	−1.84	(4)	1.60	(14)	0.5
Pt	+2.66	(4)	0.30	(8)	1.5
Absorber lattice					
LiTaO$_3$	−24.04	(30)	1.6	(2)	0.9
NaTaO$_3$	−13.26	(30)	1.0	(2)	0.9
KTaO$_3$	−8.11	(15)	1.5	(2)	0.3
TaC	+70.8	(5)	2.4	(4)	0.2

in isomer shift in going from one sample to another are reproduced in Figs. 7.5.3−5. A graphical representation of isomer shifts of the 6.2 keV γ-transition in ^{181}Ta for tantalum compounds versus ^{181}W/W (single-crystal) source and for dilute impurities of ^{181}Ta(^{181}W) in transition metal hosts (versus Ta metal as absorber), published by Kaindl et al. [7.5.15], is shown in Fig. 7.5.6. The transition energy increases in this plot from bottom to top. The range of isomer shifts for tantalum compounds is much larger than for ^{181}Ta in transition metal hosts. The isomer shift of TaC is, together with that of ^{237}Np in Li$_5$NpO$_6$ (δ = 68.9 ± 2.6 mm s^{-1} versus ^{241}Am/Th source at 4.2 K) [7.5.20], the largest one observed so far. The total range of isomer shifts (110 mm s^{-1}) in ^{181}Ta (6.2 keV) corresponds to 17,000 times twice the natural width or 1,600 times the best experimental line width obtained up to now.

Kaindl et al. [7.5.15] have plotted the isomer shift results for metallic hosts versus the number of outer electrons of the 3d, 4d, and 5d metals and found the transition energy to decrease when proceeding from a 5d to a 4d and further to a 3d host metal in the same column of the periodic table. This systematic behaviour is similar

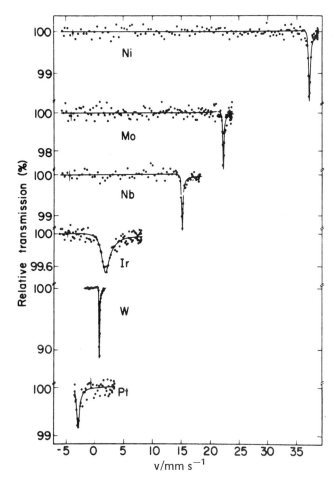

Fig. 7.5.3. ^{181}Ta (6.2 keV) Mössbauer spectra obtained with a tantalum metal absorber and sources of ^{181}W diffused into various cubic transition metal hosts. The solid lines represent the results of least-squares fits of dispersion-modified Lorentzian lines to the experimental spectra (from [7.5.15])

to that observed for isomer shifts of γ-rays of ^{57}Fe(14.4 keV) [7.5.21], ^{99}Ru(90 keV), ^{197}Au(77 keV), and ^{193}Ir(73 keV) [7.5.22]. The changes of $\Delta\langle r^2 \rangle = \langle r^2 \rangle_e - \langle r^2 \rangle_g$ for these Mössbauer isotopes are all reasonably well established. Kaindl et al. [7.5.15] have used these numbers to estimate, with certain assumptions, the $\Delta\langle r^2 \rangle$ value for ^{181}Ta (6.2 keV) and found a mean value of $\Delta\langle r^2 \rangle = -5 \cdot 10^{-2}$ fm^2 with some 50% as an upper limit of error. The negative sign of $\Delta\langle r^2 \rangle$ is in agreement with the observed variation of the isomer shift of LiTaO$_3$, NaTaO$_3$, and KTaO$_3$, as well as with the isomer shift found for TaC [7.5.15].

The temperature dependence of the transition energy of the 6.2 keV γ-rays in ^{181}Ta requires particular attention, especially in studies over a wide temperature range. In contrast to ^{57}Fe and ^{119}Sn, where the observed temperature shifts were

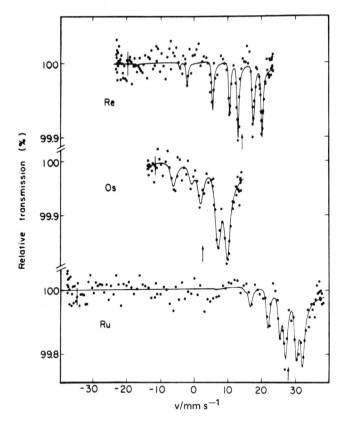

Fig. 7.5.4. Electric quadrupole split ^{181}Ta (6.2 keV) Mössbauer spectra for sources of ^{181}W diffused into hexagonal transition metals (Re, single crystal; Os, polycrystalline; Ru, single crystal). The isomer shifts relative to a tantalum metal absorber are indicated by the arrows (from [7.5.15, 18])

found to be predominantly caused by the second-order Doppler shift (SOD) (thermal red shift) [7.5.23], Kaindl and Salomon [7.5.24] have found that the 6.2 keV transition energy, emitted from ^{181}Ta as a dilute impurity in transition metal hosts, exhibits a strong temperature dependence far beyond the SOD shift. In the nickel hosts, for instance, the 6.2 keV transition energy increases with temperature with a slope which is 32 times larger and of opposite sign than the one expected from the SOD shift alone (cf. Fig. 1 of Ref. [7.5.24]). The line shift corresponds to 2.3 natural widths per degree. The slopes of the temperature shifts for W, Ta, and Pt have the same sign as the SOD shift, but are up to 8 times larger. After [7.5.24], the experimentally observed isobaric temperature variation of the line position S may be written as

$$\left(\frac{\partial S}{\partial T}\right)_P = \left(\frac{\partial S_{SOD}}{\partial T}\right)_P + \left(\frac{\partial S_{IS}}{\partial T}\right)_V + \left(\frac{\partial S_{IS}}{\partial \ln V}\right)_T \left(\frac{\partial \ln V}{\partial T}\right)_P.$$

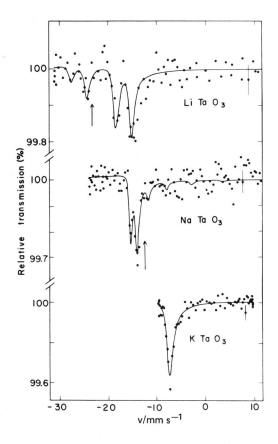

Fig. 7.5.5. ^{181}Ta (6.2. keV) Mössbauer spectra of alkali tantalates obtained at room temperature with a source of ^{181}W in W (single crystal). The centers (isomer shifts) of the quadrupole split spectra are indicated by arrows (from [7.5.15])

With the observed temperature shift data for $(\partial S/\partial T)_P$ and calculated (in the framework of the Debye model) numbers for the temperature shift of SOD and using the known thermal expansion coefficient as well as results from ^{181}Ta Mössbauer experiments under pressure, the authors of [7.5.14] were able to evaluate the true temperature dependence of the isomer shift, $(\partial S_{IS}/\partial T)$. It turned out to be $-33 \cdot 10^{-4}$ and $-26 \cdot 10^{-4}$ mm s^{-1} degree $^{-1}$ for Ta and W host metal, respectively.

Another study of the temperature dependence of the 6.2 keV Mössbauer resonance of ^{181}Ta has been carried out by Salomon et al. [7.5.25] for sources of ^{181}W/W metal and ^{181}W/Ta metal in the temperature range from 15 to 457 K. In more recent investigations Salomon et al. [7.5.26] have extended such studies of the temperature behaviour of the 6.2 keV Mössbauer transition of ^{181}Ta in tantalum metal to temperatures up to 2,300 K, which has been the highest temperature range for any Mössbauer study so far.

Enormously large isomer shift changes have been observed by Heidemann et al. [7.5.27] in tantalum metal loaded with hydrogen (α-phase of Ta-H) as a function of hydrogen concentration. Some typical spectra of their work demonstrating this striking effect are shown in Fig. 7.5.7. One also notices a substantial line broadening with increasing hydrogen concentration, which the authors explained quantitatively by

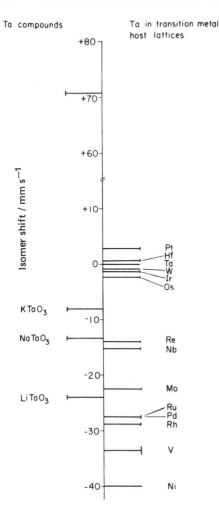

Fig. 7.5.6. Graphical representation of isomer shifts of the 6.2 keV γ-transition in ^{181}Ta for tantalum compounds and for dilute impurities of ^{181}Ta (^{181}W) in transition metal hosts (from [7.5.15])

isomer shift fluctuations as arising from varying hydrogen configurations. With a motional narrowing approach they arrived at the jump frequency and the activation energy of the hydrogen interstitials from the concentration and temperature dependence, respectively, of the line broadening.

7.5.3. Hyperfine Splitting in ^{181}Ta(6.2 keV) Spectra

The ground state as well as the 6.2 keV excited state of ^{181}Ta possess sizeable electric quadrupole and magnetic dipole moments (cf. Table 7.1 at the end of the book) which, in cooperation with the extremely narrow line width, generally yield well-resolved Mössbauer spectra even in cases with relatively weak interactions.

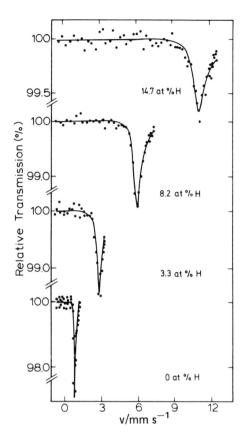

Fig. 7.5.7. ^{181}Ta (6.2 keV) Mössbauer spectra of Ta-H absorbers at 300 K with various hydrogen concentration (from [7.5.27])

a) Quadrupole Splitting

Electric quadrupole interaction will split the $I_e = 9/2^-$ excited state into 5 sublevels (with $m_I = \pm 9/2, \pm 7/2, \pm 5/2, \pm 3/2, \pm 1/2$) and the $I_g = 7/2^+$ ground state into 4 sublevels (with $m_I = \pm 7/2, \pm 5/2, \pm 3/2, \pm 1/2$), and there are a total of 11 possible E1 transitions, whose relative intensities are fixed by the relevant Clebsch-Gordan coefficients.

The broadened single line spectrum of the early measurements by Cohen et al. [7.5.4], obtained with a TaC absorber and an annealed ^{181}W/W metal source at room temperature, was assigned by the authors to the $\pm 3/2 \longleftrightarrow \pm 3/2$ transition with an estimated Q_e/Q_g ratio between 0.8 and 0.95.

Steyert et al. [7.5.5] observed quadrupole effects in their ^{181}Ta Mössbauer spectra of source/absorber combinations with "nominally" cubic environments of the ^{181}Ta atoms in both source and absorber. They ascribed this unexpected effect to interstitial impurities of O_2, N_2, and C, which likely produced the observed field gradients.

Sauer et al. [7.5.13] derived a weak quadrupole interaction from the asymmetry of a poorly resolved Zeeman split spectrum of ^{181}W in W metal versus a Ta metal absorber. They also ascribed the unexpected weak quadrupole effect to deviations from cubic symmetry at the source or absorber atom arising from either interstitial impurities or crystal defects.

Well resolved quadrupole split [181]Ta Mössbauer spectra were observed by Kaindl and co-workers with a direction oriented source of [181]W in a Re single crystal with hexagonal symmetry versus a specially treated Ta metal absorber [7.5.16], from which they derived the sign and the magnitude of the quadrupole interaction and the quadrupole moment ratio $Q_e/Q_g = 1.333 \pm 0.010$. In further investigations they used sources of [181]W in the hexagonal transition-metals Hf (single crystal), Os and Ru (both polycrystalline) versus a single-line absorber of Ta metal foil (4 mg/cm^2) [7.5.18] (cf. Fig. 7.5.4), and a [181]W/W single crystal source versus hexagonal LiTaO$_3$ and orthorhombic NaTaO$_3$ absorbers [7.5.15] (cf. Fig. 7.5.5). All these spectra were recorded with the source and the absorber kept at room temperature. In all cases dispersion-modified Lorentzians were fitted to the experimental spectra by a least-squares computer fit.

b) Magnetic Dipole Splitting

The nuclear Zeeman effect splits the 9/2$^-$ excited state into 10 sublevels and the 7/2$^+$ ground state into 8 sublevels, and a large number of allowed transitions would occur in [181]Ta Mössbauer experiments on systems with magnetic dipole interaction. The number of transitions can be reduced by applying a small magnetic field ($\sim$1–2 kOe). In fact, in all studies of magnetic dipole splitting by [181]Ta Mössbauer spectroscopy published so far, the authors have applied magnetic fields of 0.15–0.17 T (1,500–1,700 Oe) such as to magnetize the source parallel to the γ-ray direction. All the $\Delta m = \pm 0$ transitions do not show up this way, and a [181]Ta Zeeman spectrum consists only of 16 hyperfine components with $\Delta m = \pm 1$ and relative intensities given by the Clebsch-Gordan coefficients.

Sauer et al. [7.5.13] determined the gyromagnetic ratio g(9/2)/g(7/2) and the magnetic moment of the 6.2 keV level in [181]Ta in two ways, (i) from the Zeeman split velocity spectrum of a [181]W/W metal source in a longitudinal field versus a Ta

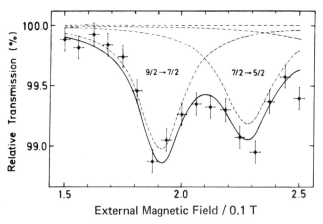

Fig. 7.5.8. Outer two components of the magnetic hyperfine spectrum of a [181]W/W source versus a Ta absorber, measured at zero velocity as a function of a longitudinal magnetic field ("Zeeman drive" experiment) (from [7.5.13])

metal absorber and (ii) in a "Zeeman drive" experiment, in which the hyperfine components $9/2 \longleftrightarrow 7/2$ and $7/2 \longleftrightarrow 5/2$ were observed at zero velocity as a function of the applied magnetic field ranging up to 0.25 T. A typical spectrum obtained in this kind of experiment is shown in Fig. 7.5.8. The Zeeman split velocity spectrum yielded $g(9/2)/g(7/2) = +1.72 \pm 0.05$ and, with the known ground state moment, the excited state moment has been found to be $\mu_e = (+5.20 \pm 0.15)\mu_N$. The "Zeeman drive" experiment gave $g(9/2)/g(7/2 = 1.62 \pm 0.09$ and $\mu_e = (4.9 \pm 0.3)\mu_N$, somewhat less accurate and less reliable because of the weak quadrupole perturbation neglected in this case. Similar experiments have been done by Kaindl and Salomon [7.5.28]. They found $g(9/2)/g(7/2) = 1.74 \pm 0.03$ and 1.76 ± 0.04 from the magnetically split velocity spectrum and a "Zeeman drive" experiment, respectively, in good agreement within experimental error with the results of [7.5.13].

Using a single-line Ta metal absorber (4 mg/cm^2) and a ^{181}W/Ni (single-crystal) source in a polarizing external field, Kaindl and Salomon [7.5.19] measured the sign and the magnitude of the supertransferred hyperfine field at the ^{181}Ta nucleus and found a value of $H = -9 \pm 1$ T at room temperature.

References

[7.5.1] Stevens, J. G., Stevens, V. E.: *Mössbauer Effect Data Index,* Covering the 1975 Literature. New York: IFI Plenum Press 1976, p. 166

[7.5.2] Steiner, P., Gerdau, E., Hautsch, W., Steenken, D.: Z. Phys. *221,* 281 (1969)

[7.5.3] Boyle, A. J. F., Hall, H. E.: Rept. Progr. Phys. *25,* 441 (1962)

[7.5.4] Cohen, S. G., Marinov, A., Budnick, J. I.: Phys. Lett. *12,* 38 (1964)

[7.5.5] Steyert, W. A., Taylor, R. D., Storms, E. K.: Phys. Rev. Lett. *14,* 739 (1965)

[7.5.6] Muir Jr., A. H., Nadler, H.: Bull. Am. Phys. Soc. *12,* 202 (1967)

[7.5.7] Sauer, C.: Z. Phys. *222,* 439 (1969)

[7.5.8] Wortmann, G.: Phys. Lett. *35A,* 391 (1971)

[7.5.9] Kaindl, G., Maier, M. R., Schaller, H., Wagner, F.: Nucl. Instr. Methods *66,* 277 (1968)

[7.5.10] Pfeiffer, L.: Nucl. Instr. Methods *140,* 57 (1977)

[7.5.11] Trammel, G. T., Hannon, J. T.: Phys. Rev. *180,* 337 (1969)

[7.5.12] Kagan, Yu. M., Afans'ev, A. M., Vojtovetskii, V. K.: Zh. Eksperim. teor. Fiz. Pis'ma Red. *9,* 155 (1969), JETP Lett. *9,* 91 (1969)

[7.5.13] Sauer, C., Matthias, E., Mössbauer, R. L.: Phys. Rev. Lett. *21,* 961 (1968)

[7.5.14] Kaindl, G., Salomon, D., Wortmann, G.: in *Mössbauer Effect Methodology,* I. J. Gruverman (ed.), Vol. 8. New York: Plenum Press 1973, p. 211

[7.5.15] Kaindl, G., Salomon, D., Wortmann, G.: Phys. Rev. *B 8,* 1912 (1973)

[7.5.16] Kaindl, G., Salomon, D., Wortmann, G.: Phys. Rev. Lett. *28,* 952 (1972)

[7.5.17] Salomon, D., Kaindl, G., Shirley, D. A.: Phys. Lett. *36 A,* 457 (1971)

[7.5.18] Kaindl, G., Salomon, D.: Phys. Lett. *40A,* 179 (1972)

[7.5.19] Kaindl, G., Salomon, D.: Phys. Lett. *42 A,* 333 (1973)

[7.5.20] Fröhlich, K., Gütlich, P., Keller, C.: J. C. S. Dalton 971 (1972)

[7.5.21] Quaim, S. M.: Proc. Phys. Soc. London *90,* 1065 (1967)

[7.5.22] Wagner, F. E., Wortmann, G., Kalvius, G. M.: Phys. Lett. *A 42,* 483 (1973)

[7.5.23] Josephson, B. D.: Phys. Rev. Lett. *4,* 341 (1960)

[7.5.24] Kaindl, G., Salomon, D.: Phys. Rev. Lett. *30,* 579 (1973)

[7.5.25] Salomon, D., Triplett, B. B., Dixon, N. S., Boolchand, P., Hanna, S. S.: J. Phys. Colloq. C6 *35,* C6-285 (1974)

[7.5.26] Salomon, D., Wallner, W., West, P. J.: Proc. Int. Conf. Mössb. Spec. 1975 (Cracow), p. 105

[7.5.27] Heidemann, A., Kaindl, G., Salomon, D., Wipf, H., Wortmann, G.: Proc. Int. Conf. Mössb. Spec. 1975 (Cracow), p. 411

[7.5.28] Kaindl, G., Salomon, D.: Phys. Lett. *32B,* 364 (1970)

7.6. Tungsten (^{180,182,183,184,186}W)

There are five isotopes of tungsten for which the Mössbauer effect has been observed. In the even A isotopes (A, mass number) the γ-resonances from the 0^+ ground state to the 2^+ excited states at 103.65 keV ($t_{1/2}$ = 1.27 ns), 100.10 keV (1.37 ns), 111.19 keV (1.28 ns), and 122.3 keV (1.01 ns) have been observed for ^{180}W, ^{182}W, ^{184}W, and ^{186}W, respectively. In ^{183}W two nuclear levels are suitable for Mössbauer measurements, viz., the 46.48 keV ($t_{1/2}$ = 0.183 ns) and the 99.08 keV ($t_{1/2}$ = 0.69 ns) levels. The nuclear data for the Mössbauer transitions are given in Table 7.1 at the end of the

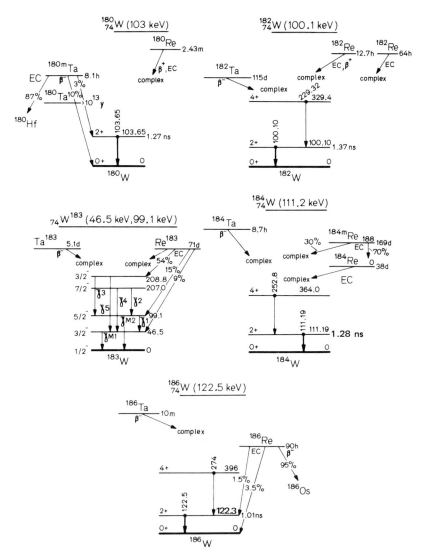

Fig. 7.6.1. Nuclear decay schemes for the tungsten isotopes ^{180,182,183,184,186}W (from [7.6.1])

book. In Fig. 7.6.1 the nuclear decay schemes and parent isotopes are depicted. The recoilless nuclear resonance absorption with the 100 keV transition in [182]W was observed shortly after Mössbauer's discovery on iridium [7.6.2−4]; physicists have been concerned so far mainly with measurements of the nuclear γ-resonance in tungsten to determine nuclear properties like magnetic dipole and electric quadrupole moments. The reason why there are only few papers devoted to applications in chemistry or solid-state physics is the small Debye-Waller factors caused by the relatively high transition energies, the short half-lives of the excited Mössbauer levels leading to broad resonance lines and therefore to poorly resolved spectra, and the small changes of the nuclear radii $\delta \langle r^2 \rangle$ in the excited 2^+ (even W isotopes) or $3/2^-$ ([183]W) rotational states causing only changes in the isomer shift of a few percent of the linewidth.

Nevertheless there has been some work in the past 6 years, mainly from research groups in Cambridge, Darmstadt and Munich, on measurements of isomer shift and quadrupole hyperfine interaction in [180,182]W under chemical aspects.

7.6.1. Practical Aspects of Mössbauer Spectroscopy with Tungsten

The source preparation for tungsten Mössbauer spectroscopy is in general cumbersome, apart from the production of [182]Ta, the parent nuclide of [182]W. The first excited rotational levels (2^+ in even W or $3/2^-$ in [183]W) can be achieved by Coulomb excitation and in beam Mössbauer measurements [7.6.5−8]. Other nuclear reactions leading to Mössbauer parent nuclei are the following:

$$^{181}Ta(d,p2n)^{180}Ta \xrightarrow[8.1h]{\beta^-} {}^{180}W \quad [7.6.9]$$

$$^{181}Ta(\gamma,n)^{180}Ta \xrightarrow[8.1h]{\beta^-} {}^{180}W \quad [7.6.10]$$

$$^{181}Ta(2n,\gamma)^{183}Ta \xrightarrow[5.1d]{\beta^-} {}^{183}W \quad [7.6.11]$$

$$^{181}Ta(n,\gamma)^{182}Ta \xrightarrow[115d]{\beta^-} {}^{182}W \quad [7.6.2-4, 11, 12].$$

The last reaction is the most important for chemical applications. [182]W is almost exclusively used for the investigation of chemical compounds. To give an impression of the resolving power of [182]W Mössbauer spectroscopy, the spectra from WS_2 powder and single-crystal measurements are shown in Fig. 7.6.2 (from [7.6.6]). On the right-hand side of the picture the angular dependence of E2 transitions for $\Delta m = \pm 2, \pm 1, 0$ hyperfine components is indicated.

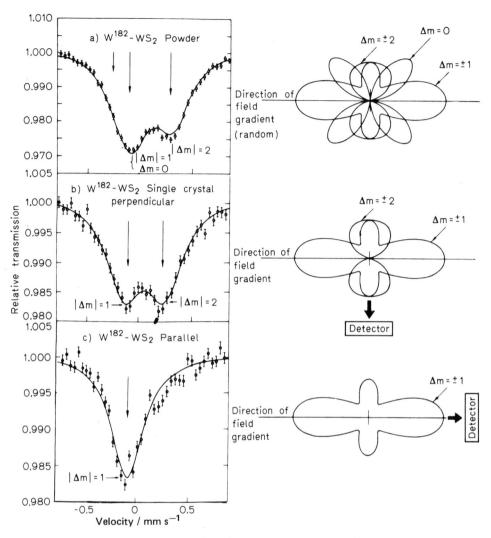

Fig. 7.6.2. Mössbauer spectrum of the $2^+ \leftrightarrow 0^+$ (100 keV) transition of ^{182}W in WS$_2$ powder and WS$_2$ single-crystals (from [7.6.6])

The ratio of the quadrupole moments in tungsten isotopes as measured by Mössbauer spectroscopy — collected from [7.6.6–11, 13] — turns out to be

$$Q_{2^+}(180)/Q_{2^+}(182)/Q_{2^+}(184)/Q_{2^+}(186)/Q_{5/2^-}(183)/Q_{3/2^-}(183) =$$
$$= (0.988 \pm 0.012)/1/(0.955 \pm 0.03)/(0.907 \pm 0.014)/(0.86 \pm 0.06)/(0.88 \pm 0.14).$$

The nuclear g_{2^+} factors for the first excited 2^+ states in ^{180,182}W have been determined by Zioutas et al. [7.6.9] to be $g_{2^+}(^{180}$W$) = 0.260 \pm 0.017$ and by Frankel et al.

[7.6.14] to be $g_{2^+}(^{182}W) = 0.22 \pm 0.02$ applying external magnetic fields up to 12.8 Tesla on ^{182}W diluted in iron. The internal magnetic field at the tungsten site in alloys of 0.5 to 5% tungsten in iron turned out to be -70.8 ± 2.5 Tesla. This field and the older values from [7.6.15], where an internal hyperfine field of 63.0 ± 1.3 Tesla from NMR data was used, lead to $g_{5/2^-}(^{183}W) = 0.307 \pm 0.015$, $g_{2^+}(^{184}W) = 0.243 \pm 0.008$ and $g_{2^+}(^{186}W) = 0.258 \pm 0.008$. An estimation of the $\delta \langle r^2 \rangle$ value in ^{182}W $(2^+ - 0^+)$ has been derived on the basis of a relativistic Hartree-Fock-Slater calculation [7.6.10], comparing the isomer shift of Rb_2WS_4 versus W-metal and WCl_6 versus K_2WCl_6: $-0.05 \cdot 10^{-4} > \delta \langle r^2 \rangle / \langle r^2 \rangle > -0.24 \cdot 10^{-4}$, which is in agreement with a value given by Wagner et al. [7.6.21] but in disagreement with $\delta \langle r^2 \rangle = 0.65 \cdot 10^{-4}$ derived by Cohen et al. [7.6.12] from the isomer shift between tungsten metal and WCl_6.

The E2/M1 mixing ratio of the 46.48 keV transition in ^{183}W has been determined by Shikazono et al. [7.6.11] to be $\delta = E2/M1 \simeq 0.5\%$.

7.6.2. Chemical Information from Debye-Waller Factor Measurements

The tungsten isotopes are very well suited for the investigation of bond properties and crystal dynamics obtainable from the Debye-Waller factor because of their high transition energies. Small changes in the mean square displacements of the vibrational amplitude will therefore lead to large changes in the Debye-Waller factor $(f = \exp(-k^2 \langle x^2 \rangle))$. Raj and Puri [7.6.15] have calculated the recoilless fraction and thermal shift (second-order Doppler shift) δ_{SOD} for ^{182}W and ^{183}W in tungsten metal by the relations

$$f = \exp \left\{ -\frac{E_\gamma^2}{2Mc^2\hbar} \frac{1}{3N} \int_0^{\omega \, max} \frac{g(\omega)}{\omega} \coth \left(\frac{\hbar\omega}{2k_B T} \right) d\omega \right\} \qquad (7.6.1)$$

and

$$\delta_{SOD} = \frac{3\hbar}{4Mc} \frac{1}{3N} \int_0^{\omega \, max} g(\omega) \, \omega \coth \left(\frac{\hbar\omega}{2k_B T} \right) d\omega , \qquad (7.6.2)$$

which hold only for monatomic cubic crystals. M is the mass of the emitting nucleus and $g(\omega)$ is the phonon frequency distribution function. An experimentally determined phonon spectrum was used in the calculations of Raj and Puri. Ruth and Hershkowitz [7.6.16] have measured the temperature dependence of f in tungsten. They found very good agreement with the calculated values of Raj and Puri [7.6.15] (Fig. 7.6.3). The experimental results of [7.6.16] could not be interpreted in terms of one Debye temperature θ_D; applying a Debye model, one found θ_D to vary from 339 ± 3 K at 29 K to 265 ± 5 K at 100 K.

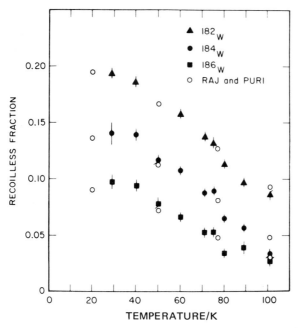

Fig. 7.6.3. Recoilless fraction as a function of temperature for ^{182}W, ^{184}W, and ^{186}W in tungsten metal (from [7.6.16])

Kaltseis, Posch, and Vogel [7.6.17] have investigated the recoil-free fraction of the 46.5 keV transition of ^{183}W in anhydrous lithium tungstate. Their results can be expressed by an effective Debye temperature of 172 ± 9 K which is in good agreement with a value of 205 ± 40 K derived from x-ray diffraction measurements of Li_2WO_4 powder.

Conroy and Perlow [7.6.18] have measured the Debye-Waller factor for ^{182}W in the sodium tungsten bronze $Na_{0.8}WO_3$. They derived a value of f = 0.18 ± 0.01 which corresponds to a zero-point vibrational amplitude of R = 0.044 Å. This amplitude is small as compared to that of beryllium atoms in metallic beryllium (0.098 Å) or the carbon atoms in diamond (0.064 Å). The authors conclude that atoms substituting tungsten in bronze may well be expected to have a high recoilless fraction.

As is well known, the recoil-free fraction of very small crystals differs markedly from that of bulk material. Roth and Hörl [7.6.19] observed a decrease of the f-factor from 0.61 to 0.57 in going from 1 μm crystals to microcrystals with a diameter of about 60 Å. Two effects will contribute to this decrease: (i) the low frequency cut-off, because the longest wavelength must not exceed the dimensions of the crystal, and (ii) high frequency cut-off caused by the weaker bonds between surface atoms.

Wender and Hershkowitz [7.6.27] used the sensitivity of the recoil-free fraction in tungsten Mössbauer spectroscopy to deduce the effect of irradiation of tungsten compounds by Coulomb excitation of the resonance levels (2^+ states of $^{182, 184, 186}$W) with 6 MeV α-particles. Whereas no effect of irradiation on the f-factors could be observed for tungsten metal in agreement with [7.6.16], a decrease of f was measured for WC, W_2B, W_2B_5, and WO_3 after irradiation.

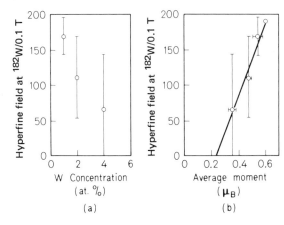

Fig. 7.6.4. Magnetic hyperfine field H_{hf} in Ni-W-alloys at the [182]W-site as a function (a) of the tungsten concentration and (b) of the average magnetic moment (from [7.6.20])

7.6.3. Chemical Information from Hyperfine Interaction

Apart from the already mentioned (Sec. 7.6.1) determination of the g-factors of [180,182]W through Mössbauer measurements with tungsten diluted in an iron foil [7.6.9, 14], where a hyperfine field at the W site of -70.8 ± 2.5 Tesla has been observed, there is only one further study of the magnetic hyperfine interaction in a tungsten alloy. Schibuya and co-workers [7.6.20] have investigated the hyperfine field H_{hf} at [182]W in Ni-W-alloys. They found a linear relation of H_{hf} on the tungsten concentration and on the average magnetic moment (Fig. 7.6.4), which can be expressed by $H_{hf} = 540 \langle \mu \rangle - 14$ T. The simple assumption leading to this relation is that the first term is due to the conduction electron polarization induced by the magnetic moment of the Ni host and that of W, and that the second term originates from the core electron polarization induced by the magnetic moment of W.

The chemical shift observed in tungsten Mössbauer spectroscopy is very small, about 1/20 of the line width for [182]W due to the small changes of $\delta \langle r^2 \rangle$ in going from the first excited rotational 2^+ state to the ground state. In spite of this difficulty there are two papers by the Munich [7.6.21] and Darmstadt [7.6.10] groups in which the isomer shift has been determined in a number of tungsten compounds. The results are presented in Fig. 7.6.5. It is obvious that the assignment of the Mössbauer isomer shift to a certain oxidation state is not unambiguous in all cases.

Maddock and co-workers have published a series of papers reporting on the measurements and interpretation of the electric quadrupole hyperfine interaction in a number of inorganic and metal-organic tungsten compounds [7.6.22–25]. The results are collected in Table 7.6.1. The electric field gradient is composed of the superposition of an electronic and lattice contribution. Whereas the latter can often be neglected, the former strongly depends on the electronic configuration, the ionicity or covalency of the chemical bond and on the geometrical arrangement of the surrounding ligands. The interplay of all these effects can be accounted for to a reasonable extent only on the basis of MO calculations of the type described in Chap. 6. This, of course, requires the knowledge of x-ray diffraction data.

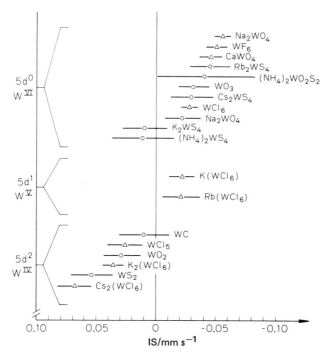

Fig. 7.6.5. Isomer shift versus formal electron configuration of tungsten compounds (from [7.6.10]). The triangles are data taken from [7.6.21], the circles from [7.6.10]

In a ^{182}W Mössbauer study of various polycrystalline W^{IV} and W^V octocyanides and in frozen solutions, Clark et al. [7.6.25] were able to determine the quadrupole coupling constant. They concluded that the EFG in the W^{IV} compounds with D_{2d} and D_{4d} symmetry arises from partial cancellation of the dominant d^2 contribution by populating the nominally empty d-orbitals through ligand-to-metal σ-bonding. The large values of the asymmetry parameter η observed for $Cd_2[W(CN)_8] \cdot 8H_2O$ and $(EtNC)_4[W(CN)_4]$ (Table 7.6.1) are indicative of distortions from D_{2d} and D_{4d} geometry, respectively.

References

[7.6.1] Stevens, J. G., Stevens, V. E.: *Mössbauer Effect Data Index,* New York: IFI Plenum 1969 to 1975

[7.6.2] Lee, L. L., Meyer-Schützmeister, L., Schiffer, J. P., Vincent, D.: Phys. Rev. Lett. *3,* 223 (1959)

[7.6.3] Bussiere de Nercy, A., Langevin, M., Spighel, M.: J. Phys. Radium *21,* 288 (1960)

[7.6.4] Kankeleit, E.: Z. Phys. *164,* 442 (1961)

[7.6.5] Hardy, K. A., Russell, D. C., Wilenzick, R. M.: Phys Lett. *27A,* 422 (1968)

[7.6.6] Chow, Y. W., Greenbaum, E. S., Howes, R. H., Hsu, F. H. H.: Phys. Lett. *30B,* 171 (1969)

Table 7.6.1. Quadrupole coupling constants $e^2qQ/mm\ s^{-1}$, asymmetry parameter $\eta = \dfrac{V_{xx}-V_{yy}}{V_{zz}}$, and experimental line width $\Gamma_{exp}/mm\ s^{-1}$ of inorganic and metal-organic tungsten compounds

Compound	$e^2qQ/mm\ s^{-1}$	η	$\Gamma_{exp}/mm\ s^{-1}$	Reference
WO_3	-8.30 ± 0.20	0.62 ± 0.03	2.50 ± 0.20	[7.6.22]
	-8.34 ± 0.05	0.63 ± 0.20	2.42 ± 0.10	[7.6.26]
$FeWO_4$	-8.80 ± 0.20	0.54 ± 0.05	2.76 ± 0.20	[7.6.22]
Na_2WO_4	$-$	$-$	2.76 ± 0.20	[7.6.22]
$W(CO)_6$	$-$	$-$	2.60 ± 0.20	[7.6.22]
$2H\text{-}WSe_2$	9.38 ± 0.20	$-$	2.34 ± 0.20	[7.6.23]
$W_{0.92}Ta_{0.08}Se_2$	9.09 ± 0.20	$-$	2.32 ± 0.20	[7.6.23]
$W_{0.5}Ta_{0.5}Se_2$	9.61 ± 0.20	$-$	2.57 ± 0.20	[7.6.23]
WS_3	-7.72	$-$	3.43	[7.6.24]
W_2N	± 2.25	$-$	3.39	[7.6.24]
W_2C	7.38	$-$	3.20	[7.6.24]
$SiO_2, 12WO_3, 26H_2O$	-18.13	0.42	3.60	[7.6.24]
$Na_2W_2O_7(2\ sites)$	10.29	0.95	2.31	[7.6.24]
	-3.42	0.95	2.24	
$Li_2W_2O_7(2\ sites)$	-11.66	0.83	2.78	[7.6.24]
	$+5.95$	0.71	2.78	
$Ba_3Fe_2SO_9$	± 3.15	0	3.30	[7.6.24]
$WOCl_4$	0	0	3.52	[7.6.24]
Me_2SnWO_4	-9.32	0.71	2.81	[7.6.24]
$(NH_4)_2WS_4$	-8.31	0.76	2.88	[7.6.24]
$K_3[W_2Cl_9]$	8.89	0	2.24	[7.6.24]
$Et_4N[WCl_6]$	-12.58	0	3.95	[7.6.24]
$Et_4N[WCl_6]$	-12.3	0.4	4.0	[7.6.24]
$Na_{0.53}WO_3$	0	0	4.28	[7.6.24]
$Li_{0.36}WO_3$	0	0	5.25	[7.6.24]
$K_4[W(CN)_8] \cdot 2H_2O$	-16.02[a]	$-$	2.49	[7.6.25]
$K_4[W(CN)_8]$ frozen soln.	-16.48[b]	$-$	2.13	[7.6.25]
$Li_4[W(CN)_8] \cdot nH_2O$	-15.90[a]	$-$	2.59	[7.6.25]
$H_4[W(CN)_8] \cdot 6H_2O$	$+12.59$[a]	$-$	2.81	[7.6.25]
$H_4[W(CN)_8]$ frozen soln.	$+14.79$[b]	$-$	2.63	[7.6.25]
$Cd_2[W(CN)_8] \cdot 8H_2O$	$+12.80$[a]	0.82	3.00	[7.6.25]
$K_3[W(CN)_8] \cdot H_2O$	0	$-$	2.93	[7.6.25]
$Na_3[W(CN)_8] \cdot 4H_2O$	0	$-$	3.20	[7.6.25]
$Ag_3[W(CN)_8]$	0	$-$	2.42	[7.6.25]
$[Co(NH_3)_6][W(CN)_8]$	0	$-$	2.44	[7.6.25]
$(EtNC)_4[W(CN)_4]$	-8.59[a]	0.74	1.83	[7.6.25]

[a] Estimated errors 0.2 mm s^{-1}
[b] Estimated errors 0.5 mm s^{-1}

[7.6.7] Oberley, L. W., Hershkowitz, N., Wender, S. A., Carpenter, A. B.: Phys. Rev. *C 3,* 1585 (1971)

[7.6.8] Hershkowitz, N., Wender, S. A., Carpenter, A. B.: Phys. Rev. *C 3,* 219 (1972)

[7.6.9] Zioutas, K., Wolbeck, B., Perscheid, B.: Z. Phys. *262,* 413 (1973)

[7.6.10] Bokemeyer, H., Wohlfahrt, K., Kankeleit, E., Eckardt, D.: Z. Phys. *A 274,* 305 (1975)

[7.6.11] Shikazono, N., Takekoshi, H., Shoji, T.: J. Phys. Soc. Jap. *21,* 829 (1966)

[7.6.12] Cohen, S. G., Blum, N. A., Chow, Y. W., Frankel, R. B., Grodzins, L.: Phys. Rev. Lett. *16,* 322 (1966)

[7.6.13] Gedikli, A., Winkler, H., Gerdau, E.: Z. Phys. *267,* 61 (1974)

[7.6.14] Frankel, R. B., Chow, Y., Grodzins, L., Wulff, J.: Phys. Rev. *186,* 381 (1969)

[7.6.15] Raj, D., Puri, S. P.: Phys. Lett *29A,* 510 (1969)

[7.6.16] Ruth, R. D., Hershkowitz, N.: Phys. Lett. *34A,* 203 (1973)

[7.6.17] Kaltseis, J., Posch, H. A., Vogel, W.: J. Phys. C: Solid State Phys. *5,* 2523 (1972)

[7.6.18] Conroy, L. E., Perlow, G. J.: Phys. Lett. *31A,* 400 (1970)

[7.6.19] Roth, S., Hörl, E. M.: Phys. Lett. *25A,* 299 (1967)

[7.6.20] Schibuya, N., Tsunoda, Y., Nishi, A., Kunitomi, N.: J. Phys. Soc. Japan *33,* 564 (1972)

[7.6.21] Wagner, F. E., Schaller, H., Felscher, R., Kaindl, G., Kienle, P.: in *Hyperfine Interactions in Excited Nuclei,* Vol 2, 603; G. Goldring, R. Kalish (eds.). Gordon and Breach Science Publishers (1971)

[7.6.22] Bancroft, G. M., Garrod, R. E. B., Maddock, A. G.: Inorg. Nucl. Chem. Lett. *7,* 1157 (1971)

[7.6.23] Clark, M. G., Gancedo, J. R., Maddock, A. G., Williams, A. F., Yoffe, A. D.: J. Phys. C: Solid State Phys. *6,* L474 (1973)

[7.6.24] Maddock, A. G., Platt, R. H., Williams, A. F., Gancedo, R.: J. C. S. Dalton 1314 (1974)

[7.6.25] Clark, M. G., Gancedo, J. R., Maddock, A. G., Williams, A. F.: J. C. S. Dalton 120 (1975)

[7.6.26] Agresti, D., Kankeleit, E., Persson, B.: Phys. Rev. *155,* 1342 (1969)

[7.6.27] Wender, S. A., Hershkowitz, A.: Phys. Rev. *B 8,* 4901 (1973)

7.7. Osmium (^{186,188,189,190}Os)

Among the Os isotopes there are four (^{186,188,189,190}Os) for which the Mössbauer effect has been observed. In Fig. 7.7.1 the nuclear decay schemes and parent nuclei of the Mössbauer isotopes are shown. Whereas only one Mössbauer transition has been observed in each of the even Os isotopes, there are three for ^{189}Os with transition energies of 95.2, 69.6, and 36.2 keV. The relevant nuclear data for all six Mössbauer transitions are collected in Table 7.1 at the end of the book. The transition energies of the even osmium isotopes are relatively high (for Mössbauer spectroscopy), 137.2, 155.0, and 187.0 keV for ^{186}Os, ^{188}Os, and ^{190}Os, respectively, resulting in recoil-free fractions of only some tenths of a percent at low temperatures (≈ 70 K). In addition, their line widths are very large compared to expected isomer shifts, due to the small $\Delta \langle r^2 \rangle$ values in going from the 0^+ ground to 2^+ excited nuclear rotational states. Therefore, Mössbauer measurements of $^{186, 188, 190}$Os have only been devoted so far to the determination of the nuclear quadrupole moment and nuclear g-factors ($\mu_{2^+} = 2g_{2^+} \mu_N$) in the 2^+-states.

Apart from the determination of nuclear parameters, the Mössbauer transition in ^{189}Os, especially the 36.2 and 69.6 keV transitions, are suited for chemical applications. As will be shown, the 36.2 keV level, in spite of its large half-width, can well be used for the measurement of isomer shifts, whereas the 69.2 keV state should be

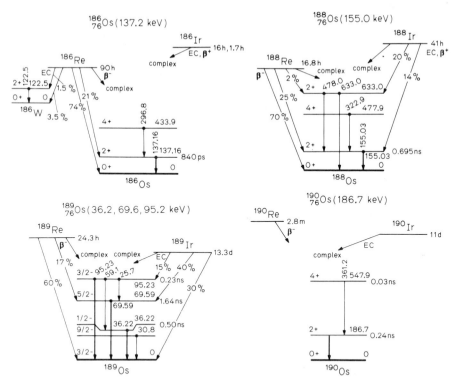

Fig. 7.7.1. Simplified nuclear decay schemes for the four osmium Mössbauer isotopes 186,188,189,190Os (from [7.7.1])

favourably taken in case of electric quadrupole or magnetic dipole interaction. Both Mössbauer levels are populated equally well by the parent isotope ^{189}Ir, and simultaneous measurement should be possible by appropriate geometrical arrangement.

7.7.1. Practical Aspects of Mössbauer Spectroscopy with Osmium

The sources for the even osmium isotopes ^{186,188}Os have been the same since the first observations of the Mössbauer effect in the scattering experiments by Morrison and co-workers [7.7.2] and Barret and Grodzins [7.7.3] for ^{186}Os and ^{188}Os, respectively. They are produced by neutron irradiation of ^{185}Re and ^{187}Re in Re metal foils making use of the reactions ^{185}Re(n, γ) ^{186}Re $\xrightarrow[90h]{\beta^-}$ ^{186}Os and ^{187}Re(n, γ) ^{188}Re $\xrightarrow[16.8h]{\beta^-}$ ^{188}Os.

Due to unresolved electric quadrupole interaction in the hexagonal Re metal lattice, the emission lines are slightly broadened by a factor of about 1.1 and 1.24 for ^{186}Os and ^{188}Os, respectively ($2\Gamma_{nat}(^{186}$Os$) = 2.42 \pm 0.08$ mm s^{-1}, $2\Gamma_{nat}(^{188}$Os$) = 2.50 \pm 0.08$ mm s^{-1}), [7.7.4]. ^{190}Os, for which the Mössbauer effect was first observed by Wagner et al. in 1972 [7.7.5], is populated via decay of ^{190}Ir as parent nucleus.

This source is produced by the reaction $^{190}Os(d, 2n)$ ^{190}Ir $\xrightarrow[11d]{EC}$ ^{190}Os with a 30 μA beam of 13 MeV deuterons. The target, which consisted in Wagner's experiment [7.7.5] of 70 wt % Cu and 30 wt % Os (enriched to 95.5% ^{190}Os), could be used as the source without further physical or chemical treatment showing no essential line broadening (Γ_{exp} = 4.0 ± 0.4 mm s^{-1}, $2\Gamma_{nat}$ = 3.67 ± 0.08 mm s^{-1}). Actually, because of the insolubility of osmium in copper, the target consisted of Os metal clusters in the host matrix. To get an impression of the resolving power, some Mössbauer spectra for the even Os isotopes are given in Figs. 7.7.2a and 7.7.2b for OsO$_2$- and OsP$_2$-absorbers, respectively (from [7.7.5]). The corresponding hyperfine fields are $|V_{zz}| \simeq 2 \cdot 10^{18}V/$cm^2, $\eta \simeq 0$ (OsO$_2$) and $|V_{zz}| \simeq 4 \cdot 10^{18}V/cm^2$, $\eta \simeq 0.74$ (OsP$_2$). It can be seen that the asymmetric hyperfine spectra – in the case of $\eta \simeq 0$ (OsO$_2$), which allow the determination of the sign of V_{zz}, – become more symmetric with increasing asymmetry parameter η. For $\eta = 1$, the spectra become completely symmetric and the sign of V_{zz} no longer influences the shape of the spectra.

The odd mass isotope ^{189}Os possesses three nuclear states at 95.2, 69.6, and 36.2 keV which are all suitable for Mössbauer effect measurements. The first observations of the resonance absorption of these three transitions were made by Gregory et al. [7.7.6] (95.2 keV), by Iha et al. [7.7.7], Persson et al. [7.7.8] (69.6 keV), and by Wagner

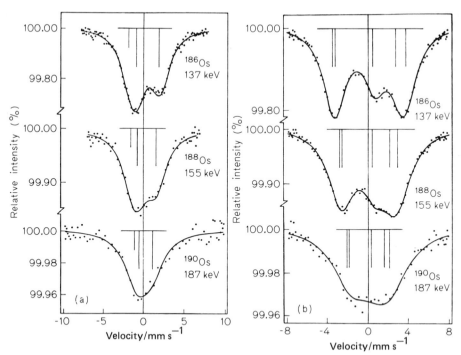

Fig. 7.7.2. Transmission Mössbauer spectra of the 137, 155 and 187 keV nuclear transitions of $^{186,188,190}Os$, taken with sources emitting an unsplit line and (a) OsO$_2$-absorber ($\eta \simeq 0$), (b) OsP$_2$-absorber ($\eta \simeq 0.74$). The curves are the results of least-squares fits. The vertical bars indicate the positions and relative intensities of the individual hyperfine components (from [7.7.5])

et al. [7.7.5] (36.2 keV). Single line sources of ^{189}Ir in Ir metal can be obtained by use of the reactions ^{189}Os(p, n) ^{189}Ir [7.7.8] or ^{189}Os(d, 2n) ^{189}Ir [7.7.4, 5, 10] and subsequent chemical separation of osmium and ^{189}Ir. The Munich group [7.7.4, 5, 10] has used noncorrosive targets with good heat conductivity containing 70 wt % Cu and 30 wt % Os enriched to 87.4% ^{189}Os, which were irradiated by a 30 μA deuterium beam of 13 MeV. After irradiation, the ^{189}Ir activity was chemically separated from the target and incorporated in Ir metal, giving nearly unbroadened emission lines since Ir metal is cubic and nonmagnetic. The ^{189}Ir source populates nearly equally well the

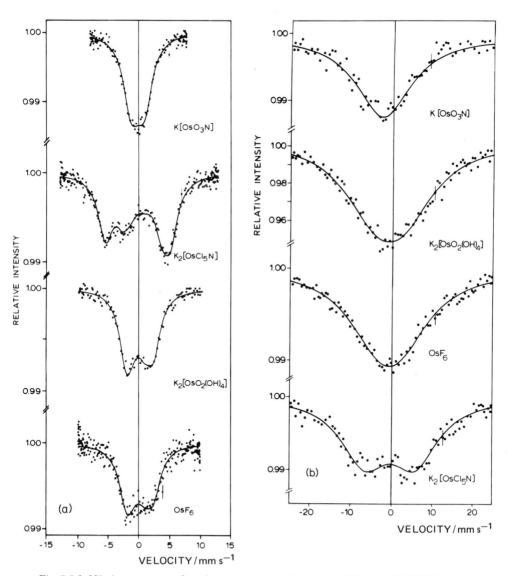

Fig. 7.7.3. Mössbauer spectra of osmium compounds obtained at 4.2 K with 69.6 KeV (a) and 36.2 keV (b) transitions of ^{189}Os with a source of ^{189}Ir in iridium metal (from [7.7.10])

three Mössbauer levels (cf. Fig. 7.7.1). However, the 69.6 keV transition is heavily masked by K_β x-rays ($E_{K_\beta} \approx 70.8$ keV) which can hardly be separated energetically, even with high resolution Ge (Li) detectors, because of the necessary strong source activities. The 95.2 keV level up to now has only been used to estimate the lifetime of this state. A value of $\tau > 0.2$ ns has been reported for the mean life of this state [7.7.6]. In Fig. 7.7.3 the quadrupole spectra for a number of compounds are given for comparison of the 36.2 and 69.6 keV Mössbauer transitions [7.7.10]. Although the line width of the former is a factor of about 7 larger, the ratio of isomer shift to line width (δ/Γ_{exp}) is a factor of approximately 4 greater, because of the larger quantity of $\Delta\langle r^2 \rangle$ for the 36 keV level (cf. Sec. 7.7.2). It is just opposite with respect to electric quadrupole and magnetic dipole interaction. Here, the 69.6 keV transition is more favorable than the 36 keV transition.

7.7.2. Determination of Nuclear Parameters of Osmium Mössbauer Isotopes

The application of Mössbauer spectroscopy in chemistry presumes the knowledge of the data of the nuclear states and transitions involved. In this section we shall describe the determination of nuclear parameters by means of Mössbauer experiments with Os nuclei.

a) Magnetic Moments and E2/M1 Mixing Parameter δ

The magnetic moment and the E2/M1 mixing parameter of the 69.6 keV level ($5/2^-$) of ^{189}Os have been determined by several authors [7.7.5, 6, 8, 11]. These measurements have been done using ^{189}Ir/Fe sources under applied external fields and single line absorbers as for example $K_2[OsCl_6]$. The weighted average of three most reliable values leads to δ = E2/M1 = +0.69 ± 0.04 and μ (69.6 keV) = 0.980 ± 0.008 μ_N, using the known magnetic moment of the ground state $\mu_{3/2-}$ = 0.6565 ± 0.0003 μ_N [7.7.12].

The magnetic moment of the 36.2 keV level ($1/2^-$) in ^{189}Os has been measured by Wagner and co-workers [7.7.9]. The source consisted of ^{189}Ir in an iron matrix ($Ir_{0.01}Fe_{0.99}$) which was produced by melting the ^{189}Ir activity with iron in an induction furnace. Applying an external magnetic field of 2.97 Tesla, the hyperfine spectrum became partly resolved leading to a ratio of $\mu(1/2^-)/\mu(3/2^-)$ = 0.302 ± 0.003. The weighted average of the internal magnetic hyperfine field of ^{189}Os in the ^{189}Ir/Fe source was found to be $-$ 111,5 ± 2.0 Tesla.

Wagner et al. [7.7.4] have also determined the magnetic moments of the first excited $I_e = 2^+$ states of ^{186}Os and ^{188}Os using ^{186}Re and ^{188}Re in $Re_{0.01}Fe_{0.99}$ sources and Os metal absorbers. The sources were polarized by an external magnetic field of 2.97 Tesla parallel to the direction of the γ-beam, so that only $\Delta m = m_e - m_g = \pm 1$ hyperfine components appeared in the Mössbauer spectrum. From their measurements they immediately obtained for the ratio of the magnetic moments of the 2^+ states $\mu_{2+}(^{188}Os)/\mu_{2+}(^{186}Os)$ = 1.08 ± 0.05. Using the value of 113 ± 2.5 Tesla for the internal hyperfine field obtained from NMR measurements which have to be corrected for the external and demagnetization field, the following magnetic moments could be derived:

$$\mu_{2^+}(^{186}\text{Os}) = 0.562 \pm 0.016 \; \mu_\text{N} \text{ and } \mu_{2^+}(^{188}\text{Os}) = 0.610 \pm 0.030 \; \mu_\text{N}.$$

b) Nuclear Quadrupole Moments

Mössbauer measurements to determine the electric quadrupole moments have been reported in [7.7.4, 5, 11]. The most recent results have been obtained by Wagner et al. [7.7.5] by measuring the quadrupole hyperfine interaction in OsO_2 and OsP_2 of the Mössbauer isotopes ^{186,188,189,190}Os. The ratios of the quadrupole moments of the $I_e = 2^+$ states in the even osmium isotopes and of the $I_e = 5/2^-$ (69.6 keV) and $I_g = 3/2^-$ states in ^{189}Os could be deduced very accurately. In Table 7.7.1 (end of the book) the experimental results [7.7.5] are given, from which the following ratios can be calculated:

$$Q_{2^+}(^{186}\text{Os})/Q_{2^+}(^{188}\text{Os})/Q_{2^+}(^{190}\text{Os})/Q_{3/2^-}(^{189}\text{Os, g. s.}) =$$
$$= (+ 1.100 \pm 0.020)/1.0/(+ 0.863 \pm 0.051)/(- 0.586 \pm 0.011)$$
and $Q_{5/2^-}(^{189}\text{Os}, 69.6 \text{ keV})/Q_{3/2^-}(^{189}\text{Os, g. s.}) = - 0.735 \pm 0.012.$

From optical hyperfine measurements a value of $Q_{3/2^-}(^{189}\text{Os, g. s.}) = (0.91 \pm 0.10)$ b has been derived [7.7.13]. Theoretical values for the quadrupole moment of ^{186}Os and experimental $I_e = 2^+$ to $I_g = 0^+$ transition probabilities led the authors of [7.7.5] to the conclusion that $Q_{2^+}(^{186}\text{Os})$ should be $-(1.50 \pm 0.10)$ b, which in turn yielded $Q_{3/2^-}(^{189}\text{Os, g. s.}) = + (0.80 \pm 0.06)$ b, which is somewhat smaller than the value from the optical measurements.

c) Change of Nuclear Charge Radii

An essential prerequisite for the determination of electron densities at the nucleus — the interesting quantity from isomer shift measurements in chemistry and solid-state physics — is the knowledge of the change of the average squared nuclear charge radii in going from the ground ($\langle r^2 \rangle_g$) to the excited ($\langle r^2 \rangle_e$) nuclear state. As already mentioned, the $\Delta\langle r^2 \rangle = \langle r^2 \rangle_e - \langle r^2 \rangle_g$ values of the 2^+ states in even Os isotopes are very small compared to the line width of the Mössbauer transition, so that there are only estimations from muonic atom spectroscopy available [7.7.14]:

$$\Delta\langle r^2 \rangle/\langle r^2 \rangle \, (^{188}\text{Os}) = - 2.37 \cdot 10^{-5}; \Delta\langle r^2 \rangle/\langle r^2 \rangle \, (^{190}\text{Os}) = - 6.05 \cdot 10^{-5}; \Delta\langle r^2 \rangle/\langle r^2 \rangle (^{192}\text{Os}) =$$
$$- 8.16 \cdot 10^{-5}.$$

Concerning the change of the nuclear charge radii in ^{189}Os, the situation is much better. Here Bohn et al. [7.7.15] have determined the ratio of $\Delta\langle r^2 \rangle_{36.2 \text{ keV}}/\Delta\langle r^2 \rangle_{69.6 \text{ keV}}$ by Mössbauer measurements with $\text{K}_2[\text{OsCl}_6]$, $\text{K}_4[\text{Os(CN)}_6]$ and OsO_4 as absorbers. Plotting isomer shifts of the 69.6 keV transition, corrected for the second-order

Table 7.7.1. Summary of results obtained for the four Os Mössbauer transitions studied. The absorber thickness d refers to the amount of the resonant isotope per unit area. The estimates of the effective absorber thickness t are based on Debye-Waller factors f for an assumed Debye temperature of $\theta = 400$ K. For comparison with the full experimental line widths at half maximum, Γ_{exp}, we give the minimum observable width $2\Gamma_{nat} = 2\hbar/\tau$ as calculated from lifetime data. The electric quadrupole interaction is described by $E_Q = (eQV_{zz}/4) \cdot (1 + 1/3\eta^2)^{1/2}$ and the asymmetry parameter η. The latter was assumed to be zero in the least-squares fits of the OsO_2 data. In the fit of the OsO_2 spectrum for ^{190}Os, the quadrupole splitting E_Q was constrained to the value given in the table. For the 69 keV transition of ^{189}Os, the results for the ratio $Q_{5/2}-/Q_{3/2}-$ of the quadrupole moments and the E2/M1 mixing parameter δ are included (from [7.7.5])

Source	Ab-sorber	d $\frac{mg}{cm^2}$	t	Γ_{exp} mm s^{-1}	E_Q mm s^{-1}	η
^{186}Re in Re	OsP_2	10	0.7	2.80 ± 0.05	-3.68 ± 0.05	0.74 ± 0.02
^{186}Re in Re	OsP_2	4	0.3	3.00 ± 0.15	-3.66 ± 0.08	0.78 ± 0.04
^{186}Re in Re	OsO_2	11	0.8	2.86 ± 0.12	$+1.91 \pm 0.04$	<0.3
^{186}Re in Pt	OsO_2	8	0.6	2.75 ± 0.13	$+1.82 \pm 0.05$	<0.3
^{186}Re in Pt	OsO_2	5	0.4	2.70 ± 0.11	$+1.87 \pm 0.04$	<0.3

a) ^{186}Os, 137.16 keV, $2^+ - 0^+$; $2\Gamma_{nat} = 2.41 \pm 0.04$ mm s^{-1}; $f(\theta = 400K) = 0.082$

b) ^{188}Os, 155.02 keV, 2^+-0^+; $2\Gamma_{nat} = 2.50 \pm 0.05$ mm s^{-1}; f $(\theta = 400$ K$) = 0.043$

188Re in 187Re	OsP$_2$	43	1.6	3.10 ± 0.10	−2.93 ± 0.05	0.74
188Re in Pt	OsO$_2$	23	0.7	3.06 ± 0.20	+1.51 ± 0.06	<0.3
188Re in Pt	OsO$_2$	8	0.3	2.80 ± 0.47	+1.61 ± 0.13	<0.3

c) ^{190}Os, 186.7 keV, 2^+-0^+; $2\Gamma_{nat} = 3.67 \pm 0.08$ mm s^{-1}; f $(\theta = 400$K$) = 0.011$

190Ir in 190Os	OsP$_2$	150	1.3	4.32 ± 0.50	−2.10 ± 0.10	0.74
				(4.04 ± 0.44)	(+2.13 ± 0.09)	
190Ir in 190Os	OsO$_2$	180	1.5	3.76 ± 0.34	+1.1	<0.3

d) ^{189}Os, 69.6 keV, $5/2^- - 3/2^-$; $2\Gamma_{nat} = 2.41 \pm 0.06$ mm s^{-1}; f $(\theta = 400$K$) = 0.53$

$Q_{5/2^-}/Q_{3/2^-} = -0.735 \pm 0.012$; $\delta = +0.685 \pm 0.025$

189Ir in Ir	OsP$_2$	16	1.2	2.86 ± 0.05	+3.83 ± 0.05	0.73 ± 0.03
189Ir in Ir	OsO$_2$	11	0.8	3.30 ± 0.10	−1.99 ± 0.06	<0.3

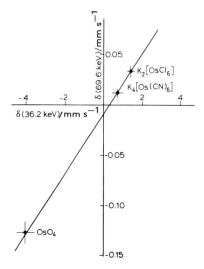

Fig. 7.7.4. Isomer shift $\delta = \delta_{exp} - \delta_{SOD}$ of the 69.6 keV Mössbauer transition versus the isomer shift of the 36.2 keV line in ^{189}Os (from [7.7.15])

Doppler shift, versus the isomer shift of the 36.2 keV transition, a straight line is obtained [Fig. 7.7.4]. The slope of this line leads to the ratio of $\Delta\langle r^2\rangle_{36.2}/\Delta\langle r^2\rangle_{69.6} = 16.6 \pm 1.3$. Wagner et al. [7.7.10] (cf. Sec. 7.7.3) derived a value of $\Delta\langle r^2\rangle_{36.2} = -2.0 \cdot 10^{-3}$ fm^2 from Mössbauer measurements on osmium compounds with Os in the oxidation states $+8$, $+6$, $+4$, $+3$, and $+2$, using the calculated electron densities from relativistic self-consistent field calculations for free Os ions. Of course, these calculations must be considered with care in the case of covalently bound Os atoms. Therefore, the $\Delta\langle r^2\rangle_{36.2}$ value should be taken only as an estimate of the order of magnitude rather than as an accurate number.

7.7.3. Inorganic and Metal Organic Osmium Compounds

To date there is only one paper — by Wagner et al. [7.7.10] — devoted to a systematic investigation of the isomer shift and quadrupole splitting in various osmium compounds. Of special interest is the comparison with similar or isoelectronic compounds of iridium and ruthenium. The appropriate Mössbauer transitions for the determination of the isomer shift and quadrupolar interaction are — for reasons quoted above — the 36.2 and 69.6 keV transitions, respectively, in ^{189}Os. In Table 7.7.2 a summary of the results from Wagner et al. [7.7.10] is given. Magnetic hyperfine interaction has been neglected in the evaluation of the Mössbauer spectra, and only single Lorentzian lines or a pure quadrupole pattern has been adjusted to the experimental data points. In Fig. 7.7.5 the isomer shift of the 36.2 keV transition is compared with the shift found for the 89 keV Mössbauer resonance of ^{99}Ru. The velocity scale has been chosen such that isoelectronic compounds lie on horizontal lines (e.g., $RuO_4 - OsO_4$; $K_2[RuCl_6] - K_2[OsCl_6]$; $(but_4N)_3[Ru(NCS)_6] - (but_4N)_3[Os(NCS)_6]$). The agreement between both elements is good except for the compounds $A(C_5H_5)_2$ ($A = Ru$, Os). As in the cases of Fe, Ru, Ir, Pt, and Au, the authors found a monotonic dependence of the isomer shift for osmium compounds on the oxidation state (Fig. 7.7.6).

Table 7.7.2. Summary of the results obtained from the Mössbauer spectra of the 36.2 and 69.5 keV γ-rays of [189]Os in absorbers of various osmium compounds containing d mg/cm² of [189]Os. W is the full experimental line width at half maximum, δ the isomer shift with respect to the source of [189]Ir metal. $E_Q = 1/4e^2qQ_{3/2}$ is the electric quadrupole coupling constant of the [189]Os ground state and V_{zz} the electric field gradient at the Os nuclei as calculated from the E_Q values with $Q_{3/2} = +(0.80 \pm 0.06)b$ (from [7.7.10l]).

Compound and formal oxidation state		36.2 keV transition			69.6 keV transition				
		d mg/cm²	Γexp mm s⁻¹	δ mm s⁻¹	d mg/cm²	Γexp mm s⁻¹	δ mm s⁻¹	E_Q mm s⁻¹	V_{zz} 10¹⁸Vcm⁻²
1. OsO$_4$	+8	20	21.9 ± 1.7	−3.64 ± 0.24	43	3.07 ± 0.08	−0.109 ± 0.011	—	—
2. K[OsO$_3$N]	+8	7	15.0 ± 0.7	−3.30 ± 0.15	24	2.84 ± 0.14	−0.154 ± 0.029	0.83 ± 0.03	0.96 ± 0.09
3. OsF$_6$	+6	15	18.7 ± 0.9	−0.78 ± 0.16	40	2.82 ± 0.18	+0.054 ± 0.050	−1.37 ± 0.04	−1.58 ± 0.14
4. K$_2$[OsCl$_5$N]	+6	5	14.2 ± 0.4	−0.47 ± 0.16	33	2.98 ± 0.10	+0.006 ± 0.021	+3.19 ± 0.04	+3.75 ± 0.31
5. K$_2$[OsO$_2$(OH)$_4$]	+6	23	18.2 ± 1.3	−0.82 ± 0.19	23	2.62 ± 0.12	−0.009 ± 0.021	−1.39 ± 0.02	−1.61 ± 0.13
6. K$_2$[OsF$_6$]	+4	8	16.9 ± 0.7	+1.64 ± 0.13	8	2.85 ± 0.14	+0.096 ± 0.042	—	—
7. K$_2$[OsCl$_6$]	+4	21	19.0 ± 0.7	+1.08 ± 0.13	21	3.14 ± 0.18	+0.049 ± 0.005	—	—
8. (NH$_4$)$_2$[OsCl$_6$]	+4	7	18.6 ± 1.5	+1.04 ± 0.27	16	3.15 ± 0.08	+0.043 ± 0.013	—	—
9. (NH$_4$)$_2$[OsBr$_6$]	+4	5	15.2 ± 1.5	+0.83 ± 0.37	21	3.52 ± 0.08	+0.073 ± 0.020	—	—
10. OsO$_2$	+4	11	16.4 ± 1.3	+1.13 ± 0.22	11	3.30 ± 0.10	+0.035 ± 0.020	−1.99 ± 0.04	−2.31 ± 0.19
11. (but$_4$N)$_3$[Os(NCS)$_6$]	+3	11	19.9 ± 1.1	+1.59 ± 0.19	36	—	—	—	—
12. K$_4$[Os(CN)$_6$]anhydr.	+2	5	16.6 ± 0.9	+0.89 ± 0.17	27	2.86 ± 0.05	+0.026 ± 0.003	—	—
13. K$_2$[Os(CN)$_5$NO] · H$_2$O	+2	7	18.8 ± 1.5	+0.59 ± 0.21	20	6.20 ± 0.44	+0.066 ± 0.037	1.10 ± 0.10	1.27 ± 0.15
14. [Os(bipy)$_3$](ClO$_4$)$_2$ · H$_2$O	+2	13	16.9 ± 0.7	+1.48 ± 0.14	21	3.36 ± 0.13	+0.077 ± 0.019	0.58 ± 0.04	0.67 ± 0.06
15. Os(C$_5$H$_5$)$_2$	+2	9	15.4 ± 1.0	+1.67 ± 0.16	15	3.19 ± 0.06	0.184 ± 0.029	+1.53 ± 0.03	+1.77 ± 0.15
16. Os metal		22	20.5 ± 0.6	−0.41 ± 0.10	65	4.30 ± 0.06	+0.008 ± 0.005	0.77 ± 0.03	0.89 ± 0.08

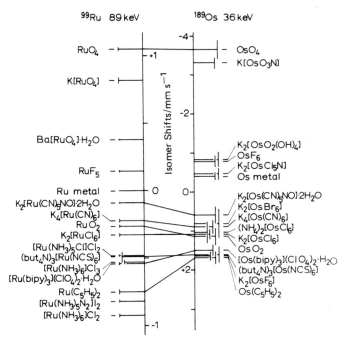

Fig. 7.7.5. Comparison of isomer shift results for compounds of Ru and Os. The velocity scales are chosen such that isoelectronic compounds of the two elements lie on horizontal lines (from [7.7.10])

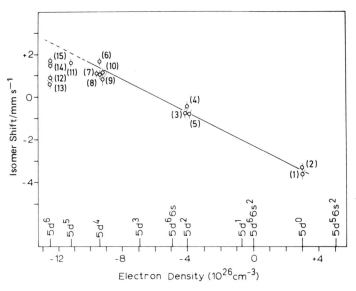

Fig. 7.7.6. Plot of the isomer shift δ of the 36.2 keV Mössbauer transition of ^{189}Os versus Dirac-Fock values for the electron density differences at the Os nuclei in free ion 5d configurations. The numbers of the data points refer to the numbering of the compounds in Table 7.7.2 (from [7.7.10])

The change of the isomer shift of d^n configurations is mainly caused by the shielding effect of the s-electrons by the varying number of d-electrons. Only in those compounds where strong backbonding effects play a concomitant role are strong deviations from this rule expected and observed. In this case d_π orbitals of the central ion and π orbitals of the ligand build up the molecular orbital with a spatial spread of the d-electrons towards the ligands, thus decreasing the shielding capability of the d-electrons. This effect can be observed in the divalent $K_4[Os(CN)_6]$ and $K_2[Os(CN)_5NO] \cdot 2H_2O$ compounds, where the Mössbauer isomer shift is close to the values observed for tetravalent compounds OsO_2 or $K_2[OsCl_6]$.

References

[7.7.1] Stevens, J. G., Stevens, V. E.: *Mössbauer Effect Data Index*. New York: IFI Plenum Press 1969–1975

[7.7.2] Morrison, R. J., Atac, M., Derbrunner, P., Frauenfelder, H.: Phys. Lett. *12*, 35 (1964)

[7.7.3] Barret, P. H., Grodzins, L.: Rev. Mod. Phys. *36*, 971 (1964)

[7.7.4] Wagner, F., Kucheida, D., Kaindl, G., Kienzle, P.: Z. Phys. *230*, 80 (1970)

[7.7.5] Wagner, F. E., Spieler, H., Kucheida, D., Kienle, P., Wäppling, R.: Z. Phys. *254*, 112 (1972)

[7.7.6] Gregory, M. C., Robinson, B. L., Iha, S.: Phys. Rev. *180*, 1158 (1969)

[7.7.7] Iha, S., Owens, W. R., Gregory, M. C., Robinson, B. L.: Phys. Lett. *25 B*, 115 (1967)

[7.7.8] Persson, B., Blumberg, H., Bent, M.: Phys. Rev. *174*, 1509 (1968)

[7.7.9] Wagner, F., Kaindl, G., Bohn, H., Biebl, U., Schaller, H., Kienle, P.: Phys. Lett. *28 B*, 548 (1969)

[7.7.10] Wagner, F. E., Kucheida, D., Zahn, U., Kaindl, G.: Z. Phys. *266*, 223 (1974)

[7.7.11] Kucheida, D., Wagner, F., Kaindl, G., Kienle, P.: Z. Phys. *216*, 346 (1968)

[7.7.12] Schrenk, A., Zimmermann, G.: Phys. Lett. *16A*, 258 (1968)

[7.7.13] Himmel, G.: Z. Phys. *211*, 68 (1968)

[7.7.14] Walter, H. K.: Nucl. Phys. *A 234*, 504 (1974)

[7.7.15] Bohn, H., Kaindl, G., Kucheida, D., Wagner, F. E., Kienle, P.: Phys. Lett. *32B*, 346 (1970)

7.8. Iridium (^{191}Ir, ^{193}Ir)

There are two iridium isotopes, ^{191}Ir and ^{193}Ir, suitable for Mössbauer spectroscopy. Each of them possesses two nuclear transitions with which nuclear resonance absorption has been observed. Fig. 7.8.1 shows the (simplified) nuclear decay schemes for both iridium Mössbauer isotopes; the Mössbauer transitions are marked therein with fat arrows. The relevant nuclear data known to date for the four Mössbauer transitions are collected in Table 7.1 at the end of the book.

As a matter of fact Mössbauer spectroscopy has its deepest root in the 129.4 keV transition line of ^{191}Ir, for which R. L. Mössbauer established recoilless nuclear resonance absorption for the first time, while he was working on his thesis under Prof. Maier-Leibnitz at Heidelberg [7.8.1]. But this nuclear transition is by far not the easiest one among the four iridium Mössbauer transitions to use for solid-state applications. The 73 keV line of ^{193}Ir with the lowest transition energy and the narrowest

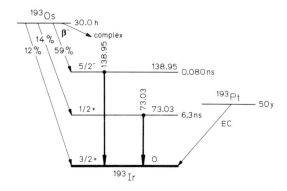

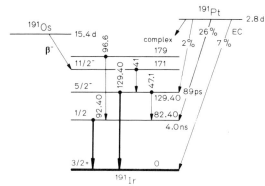

Fig. 7.8.1. Simplified decay scheme leading to the population of the four nuclear Mössbauer transitions of ^{191}Ir and ^{193}Ir (from [7.8.2])

natural line width (0.60 mm s^{-1}) fulfills best the practical requirements and is, of all four iridium transitions, therefore, most often (say in 90% of all reports published on Ir Mössbauer spectroscopy) used in studying electronic structures, bond properties, and magnetism.

7.8.1. Practical Aspects of ^{193}Ir (73 keV) Mössbauer Spectroscopy

The most popular precursor to feed the 73 keV level of ^{193}Ir is ^{193}Os ($t_{1/2}$ = 31 h). This isotope is produced by neutron activation, ^{192}Os (n, γ)^{193}Os. Most laboratories use enriched (> 95%) ^{192}Os metal as the target without further treatment after neutron irradiation. The only disadvantage in working with this source is the fact that osmium metal has hexagonal symmetry and therefore nonzero EFG at the Ir nucleus, which causes an electric quadrupole splitting of the I_g = 3/2 ground state. As a consequence, the ^{193}Os/Os metal source emits an unresolved quadrupole doublet; the two overlapping emission lines have equal intensity and are separated by $\Delta E_Q = 0.48 \pm 0.02$ mm s^{-1} [7.8.3]. This splitting must be considered in the fitting procedures of measured spectra. Wagner and Zahn [7.8.3] have also tested an Os$_{0.01}$P$_{0.99}$ alloy as a source. Although it emits a single unsplit line, it is not as convenient to work with as the metallic osmium source, because the preparation involves induction melting of the activated ^{193}Os with platinum in argon atmosphere. Another single-line source for ^{193}Ir Mössbauer spectroscopy has been described by Davies, Maddock, and Williams [7.8.4]. They use an Os$_{0.05}$Nb$_{0.95}$ alloy with 98.7% enriched ^{192}Os, which yields a single emission line with a line width of 0.64 mm s^{-1} (4.2 K) without any post-irradiation annealing.

The ^{193}Ir Mössbauer experiments are usually carried out in transmission geometry with both source and absorber kept at liquid helium temperature, and using a Ge(Li) diode or a 3 mm NaI(Tl) crystal to detect the 73 keV γ-rays. The absorbers typically contain 50–500 mg/cm^2 of natural iridium. The isomer shifts are generally given with respect to iridium metal (the isomer shift between ^{193}Os/Os and Ir metal is $-(0.540 \pm 0.004)$ mm s^{-1} at 4.2 K ([7.8.3]).

7.8.2. Coordination Compounds of Iridium

After a first brief report by Thomson et al. [7.8.5] on isomer shifts of the 73 keV γ-rays of ^{193}Ir in Ir metal, IrO$_2$, K$_2$[IrCl$_6$], and IrCl$_4$, it was only in 1967 that, at nearly the same time, the first extended papers from an Israeli group [7.8.6] and the Munich laboratory [7.8.7, 8] on applications of Mössbauer spectroscopy to intermetallic and coordination compounds of iridium appeared in the literature.

Atzmony et al. [7.8.6] have observed that the ^{193}Ir (73 keV) isomer shifts of the tetravalent Ir compounds IrO$_2$, K$_2$[IrCl$_6$], (NH$_4$)$_2$[IrCl$_6$], and IrCl$_4$ relative to the trivalent Ir compounds IrI$_3$, K$_3$[IrCl$_6$], and IrCl$_3$ are positive. They assume that in tetravalent Ir compounds ([Xe]4f^{14}5d^5 for Ir^{4+}) with one 5d electron less than in trivalent Ir compounds ([Xe]4f^{14}5d^6 for Ir^{3+}) the external (5s) electrons are less effectively shielded and that therefore the s-electron density $|\psi_s(o)|^2$ at the Ir nucleus is larger than in the trivalent Ir compounds. From this they conclude that the sign of $\Delta\langle r^2\rangle$, the difference in the mean-square radius of the charge distribution between the 73 keV level and the ground state of ^{193}Ir, is positive. An absolute value of $\Delta\langle r^2\rangle$ (^{193}Ir, 73 keV) has been derived by Wagner et al. [7.8.8] from the isomer shifts measured with metallic sources of ^{191}Os(14.6 h), ^{191}Pt(3.0 d), and ^{193}Os(31 h), and natural Ir metal as absorber.

Wagner and Zahn [7.8.3] have reported on an extensive study of a variety of hexacoordinated trivalent and tetravalent iridium compounds as well as on IrF$_5$ and IrF$_6$. Their observed isomer shifts (relative to Ir metal) are summarized in Fig. 7.8.2. The isomer shifts in halogen compounds of Ir nearly increase with the formal oxidation state of Ir, but deviations are observed for the complexes with CN$^-$ and NO$_2^-$ as ligands. The authors interpret the observations qualitatively on the basis of the shielding effect of the varying effective number of 5d electrons, paying attention to the fact that 5d$^{n_{eff}}$ not only changes by altering the Ir oxidation state but also by

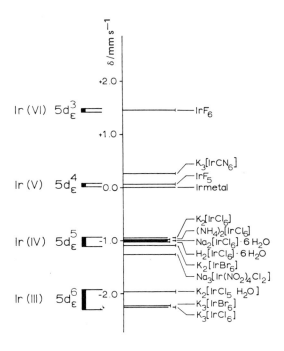

Fig. 7.8.2. ^{193}Ir (73 keV) isomer shifts as a function of oxidation state of iridium (from [7.8.3])

backbonding (in complexes with CN^- and NO_2^- ligands). They are also aware of the influence of varying σ-bond covalency (direct 6s contribution to $|\psi_s(o)|^2$ by σ-donation from the ligands) which accompanies both changes in the oxidation state of iridium and back-donation ($\pi_{M \to L}$) effects in order to fulfill the requirements of Pauling's electroneutrality principle. A quantitative interpretation of the isomer shifts observed so far in iridium compounds is only possible in connection with molecular orbital calculations, which should yield effective $5d^\alpha 6s^\beta 6p^\gamma$ configurations from the individual covalencies of the a_{1g}, e_g, and t_{1u} orbitals. Though highly desirable, such a treatment has not yet been performed for iridium compounds.

The compounds containing the hexahalogen complexes $[IrCl_6]^{2-}$, $[IrBr_6]^{2-}$ with Ir(IV) formal oxidation state ($5d^5$) and iridium hexafluoride with hexavalent iridium ($5d^3$) posses an odd number of electrons and have therefore Kramers degenerate electronic levels with an effective spin of S = 1/2 and S = 3/2 in the cases of $5d^5$ and $5d^3$, respectively. These compounds are therefore expected to order magnetically at sufficiently low temperatures, as indeed has been established by susceptibility and heat capacity measurements. The Néel temperatures occur around 2–3 K for the $5d^5$ compounds and around 8 K in IrF_6. Wagner and Zahn [7.8.3] have measured the Mössbauer hyperfine spectra of such compounds between 0.5 and 4.2 K. Some representative spectra of $K_2[IrCl_6]$, taken with the single-line source $^{193}Os_{0.01}Pt_{0.99}$ for better resolution, are shown in Fig. 7.8.3. Using a single-line source the magnetic hyperfine pattern is composed of eight lines, arising from the M1/E2 transitions between the sublevels of the excited and the ground nuclear states as depicted in Fig. 7.8.4 for the general case of a combined magnetic dipole and electric quadrupole interaction. The spectra of Fig. 7.8.3 show nicely the gradual decrease of the magnetic

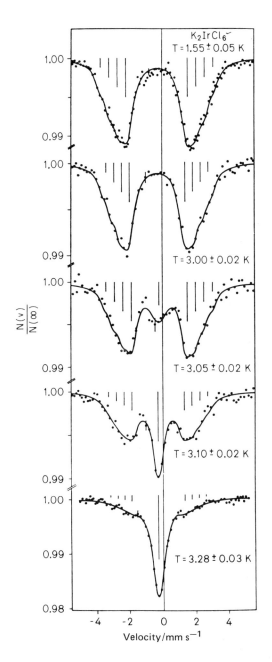

Fig. 7.8.3. Mössbauer spectra of K$_2$[IrCl$_6$] at various temperatures, taken with the single-line source ^{193}Os$_{0.01}$Pt$_{0.99}$. The positions and relative intensities of the resonance lines are indicated by vertical bars (from [7.8.3])

hyperfine pattern with increasing temperature, in favour of a single center line from the paramagnetic state of K$_2$[IrCl$_6$], due to the gradual decrease of the effective field as a consequence of the thermally induced increase in electronic spin fluctuation rate. An example for a well-resolved ^{193}Ir (73 keV) magnetic hyperfine pattern is the spectrum of IrF$_6$ at 4.18 K shown in Fig. 7.8.5 from the work of Wagner and Zahn [7.8.3].

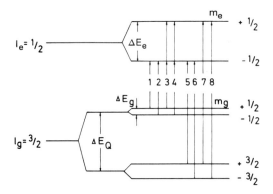

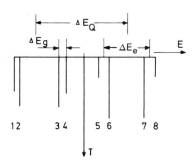

Fig. 7.8.4. Effect of combined magnetic dipole and electric quadrupole interaction on the nuclear excited (e) and ground (g) states of the 73 keV transition in ^{193}Ir. The relative intensities, given in the bar diagram, correspond to a E2/M1 mixing ratio of $\delta^2 = 0.340$ in case of absence of preferred orientation. ΔE_e = magnetic splitting of the excited state ($I_e = 1/2$), ΔE_g = magnetic splitting of the ground state ($I_g = 3/2$), ΔE_Q = electric quadrupole splitting. The relatively large magnetic splitting of the excited state reflects the fact that the g factor of the 73 keV level is about nine times larger than that of the $3/2^+$ ground state (from [7.8.7])

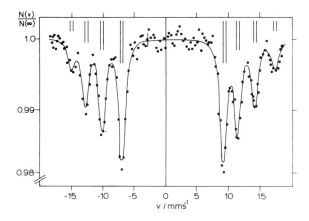

Fig. 7.8.5. Mössbauer spectrum of IrF_6 at 4.18 K, taken with a ^{193}Os/Os source. The positions and relative intensities (in accordance with an E2/M1 mixing ratio of $\delta^2 = 0.311 \pm 0.006$ [7.8.9]) are indicated by the vertical bars (from [7.8.3])

Here the eight hyperfine resonances are an unresolved quadrupole doublet each, due to the quadrupole interaction in the hexagonal ^{193}Os in Os metal source. The authors interpret the hyperfine fields in terms of core polarization, orbital and spin-dipolar contributions.

In another paper from the Munich laboratory, Rother, Wagner, and Zahn [7.8.10] have reported on studies of chemical consequences of the β^- decay of ^{193}Os in osmium compounds by the recoilless resonance absorption of the 73 keV γ-rays of ^{193}Ir. They have looked into a number of ^{193}Os labelled osmium compounds, like $K_2[OsO_4] \cdot 2H_2O$, OsO_4, $K_2[OsCl_6]$, $K_2[OsBr_6]$, $(NH_4)_2[OsCl_6]$, $(NH_4)_2[OsBr_6]$ and $Os(C_5H_5)_2$, which they have synthesized by standard methods either in the presence of ^{193}Os activity or as inactive compounds with subsequent neutron activation. These Mössbauer sources were then measured at 4 K versus Ir metal as absorber. The prominent resonance lines of the emission spectra were assigned to different charge states on the basis of the measured isomer shift data and in comparison with the isomer shift/oxidation state scale for ^{193}Ir (73 keV) [7.8.3]. This way the authors found that in the case of OsO_4, $K_2[OsO_4] \cdot 2H_2O$ and $Os(C_5H_5)_2$ the ^{193}Ir daughter species were isoelectronic with the parent compounds, which implies that species with iridium in the valence states +9 and +7 with isomer shifts of -3.78 and -2.14 mm s^{-1}, respectively (compare with Fig. 7.8.2), and $Ir(C_5H_5)_2^+$ must have been produced as a consequence of the β^- decay and have existed with lifetimes at least in the order of the lifetime of the ^{193}Ir (73 keV) excited state. In case of the Os(IV) hexahalides, however, Ir(IV), Ir(III), and possibly Ir(II) species were observed, which the authors understand on the basis of a fast electronic recombination mechanism. No major influence of the ^{192}Os(n, γ) ^{193}Os reaction and of post-irradiation annealing on the pattern of the emission spectra could be observed.

An interesting effect, called hyperfine anomaly[11], has been observed by Wagner and Zahn [7.8.3] and by Perlow et al. [7.8.11] for the 73 keV transition of ^{193}Ir, which apparently has favourable characteristics to establish this phenomenon, in various Ir compounds.

Wickman and Silverthorn [7.8.12] have investigated bond properties in molecular adducts of the planar Vaska type compound trans-bis(triphenyl-phosphine)iridium carbonyl chloride, $IrCl(CO) ((C_6H_5)_3P)_2$, with small molecules such as H_2, O_2, Cl_2, I_2, CH_3I, HCl employing the ^{193}Ir (73 keV) Mössbauer method. They essentially observed a decrease of the isomer shift in the following series of adduct molecules XY: $H_2 > HCl > CH_3J > O_2 > I_2 > Cl_2$, which they explained by variations of the electron population in the d^2sp^3 hybrid orbitals owing to decreasing relative

---

[11] The observed ratio of the nuclear magnetic moments μ_e/μ_g is generally taken to be constant for the nuclear states of a given Mössbauer transition. This, however, is not necessarily so, because there may be slight differences in the interactions between the nuclear magnetic moment (with radial distribution due to orbital and spin motions of unpaired nucleons) and an applied magnetic field (which is uniform over the nuclear magnetization distribution) on the one hand, and with an internal hyperfine field (which may also have radial distribution) on the other hand. Such differences will be reflected in differences of the ratio μ_e/μ_g observed for the two cases of interaction. This is called "hyperfine anomaly".

σ-donation power on going from H_2 to Cl_2 to be the predominant effect. π-back-donation from the filled t_{2g} orbitals (the adducts in question are formally Ir(III) systems) to these ligand molecules is out of discussion. There is, however, a π-effect as indicated by the observed trend in the i.r. carbonyl frequencies. With the exception of the O_2 adduct, which differs from the others in that the O_2 bond is partially retained in forming the adduct, whereas in the other ligand molecules XY of the above series the bond is ruptured and the constituents form bonds only to the iridium atom [7.8.13], the carbonyl frequency increases on going from H_2 to Cl_2. This apparently implies that the tendency of charge delocalization from the filled t_{2g} orbitals to the π carbonyl orbital decreases in the same order of XY ligands, which in turn influences the electron density $|\psi_s(o)|^2$ at the iridium nucleus via shielding and thus alters the isomer shift in the same direction as the σ-donation effect. Both effects are superimposed constructively; this means that the isomer shift varies as a function of $\sigma(XY \rightarrow Ir) - \pi(Ir \rightarrow CO)$. It is assumed that the two effects are concomitant (synergic) to fulfill the requirements of the electroneutrality principle, e.g., H_2 with the strongest relative σ-donation power is expected to cause the largest loss of t_{2g} charge by back-donation into the π (CO) orbital.

In a similar study of a series of $((C_6H_5)_3P)_2Ir(CO)X$ complexes (X = Cl, Br, N_3, SCN; NO^+) by x-ray photoelectron and ^{193}Ir (73 keV) Mössbauer spectroscopy, Holsboer, Beck, and Bartunik [7.8.14] found that substitution of Cl^- by other halides and pseudohalides in the planar trans-$(PPh_3)_2Ir(CO)X$ complexes affects only slightly the total electron density and the EFG at the Ir nucleus. There is, however, a distinct influence of the variation of the anionic X^- ligands on the IR stretching frequency of the CO bond [7.8.15]. The changes in the electron density at the Ir nucleus caused by the variation of X are apparently nearly compensated by the concomitant $\pi(Ir \rightarrow CO)$ effect. The quadrupole splittings for the $(PPh_3)_2Ir(CO)X$ complexes (X = Cl, Br, N_3, SCN) are among the largest ones observed in ^{193}Ir (73 keV) Mössbauer spectroscopy.

In another publication Holsboer, Beck, and Bartunik [7.8.16] have investigated bond properties in the five-coordinated Ir(I) complexes $(PPh_3)_3Ir(CO)H$, $(PPh_3)_3Ir(CO)CN$, and the tetracyanoethylene, fumaronitrile and acrylonitrile adducts of $(PPh_3)_2Ir(CO)Cl$ by x-ray photoelectron and ^{193}Ir Mössbauer spectroscopy.

Ginsberg, Cohen, DiSalvo, and West [7.8.17] have applied ^{193}Ir (73 keV) Mössbauer spectroscopy and magnetic susceptibility measurements to the one-dimensional conductor iridium carbonyl chloride, $Ir(CO)_3Cl$. By carbon and chlorine analyses they confirmed the earlier report by Krogmann et al. [7.8.18] that this compound is partially oxidized. The ^{193}Ir Mössbauer spectra measured in the temperature range from 1.8 to 35 K, however, are not in accordance with Krogmann's suggested formulation of $Ir(CO)_{3-x}Cl_{1+x}$ (x = 0.07). They rather prefer the formulation $Ir(CO)_3Cl_{1+x}$ (x = 0.10 ± 0.03), because the ^{193}Ir (73 keV) Mössbauer spectra exhibit no line attributable to a species other than partially oxidized $Ir(CO)_3Cl$ units. Ginsberg et al. presume that the "extra" chloride occurs in disordered interchain positions. The results from both susceptibility and Mössbauer measurements imply that the charges arising from partial oxidation of $Ir(CO)_3Cl$ chains are not localized on individual atoms.

Another paper dealing with ^{193}Ir Mössbauer studies of planar Ir(I) compounds with metal-metal bonds has been published by Aderjan, Keller, Rupp, Wagner, and Wagner [7.8.19]. They investigated chelato-dicarbonyl-iridium(I) compounds of composition $Ir(CO)_2 L$ and $Ir(CO)_2 L'Cl$ (L = singly charged bidentate organic ligand;

L' = neutral monodentate organic base) and found that the formation of columnar stacks (one-dimensional chains) with metal-metal bonds results in a reduction of both the magnitude of the quadrupole splitting and the isomer shift. They interpreted these findings in terms of a model which assumes that in the columnar stacks a wide d_{z2} band, overlaps with a narrow energetically higher d_{x2-y2} band, and that the partial filling of the antibonding d_{x2-y2} band would then destabilize the bonds with σ-donating and π-accepting ligands.

Ginsberg, Koepke, Cohen, and West [7.8.20] have explored the dependence of ^{193}Ir (73 keV) isomer shifts on the oxidation state in a set of closely related low-valent iridium carbonyl complexes, $(C_6H_5)_4As[Ir(CO)_2Cl_2]$, $Ir(CO)_3Cl_{1.1}$, $K_{0.6}[Ir(CO)_2Cl_2]$ · $0.5H_2O$, and $[(C_6H_5)_4As]_2[Ir(CO)_4Cl_2]$, with formal iridium oxidation states + 1.0, + 1.1, + 1.4, and + 2, respectively. In contradiction to the generally observed increase in isomer shift with increasing oxidation state (cf. Fig. 7.8.2), the measured isomer shifts here decrease with increasing formal oxidation state and are therefore opposite in direction to what is expected for decreased shielding by removal of 5d electrons. The authors believe that a combination of two effects may account for the phenomenon: (i) the increase in electron density $|\psi(o)|^2$ expected from 5d electron removal will largely be cancelled by an accompanying decrease in $\pi(5d, t_{2g}) \rightarrow \pi^*(CO)$ backbonding; (ii) the electronic charge removed in going down the series from + 1.0 to + 2.0 will contain a significant 6s component. A more satisfactory answer to this problem could be obtained from MO calculations.

A study of six-coordinate pentammine complexes of iridium, $[Ir(NH_3)_5X]Y_n$ ($X = I^-$, Cl^-, NO_2^-, $HCOO^-$, CH_3COO^-, NCS^-, and NH_3; $Y = Cl^-$, ClO_4^-), of $(but_4N)_3$ $[Ir(SCN)_6]$, and of the cis and trans isomers of $[IrCl_4pyr_2]$, $C_5H_5NH[IrCl_4pyr_2]$, and $[IrCl_2pyr_4]Cl$ (pyr = pyridine) has been performed by Wagner, Potzel, Wagner, and Schmidtke [7.8.21]. Some of their measured spectra are shown in Fig. 7.8.6. The spectra are more or less resolved quadrupole doublets with each line being an unresolved doublet by itself due to the quadrupole splitting in the source. The ratio of the quadrupole splittings for the trans and cis isomers of the pyridine complexes has been observed to be about 2:1 (cf. upper two spectra of Fig. 7.8.6), as has been predicted on the basis of a point charge model [7.8.22, 23] and also observed, e.g., in low-spin iron(II) complexes [7.8.22]. The isomer shift tendencies are understood in terms of varying powers of σ-donation and π-backbonding. The authors have attempted to interpret the isomer shifts in the framework of the angular overlap model [7.8.24–26], which enables one to make separate estimates of the strength of σ and π bonds.

Two more publications on ^{193}Ir (73 keV) Mössbauer spectroscopy of complex compounds of iridium by Williams, Maddock et al. [7.8.27, 28] deserve particular attention. In the first of their papers [7.8.27], which deals with iridium(III) complex compounds, they have shown that the partial isomer shift and partial quadrupole-splitting treatments are not very satisfactory for Ir(III) complexes. They have attributed this to the picture of localized bonding, which is the basis of this treatment, otherwise called "additive model" [7.8.29], to be less true for Ir(III) than for Fe(II) complexes. The idea of interligand interactions in Ir(III) compounds playing a considerable role has been supported by NMR data recorded by these authors. Their second paper deals with four-coordinate formally Ir(I) complexes. They also observed, like other authors on similar low-valent iridium compounds [7.8.20], only small differences in the isomer shifts, which they attributed to the interaction between the metal-ligand bonds leading to compensation effects. Their interpretation is supported by changes in the ^{31}P NMR data of the phosphine ligands and in the frequency of the carbonyl stretching vibration.

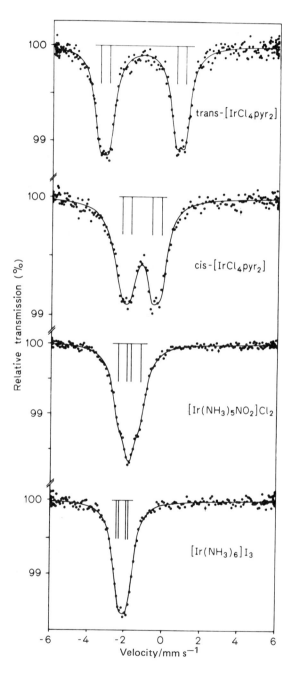

Fig. 7.8.6. Mössbauer spectra of some hexacoordinated ammine and pyridine complexes of trivalent iridium taken at 4.2 K with a source of ^{193}Os in Os metal. The stick spectra indicate the positions and relative intensities of the individual resonance lines (from [7.8.21])

7.8.3. Intermetallic Compounds and Alloys of Iridium

Heuberger, Pobell, and Kienle [7.8.7] and an Israeli group [7.8.6] have, independently from each other and at nearly the same time, measured the hyperfine splitting and the isomer shift of the ^{193}Ir (73 keV) Mössbauer transition in intermetallic compounds of iridium with rare earths, RIr_2. The systems studied by the former group included R = Sm, Gd, Tb, Dy, Ho, and Er; those of the latter research group contained R = Nd, Pr, Sm, Gd, Tb, Dy, and Ho. The intermetallic compounds RIr_2 belong to the family of fcc Laves phases. They are magnetically ordered with varying Curie temperatures below 100 K. Above the Curie temperature T_C these phases exhibit a quadrupole doublet resulting from a quadrupole interaction due to the nonzero EFG at the noncubic Ir site. Below T_C the resonances are generally split into four lines by both electric qua-

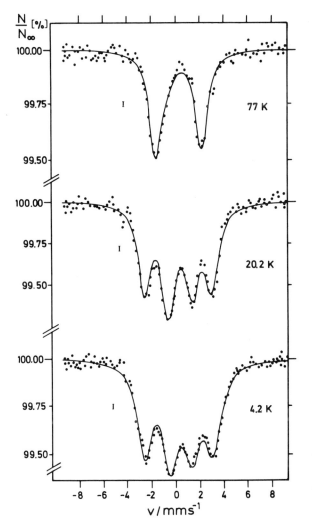

Fig. 7.8.7. ^{193}Ir (73 keV) Mössbauer spectra of $TbIr_2$ at various temperatures, taken with a ^{193}Os/Os source kept at the absorber temperature (from [7.8.7])

drupole and magnetic hyperfine interaction, and into two broad lines only, if the electric quadrupole interaction is practically zero; the magnetic splitting of the $I = 3/2$ ground state has never been resolved in these systems. The hyperfine structure of the ^{193}Ir (73 keV) transition of mixed E2/M1 multipolarity is depicted in Fig. 7.8.4, together with a bar spectrum indicating the relative positions and intensities of the eight (E2 + M1) allowed transitions. Typical spectra of TbIr$_2$ with pure quadrupole interaction (above $T_C = 45$ K) and combined quadrupole and magnetic hyperfine interaction (at $T < T_C$) are shown in Fig. 7.8.7.

The main object of these studies was to investigate the dependence of the effective magnetic field acting on the nucleus of the nonmagnetic Ir atom on the properties of the rare earth constituents, and to learn this way about the mechanism of the formation of the hyperfine fields at the Ir nuclei induced by the magnetic rare earth neighbour atoms. Essentially it has been found that the saturation field at the ^{193}Ir nucleus in RIr$_2$ is a linear function of $(g - 1)J$, i.e., of the projection of the spin of the rare earth atoms onto their total angular momentum. This in turn implies that the induced fields predominantly originate from a spin polarization of the s conduction electrons by their interactions with the 4f electrons of the R atoms (s–f interaction). The polarized conduction electrons are believed to contribute to the effective field H_{eff} at the Ir nucleus mainly through two mechanisms: (i) direct contribution through the Fermi contact interaction, and (ii) indirect contribution of the non-s conduction electrons polarized by the s conduction electrons and of the core electrons of the iridium atom polarized by overlap with the polarized conduction electrons. These effects lead to polarized core s-electrons which build up the field contributions at the nucleus by Fermi contact interaction. Both mechanisms (i) and (ii) depend on the spin of the rare earth atoms and the number of s conduction electrons, and the Kasuya-Yosida theory [7.8.30] on conduction electron polarization should apply. This could be confirmed in the studies of the RIr$_2$ systems by both research groups.

The same Israeli group [7.8.6] has also measured the magnetic hyperfine interaction in the alloy Ir$_{0.01}$Fe$_{0.99}$. The spectrum taken at 4.2 K with a ^{193}Os/Os metal source looks very similar to the spectrum of IrF$_6$ shown in Fig. 7.8.5. The supertransferred field at the nuclear site of the "nonmagnetic" iridium induced by the magnetic iron atoms of the host lattice is so large that all eight (E2 + M1) allowed transitions (cf. Fig. 7.8.4) are observed and the magnetic splitting of the ground state is well resolved. Some broadening of the eight resonance lines is due to the quadrupole interaction intrinsic to the Os metal source.

Mössbauer et al. [7.8.31] studied Ir-Fe and Ir-Ni alloy systems over the whole composition range by means of ^{193}Ir (73 keV) and ^{57}Fe nuclear resonance absorption. They found that the magnetic hyperfine field at the Ir nuclei in Ir-Ni alloys decreases approximately linearly with the Ir concentration from -46.0 T at 4.2 K in very dilute alloys to zero at about 20 at.-% Ir. Some representative ^{193}Ir (73 keV) spectra from their work, which demonstrate nicely the gradual decrease of the magnetic hyperfine pattern with increasing Ir concentration in favour of a center absorption line characteristic of nonmagnetic Ir metal, are shown in Fig. 7.8.8. The concentration dependence of the hyperfine fields has been discussed in terms of a rigid 3d band model combined with local shielding. The suggestion of conduction electron polarization being the primary source of the magnetic hyperfine fields at dilute Ir impurities in the

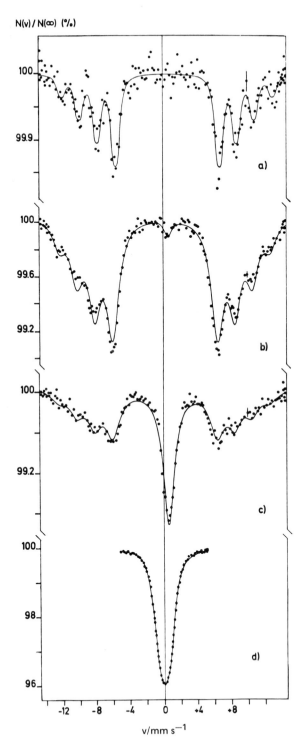

Fig. 7.8.8. ^{193}Ir (73 keV) Möss-bauer spectra of Ir_xFe_{1-x} al-loys containing x = 0.7 at.-% (a), 13.5 at.-% (b), 16.2 at.-% (c), and 53.6 at.-% (d) iridum. The spectra were recorded at 4.2 K using a ^{193}Ir/Os metal source (from [7.8.31])

ferromagnetic host lattices is supported by the observation that the hf fields are approximately proportional to the host magnetic moment.

Vogl et al. [7.8.32] have employed the Mössbauer effect in ^{193}Ir (73 keV) and ^{197}Au to study radiation damage in α-iron, which was doped with 0.6 at.-% Os and 4 at.-% Pt, respectively, and irradiated in a nuclear reactor at 4.6 K. Thermal neutron capture produces the precursor nuclides ^{193}Os and ^{197}Pt, respectively, for the Mössbauer isotopes ^{193}Ir and ^{197}Au and simultaneously could lead to radiation damage. The influence of annealing on the average magnetic field and on line broadening was studied.

Salomon and Shirley [7.8.33] have performed Mössbauer-source experiments on Fe(^{193}Os) alloys with less than 0.1 at.-% osmium against a metallic Ir absorber. Their spectra were well-resolved magnetic 8-line patterns, which the authors could satisfactorily analyze using the Hamiltonian

$$\hat{H} = -g\mu_N \, \hat{H}_z\hat{I}_z + \frac{eV_{zz}Q}{4I(2I-1)} \, [3\hat{I}_z^2 - I(I+1)].$$

Their results show that quadrupole coupling persists in domains at high dilution, for which a reasonably accurate magnitude and the sign may be determined, even in more concentrated systems and even in the presence of substantial solute-solute induced magnetic and quadrupole perturbations. The derived hf field at Ir was 148.1 T.

A Japanese group [7.8.34] recently measured the hyperfine field at Ir nuclei in $Fe_{0.7}Pt_{0.3-x}Ir_x$ ($0.03 \leqslant x \leqslant 0.2$) using the ^{193}Ir (73 keV) Mössbauer method, in order to understand the mechanism of the moment decrease in these alloys. The isomer shift was found to increase with increasing x, which could be rationalized by the contraction of the lattice volume. The observed decrease of the hf field in the region of small x is consistent with the decrease of the average magnetic moment.

References

[7.8.1] Mössbauer, R. L.: Z. Phys. *151*, 124 (1958); Naturwiss. *45*, 538 (1958); Z. Naturforsch. *14A*, 211 (1959)

[7.8.2] Stevens, J. G., Stevens, V. E. (eds.): Mössbauer Effect Data Index, Vols. 1972, 1975. New York: Plenum Press 1973, 1976

[7.8.3] Wagner, F., Zahn, U.: Z. Phys. *233*, 1 (1970)

[7.8.4] Davies, G. J., Maddock, A. G., Williams, A. F.: J. C. S. Chem. Comm. 264 (1975)

[7.8.5] Thomson, J. O., Werkheiser, A. H., Lindauer, M. W.: Rev. Mod. Phys. *36*, 357 (1964)

[7.8.6] Atzmony, U., Bauminger, E. R., Lebenbaum, D., Mustachi, A., Ofer, S., Wernick, J. J.: Phys. Rev. *163*, 314 (1967)

[7.8.7] Heuberger, A., Pobell, F., Kienzle, P.: Z. Phys. *205*, 503 (1967)

[7.8.8] Wagner, F., Klöckner, J., Körner, H. J., Schaller, H., Kienle, P.: Phys. Lett. *25B*, 253 (1967)

[7.8.9] Wagner, F., Kaindl, G., Kienle, P., Körner, H.-J.: Z. Phys. *207*, 500 (1967)

[7.8.10] Rother, P., Wagner, F., Zahn, U.: Radiochim. Acta *11*, 203 (1969)

[7.8.11] Perlow, G. J., Henning, W., Olson, D., Goodman, G. L.: Phys. Rev. Lett. *23*, 680 (1969)

[7.8.12] Wickman, H. H., Silverthorn, W. E.: Inorg. Chem. *10*, 2333 (1971)

[7.8.13] Vaska, L.: Accounts Chem. Res. *1*, 335 (1968)

[7.8.14] Holsboer, F., Beck, W., Bartunik, H. D.: J. C. S. Dalton, 1829 (1973)

[7.8.15] Vaska, L., Peone, J.: Chem. Comm., 418 (1971)

[7.8.16] Holsboer, F., Beck, W., Bartunik, H. D.: Chem. Phys. Lett. *18*, 217 (1973)

[7.8.17] Ginsberg, A. P., Cohen, R. L., DiSalvo, F. J., West, K. W.: J. Chem. Phys. *60*, 2657 (1974)

[7.8.18] Krogmann, K., Binder, W., Hausen, H. D.: Angew. Chem. Int. Ed. *7*, 812 (1968)

[7.8.19] Aderjan, R., Keller, H. J., Rupp, H. H., Wagner, F. E., Wagner, U.: J. Chem. Phys. *64*, 3748 (1976)

[7.8.20] Ginsberg, A. P., Koepke, J. W., Cohen, R. L., West, K. W.: Chem. Phys. Lett. *38*, 310 (1976)

[7.8.21] Wagner, F. E., Potzel, W., Wagner, U., Schmidtke, H.-H.: Chem. Phys. *4*, 284 (1974)

[7.8.22] Berret, R. R., Fitzsimmons, B. W.: J. Chem. Soc. A., 525 (1967)

[7.8.23] Clark, M. G.: Mol. Phys. *20*, 257 (1971)

[7.8.24] Jørgensen, C. K., Pappalardo, R., Schmidtke, H.-H.: J. Chem. Phys. *39*, 1422 (1963)

[7.8.25] Schäffer, C. E., Jørgensen, C. K.: Mol. Phys. *9*, 401 (1965)

[7.8.26] Schmidtke, H.-H.: Theoret. Chim. Acta *20*, 92 (1971)

[7.8.27] Williams, A. F., Jones, G. C. H., Maddock, A. G.: J. C. S. Dalton, 1952 (1975)

[7.8.28] Williams, A. F., Bhaduri, S., Maddock, A. G.: J. C. S. Dalton, 1958 (1975)

[7.8.29] Bancroft, G. M.: Coord. Chem. Rev. *11*, 247 (1973)

[7.8.30] Kasuya, T.: Progr. Theoret. Phys. (Kyoto) *16*, 45 (1956)
Yosida, K.: Phys. Rev. *106*, 893 (1957)

[7.8.31] Mössbauer, R. L., Lengsfeld, M., von Lieres, W., Potzel, W., Teschner, P., Wagner, F. E., Kaindl, G.: Z. Naturforsch. *26 a*, 343 (1971)

[7.8.32] Vogl, G., Schaefer, A., Mansel, W., Prechtel, J., Vogl, W.: phys. stat. sol. (b) *59*, 107 (1973)

[7.8.33] Salomon, D., Shirley, D. A.: Phys. Rev. B *9*, 29 (1974)

[7.8.34] Kanashiro, M., Nishi, M., Kunitomi, N., Sakai, H.: J. Phys. Soc. Japan *38*, 897 (1975)

7.9. Platinum (^{195}Pt)

Mössbauer spectroscopy with ^{195}Pt started only in 1965, when Harris, Benczer-Koller, and Rothberg [7.9.1] measured the Mössbauer absorption spectra of the 99 keV transition of ^{195}Pt in platinum metal as a function of temperature (between 20 and 100 K) and of absorber thickness and derived the temperature dependence of the Debye-Waller factor.

There are two gamma transitions in ^{195}Pt amenable to the Mössbauer effect — the 130 keV transition between the $5/2^-$ excited state and the $1/2^-$ ground state and the 99 keV transition between the first excited $3/2^-$ state and the ground state. Fig. 7.9.1 shows the simplified decay scheme of ^{195}Pt. The relevant nuclear data may be taken from Table 7.1 (at the end of the book).

It is much more difficult to observe the Mössbauer effect with the 130 keV transition than with the 99 keV transition because of the relatively high transition energy of 130 keV, the low transition probability and thus the small cross section for resonance absorption. Therefore most of the Mössbauer work with ^{195}Pt published so far has been performed using the 99 keV transition. Unfortunately, its line width is about 5 times larger than that of the 130 keV transition, and hyperfine interactions in most cases are poorly resolved. However, isomer shifts in the order of one tenth of the line width and magnetic dipole interaction, which manifests itself only in line broadening, may be extracted reliably from ^{195}Pt (99 keV) spectra.

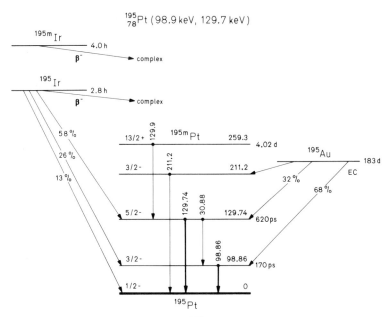

Fig. 7.9.1. Simplified decay scheme of ^{195}Pt. The two Mössbauer transitions are indicated by fat arrows (from [7.9.2])

7.9.1. Experimental Aspects

The two Mössbauer levels of ^{195}Pt, 99 and 130 keV, are populated by either the EC decay of ^{195}Au ($t_{1/2}$ = 183 d) or the isomeric transition of ^{195m}Pt ($t_{1/2}$ = 4.1 d). Only a few authors, e.g., [7.9.3, 13] reported on the use of ^{195m}Pt, which is produced by thermal neutron activation of ^{194}Pt via ^{194}Pt(n, γ) ^{195m}Pt. The source used in the early measurements by Harris, Benczer-Koller and Rothberg [7.9.1, 4] was carrier-free ^{195}Au diffused into platinum metal. Walcher [7.9.5] irradiated natural platinum metal with deuterons to obtain the parent nuclide ^{195}Au by (d, xn) reactions. After the decay of short-lived isotopes, especially ^{196}Au($t_{1/2}$ = 6.18 d), ^{195}Au was extracted with ethyl acetate, and the ^{195}Au/Pt source prepared by induction melting. Buyrn and Grodzins [7.9.3] made use of (α, xn) reactions when bombarding natural iridium with 30 MeV α-particles and used the ^{195}Au/Ir source after annealing without any further chemical or physical treatment. Commercially available sources are produced via ^{195}Pt(p, n) ^{195}Au. The most popular source matrix, into which ^{195}Au is diffused, is platinum metal, although it has the disadvantage of being a resonant matrix — natural platinum contains 33.6% of ^{195}Pt. Using copper and iridium foils as host matrices for the ^{195}Au parent nuclide, Buyrn et al. [7.9.6] observed natural line widths and reasonable resonance absorption of a few percent at 4.2 K.

^{195}Pt (99 keV) Mössbauer spectra have been taken at temperatures ranging from 4 to about 100 K for source and absorber. There is an increase in the percentage resonance effect of about one order of magnitude on going from 77 down to 20 K (cf. Fig. 3 in [7.9.1]).

Typical absorbers contain 50–700 mg/cm^2 of natural platinum. The observed experimental line widths in ^{195}Pt (99 keV) spectra range from values close to the natural width ($2\Gamma_{nat}$ (99 keV) = 16.28 mm s^{-1}) to approximately 25 mm s^{-1}. With respect to the line width the 130 keV transition with a natural width of $2\Gamma_{nat}$ (130 keV) = 3.40 mm s^{-1} seems to be more favourable for the study of hyperfine interaction in platinum compounds; in fact, experimental line widths of 3.4 ± 0.4 [7.9.7] and 3.5 ± 0.7 mm s^{-1} [7.9.8] have been measured. The considerably higher energy resulting in a much smaller recoilless fraction and the lower probability for the population of the 130 keV excited state, however, make the use of the 130 keV transition less favourable for ^{195}Pt Mössbauer spectroscopy than the 99 keV transition.

The 99 keV gamma quanta are usually counted with NaI(Tl) scintillation counters or Ge(Li) diodes in transmission geometry. A Cd absorber should be used to reduce the background counting rate of the K x-rays and to avoid pile-up of the different x- and γ-rays (cf. Fig. 4 in [7.9.4]).

In the only report on the Mössbauer scattering technique Atac, Debrunner and Frauenfelder [7.9.9] recorded Mössbauer spectra of platinum in iron, nickel, and cobalt at 29 K counting the resonantly scattered 99 keV γ-quanta.

7.9.2. Platinum Compounds

First applications of ^{195}Pt (99 keV) Mössbauer spectroscopy to platinum compounds have been attempted by Agresti, Kankeleit, and Persson [7.9.10]. For Pt(II) and Pt(IV) oxides and chlorides at 4.2 K, they measured the following isomer shifts (with respect to the source of ^{195}Au diffused into Pt foil): PtO, −0.34(11); PtO$_2$, −0.40(8); PtCl$_2$, −0.1(2); PtCl$_4$, −0.3(3) mm s^{-1}. Although the errors given here are relatively large, it seems as though the isomer shift tends to decrease with increasing oxidation state. It was, however, too early to draw from these results any definite conclusion with respect to the sign of $\Delta\langle r^2\rangle$ of the 99 keV transition.

This became possible by the thorough ^{195}Pt Mössbauer study of a number of Pt(II) and Pt(IV) complex compounds by Walcher [7.9.5]. The isomer shifts observed with the 99 keV transition at 4.2 K are given (with respect to metallic Pt) in Table 7.9.1. Comparing these data with isomer shifts of isoelectronic compounds of iridium and gold, Walcher concluded that the change in the mean-square nuclear radius on going from the 99 keV state to the ground state is negative, $\Delta\langle r^2\rangle_{99} < 0$. From the observed difference in isomer shift of [PtCl$_4$]$^{2-}$ and [PtCl$_6$]$^{4-}$ and from the estimated difference in electron density at the Pt nucleus for 5d^8 and 5d^6 electron configurations, respectively, using results from relativistic free-ion Hartree-Fock-Slater calculations, Walcher derived the value of $\Delta\langle r^2\rangle_{99}/\langle r^2\rangle = -(1.6^{+4.4}_{-0.9}) \cdot 10^{-4}$.

Walcher's isomer shift data of Table 7.9.1 may be interpreted qualitatively in a similar way as described for gold (Sec. 7.10) or iridium (Sec. 7.8). The difference in isomer shift on going from Pt(II) to Pt(IV) in compounds with the same ligands is predominantly due to the changing number of 5d electrons, 5d^8 for Pt(II) and 5d^6 for Pt(IV). This causes less shielding of the 5s electrons from the nuclear charge and, along with it, lower electron density at the nucleus and, with $\Delta\langle r^2\rangle_{99} < 0$, a more positive isomer shift in Pt(II) than in Pt(IV) compounds. Moreover, within a given oxida-

Table 7.9.1. ^{195}Pt (99 keV) isomer shifts δ in
platinum (II) and platinum (IV) complex compounds at 4.2 K (relative to platinum metal,
from [7.9.5])

Compound	δ/mm s^{-1}
$K_2[Pt(CN)_4]$	−1.7 (1)
$K_2[PtBr_4]$	+0.85 (10)
$K_2[PtCl_4]$	+0.85 (10)
$K_2[Pt(CN)_6]$	−2.1 (4)
$K_2[Pt(NO)_4Cl_2]$	−0.85 (10)
$K_2[PtBr_6]$	−0.02 (7)
$K_2[PtCl_6]$	−0.22 (7)

tion state, differences in bond properties are extractable from the isomer shifts of
Table 7.9.1. Substituting halide for cyanide as ligand in either Pt(II) or Pt(IV) complexes, d-electron delocalization from the metal ion into the antibonding π-orbitals
of the CN^- ligand becomes effective, resulting in a decrease of the shielding and thus
an increase in electron density at the nucleus and therefore a decrease in isomer shift.
Even a gradual change in π-backbonding power of the ligand sphere is evident from
the observed isomer shifts. Although NO^+ is known to be a stronger π-backbonding
ligand than CN^-, the average π-back-donation power of the total ligand sphere is
most probably larger in case of $K_2[(Pt(CN)_6]$ than in $K_2[Pt(NO)_4Cl_2]$ (which still
contains two Cl^- ligands which are not capable of π-backbonding), and one therefore
finds the latter complex placed between $K_2[Pt(CN)_6]$ and $K_2[PtCl_6]$ on the isomer
shift scale [7.9.5]. Walcher did not observe any quadrupole interaction in compounds
with nonzero EFG, which is not surprising in view of the very large line width. Buyrn
et al. [7.9.6] have pointed out that electric quadrupole interactions of usual strengths
hardly affect the ^{195}Pt (99 keV) spectrum.

Further studies of platinum compounds using the 99 keV transition in ^{195}Pt have
been performed by Rüegg and co-workers. In addition to measurements on PtI_2, they
remeasured PtO_2, $PtCl_4$, and $PtCl_2$ [7.9.8, 11]. Their isomer shift data, which are
given in Table 7.9.2 together with data of the line width and percentage resonance
effect, are considerably more accurate than and for the latter two compounds quite
different from those observed earlier by Agresti et al. [7.9.10]. The more accurate
isomer shift data by Rüegg et al. now allow a clear distinction to be made between
Pt(II) and Pt(IV) compounds similarly to what has been discussed above in connection with the work by Walcher [7.9.5], viz., isomer shifts of Pt(II) compounds are
clearly more positive than those of Pt(IV) compounds. The difference in isomer shift
between $PtCl_2$ and PtI_2 most probably originates from a more effective σ-donation
of s-like charge to the metal ion in case of PtI_2 due to the smaller electronegativity of
iodine (2.5) as compared to chlorine (3.0). The experimental line width is in the case
of $PtCl_4$ practically identical to and in the other compounds only slightly above twice

Table 7.9.2. Isomer shift δ (with respect to the source of ^{195}Au in Pt metal at 4.2 K), line width Γ, and percentage resonance effect f from ^{195}Pt (99 keV) Mössbauer measurements on platinum compounds at 4.2 K by Rüegg and co-workers [7.9.8, 11, 12]

Compound	δ [a] $\overline{\text{mm s}^{-1}}$	Γ [b] $\overline{\text{mm s}^{-1}}$	f $\overline{\%}$
PtO$_2$	−0.34 (2) [−0.40 (8)]	19.0 (8)	6.5 (21)
PtCl$_4$	0.03 (5) [−0.3 (3)]	16.6 (5)	1.7 (6)
PtCl$_2$	0.87 (6) [−0.1 (2)]	20.0 (11)	1.05 (45)
PtI$_2$	0.58 (6)	19.2 (8)	1.35 (50)
K$_2$[Pt(CN)$_4$]Br$_{0.30}$ · 3H$_2$O	−1.68 (4)	16.8 (3)	7.5 (20)
K$_2$[Pt(CN)$_4$] · 3H$_2$O	−1.13 (9)	18.7 (15)	1.4 (5)
K$_2$[Pt(CN)$_4$Br$_2$]	−1.45 (6)	16.9 (3)	7.3 (20)

[a] Data given in square brackets from [7.9.10]
[b] Data from extrapolation to zero absorber thickness

the natural line width ($2\Gamma_{nat} = 16.28$ mm s^{-1}). The relatively small percentage resonance effects require measuring times on the order of one week. A representative ^{195}Pt (99 keV) Mössbauer spectrum, that of PtO$_2$ taken at 4.2 K [7.9.11], is shown in Fig. 7.9.2.

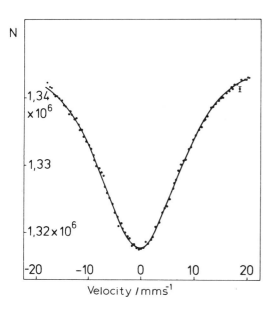

Fig. 7.9.2. ^{195}Pt (99 keV) Mössbauer spectrum of PtO$_2$ taken at 4.2 K with a source of ^{195}Au in platinum at 4.2 K (from [7.9.11])

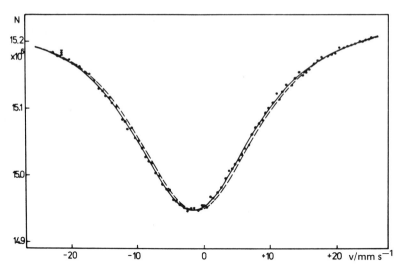

Fig. 7.9.3. [195]Pt (99 keV) Mössbauer spectrum of the one-dimensional conductor $K_2[Pt(CN)_4]Br_{0.30} \cdot 3H_2O$ at 4.2 K (source: [195]Au in platinum at 4.2 K). The solid line represents a single Lorentzian line fitted to the measured spectrum. The dashed line represents the best fit using a sum of two Lorentzian lines with an intensity ratio of 85:15 and with the isomer shifts of the spectra of $K_2[Pt(CN)_4] \cdot 3H_2O$ and $K_2[(Pt(CN)_4Br_2]$ (from [7.9.12])

In another investigation using the [195]Pt (99 keV) Mössbauer effect, Rüegg, Kuse, and Zeller [7.9.12, 11] attempted to elucidate the problem about the charge localization in the one-dimensional conductor $K_2[Pt(CN)_4]Br_{0.30} \cdot 3H_2O$. Optical properties of this compound indicate metallic behaviour. The d_c conductivity, along the Pt-Pt chains, is thermally activated below 200 K. At 4.2 K, however, the compound is an excellent insulator. All models which have been proposed to describe the one-dimensional conductors presume that the insulating state at 0 K is indicative of a certain localization of the conduction electrons. Rüegg et al. showed that the Mössbauer spectrum of $K_2[Pt(CN)_4]Br_{0.30} \cdot 3H_2O$, taken at 4.2 K (cf. Fig. 7.9.3), could be fitted ideally to a single Lorentzian absorption line. The fit assuming a sum of two Lorentzian lines with an intensity ratio of 85:15 according to the possible formation of a double salt $\{0.85\ K_2[Pt^{II}(CN)_4] + 0.15\ K_2[Pt^{IV}(CN)_4Br_2] \cdot 3H_2O\}$ (which is chemically equivalent to $K_2[Pt(CN)_4]Br_{0.30} \cdot 3H_2O$) gave worse results. The authors therefore concluded that charge is not localized on an atomic scale. The relatively high electron density obtained from the isomer shift of the one-dimensional conductor system as compared to the insulating compounds $K_2[(Pt(CN)_4] \cdot 3H_2O$ and $K_2[Pt(CN)_4Br_2]$ (cf. Table 7.9.2) appears to be a consequence of the strong metal-metal bond.

7.9.3. Metallic Systems

Much of the ^{195}Pt Mössbauer work performed so far has been devoted to studies of platinum metal and alloys in regard to nuclear properties (magnetic moments and lifetimes) of the excited Mössbauer states of ^{195}Pt, lattice dynamics, electron density and internal magnetic field H^{int} at the nuclei of Pt atoms placed in various magnetic hosts. Observed changes in the latter two quantities, $|\psi(o)|^2$ and H^{int}, within a series of platinum alloys are particularly informative about conduction electron delocalization and polarization.

In ^{195}Pt (99 keV) backscattering experiments, the only ones communicated so far, Atac, Debrunner, and Frauenfelder [7.9.9] observed magnetic hyperfine interaction in ferromagnetic alloys of natural platinum with iron, cobalt and nickel ($Pt_{0.3}$ $Fe_{0.7}, Pt_{0.07}Co_{0.93}, Pt_{0.07}Ni_{0.93}$). Typical spectra reproduced from their work are shown in Fig. 7.9.4. The spectra are unresolved due to the large natural line width of the 99 keV transition. Nevertheless, by fitting a Zeeman pattern consisting of six lines of equal width and with an intensity ratio appropriate for a pure dipole transition, reliable values for the magnitude and even the sign (from the angular correlation pattern observed with a polarized scatterer) of the internal field at the Pt nuclei could be evaluated (at 29 K) to be -124 T in $Pt_{0.3}Fe_{0.7}$, -77 T in $Pt_{0.07}Co_{0.93}$, and -23 T in $Pt_{0.07}$ $Ni_{0.93}$ (error in H^{int} about 10%). From the approximate constancy of $|H^{int}|/\mu_B$, where μ_B is the magnetic moment in Bohr magnetons of the host atoms, the authors concluded that the internal field at the Pt nuclei is predominantly due to conduction electron polarization. Buyrn, Grodzins, Blum, and Wulff [7.9.6] arrived at the same conclusion in their study of the ^{195}Pt (99 keV) Mössbauer effect in Pt_xFe_{1-x} alloys ($0.03 \leqslant x \leqslant 0.5$). By applying an external magnetic field they also found the nuclear g factor ratio to be $g(3/2)/g(1/2) < 0$. (Longitudinal polarization eliminates the $\Delta m = 0$ transitions and the intensity pattern should be $3 : 1 : 1 : 3$ for $g(3/2)/g(1/2) < 0$ but $1 : 3 : 3 : 1$ for $g(3/2/g(1/2) > 0$). H^{int} was found to be nearly constant, about 126 T (at 4.2 K), over a range of composition from 3 to 30 at. % Pt.

Agresti et al. [7.9.10] investigated alloys of 3 at. % Pt in iron, cobalt and nickel at 4.2 K in transmission geometry. They applied a small magnetic field (0.25 T) parallel and perpendicular to the γ-ray direction modifying the intensities of the transition between the nuclear magnetic sublevels and thereby increasing the obtainable information as compared to measurements using unpolarized absorbers. Their results concerning the magnitude of H^{int} are in satisfactory agreement with those measured in scattering geometry by Atac et al. [7.9.9]. Agresti et al. could also extract isomer shifts from their recorded spectra. From the observed decrease of δ with decreasing electronegativity (cf. Fig. 7.9.5) they concluded that the mean-square radius of the ^{195}Pt nucleus in the 99 keV state is smaller than in the ground state.

The first clear resonance effect using the 130 keV transition in ^{195}Pt was observed on a platinum metal absorber by Wilenzick et al. [7.9.7] (resonance effect 0.16%, line width 4.4 ± 0.4 mm s^{-1}) and by Wolbeck and Zioutas [7.9.13] (resonance effect 0.044%, line width 2.6 ± 1.5 mm s^{-1}). Considerably improved ^{195}Pt (130 keV) spectra were recorded by Walcher [7.9.5]. Two representative spectra are shown in Fig. 7.9.6. The spectrum of $Pt_{0.1}Fe_{0.9}$ is broadened due to magnetic hyperfine interaction. The solid line in Fig. 7.9.6b was obtained by adjusting 10 hyperfine components of

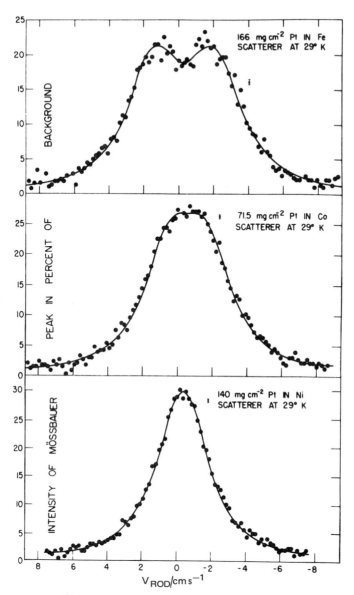

Fig. 7.9.4. ^{195}Pt (99 keV) Mössbauer spectra of Pt in ferromagnetic alloys measured in back-scattering geometry. The count rate at infinite velocity is normalized to 1 (from [7.9.9])

the 130 keV transition to the measured spectrum, with relative intensities and positions as indicated by the bars. From this fit Walcher was able to determine the magnetic moment of the 130 keV state to be $\mu = (0.81^{+0.13}_{-0.25})\mu_N$. From the isomer shift data observed for both the 99 and 130 keV transition in ^{195}Pt in various alloys (cf. Table 7.9.3), Walcher found that $\Delta\langle r^2\rangle_{130}/\Delta\langle r^2\rangle_{99} = 1.5(2)$. The same value, within experimental error, was found by Rüegg [7.9.8, 11]. Two of the ^{195}Pt Mössbauer

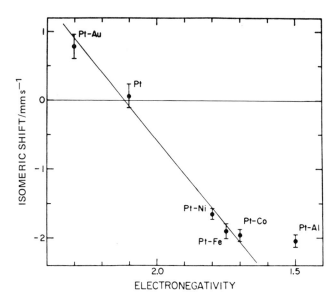

Fig. 7.9.5. ^{195}Pt (99 keV) isomer shift in platinum alloys (3 at.-% Pt) as function of electro-negativity of the host element (from [7.9.10])

spectra obtained with a Fe-Pt alloy absorber (3 at. % Pt) at 4.2 K by Rüegg are shown in Fig. 7.9.7; the upper spectrum (a) has been recorded with the 99 keV transition, the lower spectrum (b) with the 130 keV transition. The theoretical spectrum in Fig. 7.9.7a is composed of 6 hyperfine components (according to a pure M1 transi-tion), in Fig. 7.9.7b it is composed of 10 hyperfine components (according to a pure E2 transition); the individual Lorentzian lines are indicated on top of the figures.

Ianarella et al. [7.9.14] used the ^{195}Pt (99 keV) Mössbauer effect to study the influence of hydrogen loading of palladium. As in similar experiments using the Möss-bauer effect in ^{197}Au, ^{193}Ir, and ^{99}Ru they observed a decrease of the electron den-sity at the nuclei of the Mössbauer impurity on hydrogenation, but they could not decide whether the effect is due to the volume expansion of the palladium lattice or to the filling of the host conduction band with electrons donated by the hydrogen.

Table 7.9.3. Isomer shift data observed with the 99 and the 130 keV transi-tion in ^{195}Pt in various alloys at 4.2 K (relative to Pt metal)

Alloy	δ_{99}/mm s^{-1}	δ_{130}/mm s^{-1}
$Pt_{0.1}Fe_{0.9}$	−1.52 (7)	−1.70 (14)
$Pt_{30}Cu_{70}$	−0.72 (10)	−0.72 (13)
$CePt_2$	−1.34 (11)	−1.61 (15)

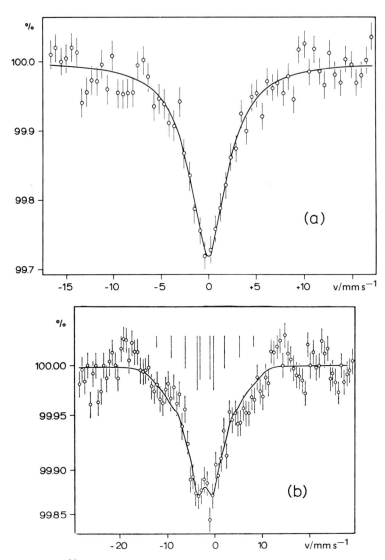

Fig. 7.9.6. ^{195}Pt (130 keV) Mössbauer spectra of (a) natural Pt metal (700 mg/cm^2) and (b) $Pt_{0.1}Fe_{0.9}$ taken at 4.2 K with a ^{195}Au/Pt metal source (from [7.9.5])

References

[7.9.1] Harris, J. R., Benczer-Koller, N., Rothberg, G. M.: Phys. Rev. *137*, A 1101 (1965)
[7.9.2] Stevens, J. G., Stevens, V. E. (eds.): *Mössbauer Effect Data Index,* Covering the 1975 Literature. New York: IFI Plenum Press 1976
[7.9.3] Buyrn, A., Grodzins, L.: Phys. Lett. *21*, 389 (1966)
[7.9.4] Harris, J. R., Rothberg, G. M., Benczer-Koller, N.: Phys. Rev. *138*, B 554 (1965)
[7.9.5] Walcher, D.: Z. Phys. *246*, 123 (1971)
[7.9.6] Buyrn, A., Grodzins, L., Blum, N. A., Wulff, J.: Phys. Rev. *163*, 286 (1967)
[7.9.7] Wilenzick, R. M., Hardy, K. A., Hicks, J. A., Owen, W. R.: Phys. Lett. *29A*, 678 (1969)

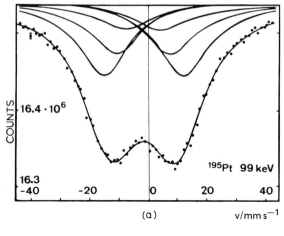

(a)   v/mm s^{-1}

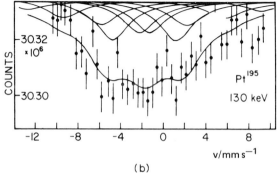

v/mm s^{-1}

(b)

Fig. 7.9.7. ^{195}Mössbauer spectra obtained with a Fe-Pt absorber (3 at.-% Pt) at 4.2 K using (a) the 99 keV transition and (b) the 130 keV transition of a ^{195}Au/Pt metal source (from [7.9.8])

[7.9.8] Rüegg, W., Launaz, J. P.: Phys. Lett. *37A*, 355 (1971)
[7.9.9] Atac, M., Debrunner, P., Frauenfelder, H.: Phys. Lett. *21*, 699 (1966)
[7.9.10] Agresti, D., Kankeleit, E., Persson, B.: Phys. Rev. *155*, 1339 (1967)
[7.9.11] Rüegg, W.: Helv. Phys. Acta *46*, 735 (1974)
[7.9.12] Rüegg, W., Kuse, D., Zeller, H. R.: Phys. Rev. B *8*, 952 (1973)
[7.9.13] Wolbeck, B., Zioutas, K.: Z. Naturforsch. *25a*, 1779 (1970)
[7.9.14] Ianarella, L., Wagner, F. E., Wagner, U., Danon, J.: J. Physique Collq. *35*, C6–517 (1974)

7.10. Gold (^{197}Au)

Recoilless nuclear resonance of the 77.34 keV transition in ^{197}Au was first reported in 1960 using gold foil absorbers and both ^{197}Pt/Pt and ^{197}Hg/Hg sources at 4.2 K [7.10.1]. During the first decade after its discovery the ^{197}Au resonance was applied almost exclusively to metallic systems, mostly in studies of the band structure and magnetic hyperfine structure of gold as a function of various host metals. Chemical applications were started by the research group of Roberts, later on by Shirley's group and Mössbauer's laboratory. Particularly the work from the latter two groups demonstrates the usefulness of the technique in distinguishing oxidation states and bond properties of gold in various surroundings.

7.10.1. Practical Aspects

The 77.34 keV level may be populated by both decaying ^{197}Pt and ^{197}Hg precursor nuclides as shown in the decay scheme of Fig. 7.10.1. ^{197}Hg, however, is hardly used because of the relatively low recoil-free fraction in gold amalgam. ^{197}Pt in platinum metal foil, activated by thermal neutron irradiation of natural or enriched platinum metal in a nuclear reactor according to ^{196}Pt(n, γ) ^{197}Pt, is the preferred source for ^{197}Au Mössbauer spectroscopy. It can be reirradiated repeatedly.

The relatively high transition energy of 77.34 keV requires cooling of both source and absorber; most experiments are therefore carried out at temperatures between 77 and 4 K in transmission geometry. With the recently developed current integration method, however, higher measuring temperatures are accessible. Diodes like Ge(Li) are most suitable as detectors.

The data of the nuclear moments and other quantities relevant for ^{197}Au Mössbauer spectroscopy are collected in Table 7.1. The spectra taken of chemical compounds are generally very simple, singlets in the absence and quadrupole doublets in the presence of electric quadrupole interaction, with line widths only slightly (some 10–20%) above the natural line width of 1.882 mm s^{-1}. Measured isomer shift changes and quadrupole splittings are several times the experimental line widths and allow the conclusive distinction of chemical species to be made. Magnetic hyperfine splitting has so far been observed only in a few metallic systems with magnetic fields up to approximately 140 T, presumably due to conduction electron polarization by the magnetic host lattice. Here, the E2/M1 mixing ratio of $\delta^2 = 0.11$ for the multipolarity of the γ-radiation determines the relative intensities of the eight E2 + M1 allowed transition lines.

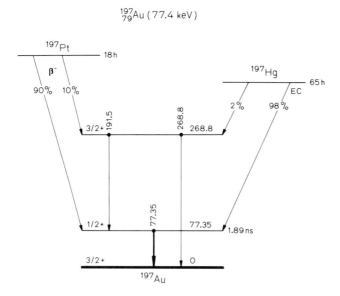

Fig. 7.10.1. Decay scheme of ^{197}Au [from 7.10.2]

7.10.2. Inorganic and Metal Organic Compounds of Gold

Mössbauer studies on gold compounds were first reported by Roberts et al. [7.10.3] and Shirley et al. [7.10.4, 5]. They observed rather large isomer shift changes in some simple gold(I) and gold(III) halides and halogeno-complexes such as AuX (X = Cl, Br, I), AuX_3 (X = F, Cl), $K[AuX_4]$ (X = F, Cl, Br), and $K[Au(CN)_2]$.

In 1970 Charlton and Nichols [7.10.6] published their results of a ^{197}Au Mössbauer study of 31 gold complexes, in which they intended to find a general correlation between Mössbauer parameters and chemical features. They investigated the series

$LAuCl$, $LAuCl_3$ (L = Ph_3P, $(C_6F_5)Ph_2P$, Ph_3As, C_5H_5N, Me_2S, and p-tolylisocyanide)

Ph_3PAuX (X = I, Br, Cl, N_3, OCOMe, CN, and ME)

$M[AuX_4]$ (X = F, Cl, Br, and I; M = H, Na, K, NH_4, Cs)

and discussed the variations in isomer shift and quadrupole splitting in terms of sp and dsp^2 hybridization of the gold atom. Some general points could be made by these authors.

1. The range of isomer shifts and quadrupole splittings is larger than that for tin and iron.

2. The range of isomer shifts is larger for aurous than for auric compounds, most probably due to the larger amount of s-character in sp than in dsp^2 hybrid orbitals, and also to the smaller variety of ligands in the auric compounds under study.

3. The quadrupole splittings of the aurous compounds are generally larger than in the auric compounds.

4. A positive correlation between isomer shift and quadrupole splitting exists for both the aurous and auric series. Moreover, in the LAuCl series they found an increasing isomer shift and thus increasing donation of charge by the ligand into the sp hybrid orbital in the following order of ligands: $Me_2S < C_5H_5N < Ph_3As < (C_6F_5)$ $Ph_2P \simeq Ph_3P$. A similar order has been found for the isomer shift of the series $LAuCl_3$: p-tolylisocyanide $< Me_2S < C_5H_5N < Ph_3P$. In the series Ph_3PAuX the order of increasing isomer shift is I < Br < Cl < OCOMe $\simeq N_3$ < CN < Me, which parallels the spectrochemical series. With the exception of fluoride, the isomer shift order also parallels the spectrochemical series in the AuX_4^- complexes (X = F, Cl, Br, I).

More systematic investigations of gold(I) and gold(III) compounds using the ^{197}Au Mössbauer effect were performed at nearly the same time by Faltens and Shirley [7.10.7] and by Mössbauer's group in Munich [7.10.8]. Both groups observed for each oxidation state a linear correlation between the isomer shift and the quadrupole splitting. It turned out that it is not possible to determine the oxidation state of gold on the basis of the isomer shift or the quadrupole splitting alone, but together they appear to determine it uniquely. This may be seen from the δ-ΔE_Q-diagrams by Faltens and Shirley [7.10.7] shown in Fig. 7.10.2 as well as from that by Mössbauer et al. [7.10.8] reproduced in Fig. 7.10.3.

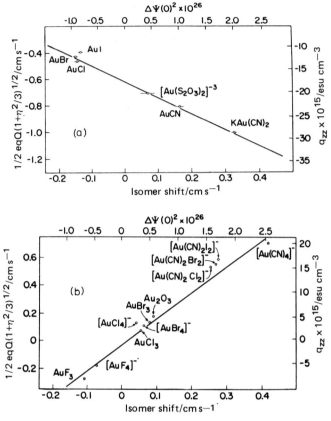

Fig. 7.10.2. Linear relation of the quadrupole splitting $\Delta E_Q = (1/2)$ eqQ $(1 + \eta^2/3)^{1/2}$ and the isomer shift δ for aurous (a) and auric (b) compounds. Also included is a correlation with the relative change in electron density at the gold nucleus, $\Delta|\psi(o)|^2$, as derived from Dirac-Fock atomic structure calculations for several electron configurations of gold. An approximate scale of the electric field gradient q_{zz} (in the principal axes system) is given on the right-hand ordinate (from [7.10.7])

Faltens and Shirley interpreted their results for the linear gold(I) and the square-planar gold(III) complexes in terms of 6s6p and $5d6s6p^2$ hybridization, respectively, with varying degree of ionic character. Employing simple linear combinations of atomic orbitals, such as

$$\psi_{\pm z}(AuL_2) = (\alpha/2) [(6s) \pm (6p_z)] \mp (1 - \alpha^2)^{1/2} \psi(L)$$

for the linear sp-bonded aurous complexes, and

$$\psi_{\pm x}(AuL_4) = \frac{1}{2} \beta [(6s) \pm \sqrt{2} (6p_x) - (5d_{x^2 - y^2})] \mp (1 - \beta^2)^{1/2} \psi(L)$$

$$\psi_{\pm y}(AuL_4) = \frac{1}{2} \beta [(6s) \pm \sqrt{2} (6p_y) + (5d_{x^2 - y^2})] \mp (1 - \beta^2)^{1/2} \psi(L)$$

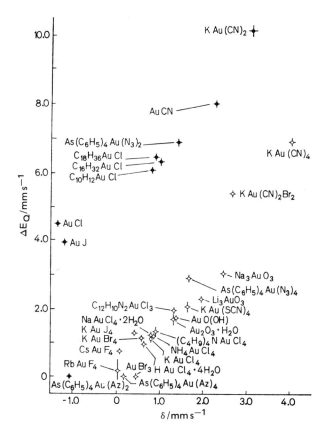

Fig. 7.10.3. Quadrupole splitting QS versus isomer shift δ in gold (I) and gold (III) compounds (from [7.10.8])

for the dsp^2-bonded square-planar complexes, and using Dirac-Fock wave functions for gold calculated by Mann they evaluated $\Delta\psi^2(o)$ values for gold(I) and gold(III) states, for which α^2, $\beta^2 = 0$ and $1/2$, and correlated these with the isomer shift (cf. Fig. 7.10.2). They also made estimates on the magnitude of $\langle r^{-3}\rangle_{5d}$ and $\langle r^{-3}\rangle_{6p}$ and the size and sign of the electric field gradient. The nuclear factor $\delta R/R$ has been concluded by these authors to be $+ (2.5 \pm 0.6) \cdot 10^{-4}$.

Some representative ^{197}Au transmission spectra taken from the work of Mössbauer et al. [7.10.8] are shown in Fig. 7.10.4. On the basis of LCAO—MO theory these authors have interpreted qualitatively the variations in isomer shift and quadrupole splitting by covalency effects in the molecular orbitals and synergic coupling of σ-donation and π-acceptance capacity of the ligands. An increase in the isomer shift with increasing rank of the ligands in the spectrochemical series is observed in both gold(I) and gold(III) complexes. The position in the spectrochemical series is known to increase with σ-donor and π-acceptor strength, while it decreases with π-donor strength.

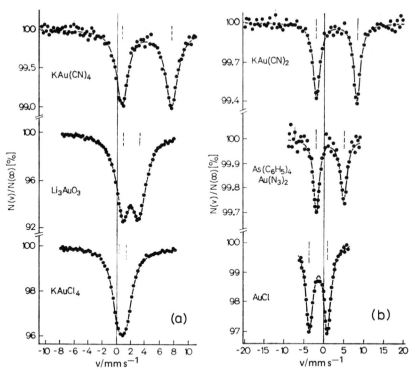

Fig. 7.10.4. ^{197}Au Mössbauer spectra of some gold (III) (a) and gold (I) (b) compounds. ^{197}Pt/Pt source and absorbers were kept at 4.2 K (from [7.10.8])

It has been found that the isomer shift in gold compounds will mainly depend on the atomic 6 s and 5 d populations of gold in the molecular orbitals. Positive contributions to the electron density at the gold nucleus arise from the $6s(a_{1g})$ populations, whereas a decrease of the electron density is caused by the atomic 5 d populations due to the shielding effect. The 6 p populations are less significant here. One has found [7.10.9] that one 5 d electron yields a screening of 25% of one 6 s electron, and one 6 p electron only 10% of one 6 s electron. The direct contribution to the electron density by 6 p electrons through relativistic effects is only about 5%. In view of this, it is expected that in gold complexes with appreciable π-donation (e.g., halides as ligands) π-charge can only be donated into the $6p_{x,y}(e_u, D_{4h})$, $6p_z(a_{2u}, D_{4h})$, and $6p_{x,y}(e_{1u}, D_{\infty h})$ metal orbitals, respectively, because the $5d_{xz,yz}(e_g, D_{4h})$ and $5d_{xz,yz}(e_{1g}, D_{\infty h})$ metal orbitals are fully occupied; therefore the isomer shift is hardly influenced by π-donation. The interpretation of the isomer shift of gold compounds can therefore be based mainly on σ-donation and π-back-donation properties of the ligands.

In square-planar gold(III) complexes, the metal orbitals $5d_{x^2-y^2}(b_{1g})$, $5d_{z^2}(a_{1g})$, $6s(a_{1g})$ and $6p_{x,y}(e_u)$ are suitable for σ-bonding. In the absence of π-backbonding the isomer shift depends on the populations of the b_{1g} and a_{1g} −MOs; their d-parts shield s-electrons, whereas the $6s(a_{1g})$ part contributes positively to the total electron density. The experimental data available so far show that the isomer shift increases with increasing σ-covalency and thus reflects the dominating role of the $6s(a_{1g})$ population. In the presence of π-backbonding (e.g., CN^- as ligand) charge will be delocalized from the $5d_{xz,yz}(e_g)$ and $5d_{xy}(b_{2g})$ metal orbitals into π^*-antibonding orbitals of the ligand. This gives rise to the further increase of the electron density at the nucleus owing to decreasing shielding effect and hence additional increase in isomer shift. Similar considerations hold for linear gold(I) complexes, where the metal orbitals $5d_{z^2}(a_{1g})$, $6s(a_{1g})$, and $6p_z$ (a_{2u}) are suitable for σ-bonding, whereas the orbitals $5d_{xz,yz}$ (e_{1g}) are engaged in π-backbonding, and the $6p_{x,y}(e_{1u})$ orbitals are employed in π-donor bonds.

Any change in the populations of the MOs of the gold ion will also have its bearing on the electric field gradient V_{zz} and thus on the quadrupole splitting

$$\Delta E_Q = \frac{1}{2}\ eQ\ (3/2)\ V_{zz}\ (1 + \eta^2/3)^{1/2}.$$

In Au(III) compounds there is a "hole" in the antibonding $b_{1g}(\sigma^*)$ $(5d_{x^2-y^2})$ –MO, which produces a negative contribution to the EFG. The contributions from the various $6p$ populations have opposite signs: the e_u-MOs contribute positively, the a_{2u}-MO contributes negatively. As the σ-overlap and thus σ-bonding is usually larger than π-overlap, it is expected that the contributions to the EFG are mainly determined by the $6p_{x,y}$ populations rather than the $6p_z$ population. With increasing σ-donor properties of the ligands, the $5d$-"hole" will be more and more filled and the population in the $e_u(\sigma)$-MOs will increase, and this way the EFG will become more and more positive. π-back-donation reduces the atomic population of the $b_{2g}(\pi^*)$ $(5d_{xy})$ and the $e_g(\pi)$ $(5d_{xz,yz})$-MOs. These MOs contribute to the EFG with opposite signs.

In Au(I) complexes the atomic populations in the $a_{2u}(6p_z)$- and $a_{1g}(5d_{z^2})$-MOs both contribute negatively to the EFG. So increasing σ-donor capacity of the ligand will increase the magnitude of the presumably negative EFG. The concomitant synergic decrease in the population of the $e_{1u}(6p_{x,y})$-MOs will reduce the positive contribution of the $6p_{x,y}$ electrons and therefore further enhance the total (negative) EFG. d_π-backbonding effects decrease the population of the $e_{1g}(\pi^*)$ $(5d_{xz,yz})$-MOs and thus cause positive contributions to the total EFG. All these trends are well supported by experimental results.

A nice extension to ^{197}Au isomer shift studies of gold compounds has been contributed by Kaindl, Leary and Bartlett, who investigated the recently synthesized [7.10.10] unusual gold(V) complex fluorides $A^+[AuF_6]^-$ with $A^+ = Xe_2F_{11}^+$, XeF_5^+, and Cs^+ [7.10.11]. Structural data indicate an octahedral $[AuF_6]^-$ anion; accordingly the Mössbauer spectra consist of a single line in each case with approximately twice the natural line width (after correction for line broadening caused by finite absorber thickness). The spectrum of $Xe_2F_{11}^+[AuF_6]^-$ is shown as an example in Fig. 7.10.5. The isomer shift data are almost identical with each other supporting the existence of similar $[AuF_6]^-$ anions in all of the studied quinquevalent gold compounds. The magnitude of the shifts is well above the range of isomer shifts of gold(III) compounds. This may be seen from Fig. 7.10.6, where the isomer shifts of

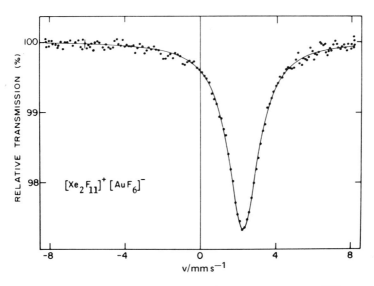

Fig. 7.10.5. Mössbauer spectrum of $Xe_2F_{11}{}^+[AuF_6]^-$ taken with a $^{197}Pt/Pt$ source at 4.2 K (from [7.10.11])

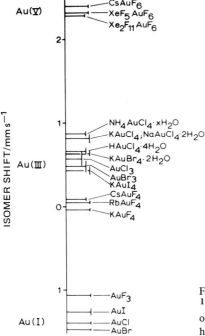

Fig. 7.10.6. ^{197}Au isomer shifts (with respect to a $^{197}Pt/Pt$ source, both source and absorbers at 4.2 K) of gold (I), gold (III) and gold (V) compounds with halogen ligands (from [7.10.11])

Table 7.10.1. ^{197}Au Mössbauer parameters observed for various gold compounds at 4.2 K. The isomer shift data refer to the used source of ^{197}Au in Pt (from [7.10.13])

Compound	Isomer shift mm s^{-1}	Quadrupole splitting mm s^{-1}
(N-Et. pip)AuCl$_4$	1.02 ± 0.02	1.4 ± 0.1
Bu$_4$N AuBr$_4$	0.92 ± 0.02	1.5 ± 0.1
KAuBr$_4$ · 2H$_2$O	0.67 ± 0.02	1.3 ± 0.1
Bu$_4$N Au(tdt)$_2$	2.99 ± 0.02	2.73 ± 0.02
Au(mnt)(dtc)	2.64 ± 0.02	2.57 ± 0.02
Bu$_4$N Au(mnt)$_2$	2.92 ± 0.01	2.33 ± 0.01
Au(H$_2$dtc)$_2$Br	2.26 ± 0.02	3.01 ± 0.04
Au(Me$_2$dtc)$_2$Br	2.04 ± 0.04	2.60 ± 0.05
Au(Et$_2$dtc)$_2$Br	2.24 ± 0.03	3.06 ± 0.05
Au(Pr$_2$dtc)$_2$Br	2.22 ± 0.02	3.02 ± 0.04
Au(Bu$_2$dtc)$_2$Br	2.20 ± 0.08	2.8 ± 0.2
Au(Ph$_2$dtc)$_2$Br	2.06 ± 0.03	2.89 ± 0.05
Bu$_4$N Au(cdc)	2.15 ± 0.05	2.71 ± 0.07
(CH$_3$)$_2$Au(Me$_2$dtc)	3.97 ± 0.01	5.07 ± 0.02
(CH$_3$)$_2$Au(Pr$_2$dtc)	3.95 ± 0.01	4.98 ± 0.01
(CH$_3$)$_2$Au(Bu$_2$dtc)	3.99 ± 0.01	5.11 ± 0.02
I$_2$Au(Bu$_2$dtc)	1.26 ± 0.03	2.10 ± 0.05
Br$_2$Au(Bu$_2$dtc)	1 47 ± 0.01	2.20 ± 0.01
BrI Au(Bu$_2$dtc)	1 28 ± 0.01	1.77 ± 0.02
Cl$_2$Au(Hep$_2$dtc)	1.35 ± 0.01	1.95 ± 0.02
Au(Et$_2$dtc)	1.70 ± 0.01	5.98 ± 0.02
Au(Pr$_2$dtc)	1.80 ± 0.01	6.39 ± 0.01
Au(Bu$_2$dtc)	1.62 ± 0.01	5.94 ± 0.01
Au(Oct · xan)	1.42 ± 0.01	6.16 ± 0.02
Au(dtp)	0.96 ± 0.01	6.09 ± 0.02
Ph$_4$P Au WS$_4$	0.86 ± 0.01	5.58 ± 0.02
Ph$_4$As Au WS$_4$	1.10 ± 0.01	5.71 ± 0.02
Ph$_4$P Au CS$_3$	1.91 ± 0.01	6.43 ± 0.01
Ph$_3$P AuBr	2.77 ± 0.01	7.35 ± 0.01
Ph$_3$P AuCl	2.93 ± 0.01	7.50 ± 0.01
Ph$_3$P AuSCN	2.85 ± 0.01	7.65 ± 0.01
Ph$_3$P AuSC(S)pip	2.79 ± 0.02	7.60 ± 0.03
Ph$_3$P Au(Ph$_2$dtc)	2.95 ± 0.02	7.79 ± 0.04
Ph$_3$P AuSeC(O)pip	3.16 ± 0.02	7.93 ± 0.02
Ph$_3$P Au SC(O)pip	3.24 ± 0.02	8.35 ± 0.03
Ph$_3$P Au(nds)	3.74 ± 0.02	8.42 ± 0.02
Ph$_3$P AuCN	3.82 ± 0.01	10.11 ± 0.01
(dmap)$_4$Au$_2$Cu$_4$I$_2$	4.00 ± 0.01	9.37 ± 0.01
(dmap)$_4$Au$_2$Cu$_4$(otf)$_2$	3.88 ± 0.01	9.14 ± 0.01
(dmap)$_4$Au$_2$Cu$_2$	4.42 ± 0.01	9.86 ± 0.01
(dmap)$_4$Au$_2$Li$_2$	5.27 ± 0.01	11.29 ± 0.01
(dmap)$_4$Au$_2$Li$_2$	5.65 ± 0.01	12.01 ± 0.01
(dmap)Au	3.69 ± 0.01	6.73 ± 0.01
(dmap)Au	4.44 ± 0.01	7.32 ± 0.01
AuCl(PPh$_3$)$_2$	1.15 ± 0.01	8.03 ± 0.01
(dmap)AuPPh$_3$	4.84 ± 0.01	10.11 ± 0.01
(dmap)AuCN(C$_6$H$_{11}$)	4.62 ± 0.01	10.35 ± 0.01
(dmap)SnAuBr	2.34 ± 0.02	6.51 ± 0.03
AuCN	2.35 ± 0.01	7.97 ± 0.01

Table 7.10.2. Abbreviations used in the chemical formulae of Table 7.10.1 (from [7.10.13])

Me = methyl, Et = ethyl, Pr = n-propyl, Bu = n-butyl, Hep = heptyl, Oct = octyl, Ph = phenyl, Bu_4N^+ = tetra-butylammonium

R_2dtc =
N, N-di-alkyl (aryl) dithiocarbamate

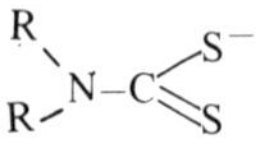

Rxan =
O-alkylxanthate

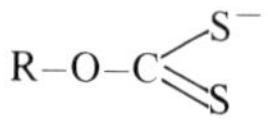

cdc =
N-cyanodithiocarbimate

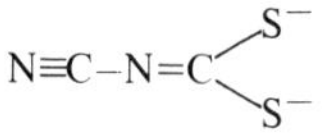

pip =
piperidyl

OC(O)pip =
N-piperidyl-carbamate

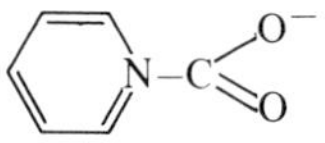

ac =
acetate

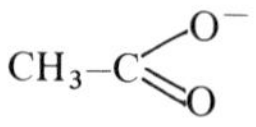

dtp =
di-alkyldithiophosphate

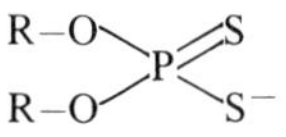

tdt =
toluene-3, 4-dithiolate

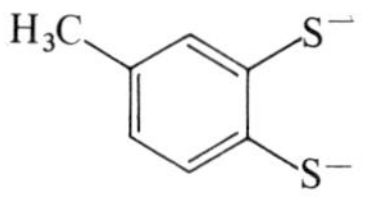

mnt =
maleonitriledithiolate

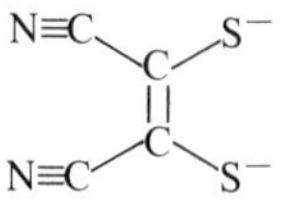

nds =
naphthalene-1, 8-dithiolate

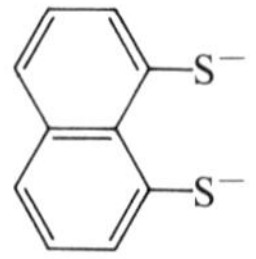

Table 7.10.2 (continued)

dmamp = 2-(dimethylamino)methyl phenyl	CH_2-NMe_2
dmap = 2-dimethylaminophenyl	NMe_2

otf = O_3SCF_3

gold(I), gold(III), and gold(V) compounds with halogen ligands are compared with each other. A decrease of the atomic 5d population is assumed on going from gold(I) to gold(III) to gold(V).

Recently Schmidbaur et al. [7.10.12] have reported results from ESCA and ^{197}Au Mössbauer studies of a series of ylide complexes of Au(I), Au(II), and Au(III). The ESCA data reveal a pronounced influence of the Au oxidation state on the Au ($4f_{7/2}$) binding energies, the differences of which are large enough to distinguish between the various oxidation states. The ^{197}Au Mössbauer spectra show a systematic decrease of the isomer shift with increasing gold oxidation state. This behaviour parallels that of the KAu(CN)$_2$ and KAu(CN)$_2$Br$_2$ or Ph$_3$PAuCl and Ph$_3$PAuCl$_3$ pairs, which is opposite to the usually observed increase of the isomer shift with increasing oxidation state of gold.

A very fine piece of work applying the ^{197}Au Mössbauer effect to gold compounds has just been completed by Viegers [7.10.13] in Trooster's laboratory in Nijmegen (Netherlands). In his thesis, Viegers studied some 50 gold compounds, AuX$_4^-$ with X = Br, Cl and different cations, bis-dithiolato Au(III) complexes, gold(I) complexes with bidentate sulphur-donor-ligands, triphenylphosphine gold(I) complexes, polynuclear organogold compounds, and some others. His data are collected in Table 7.10.1; the relevant abbreviations used in the chemical formulae are given in Table 7.10.2.

The general features observed by Viegers follow those found in earlier ^{197}Au Mössbauer effect studies by other authors [7.10.6–8]: The linear relationship between the quadrupole splitting (ΔE_Q) and the isomer shift (δ) still exists. This relation is on the average (included in this consideration are the data from [7.10.6–8]) given by

$$\Delta E_Q(Au^{III}) = 1.74\,\delta - 0.22 \text{ mm s}^{-1}$$

$$\Delta E_Q(Au^I) \;\; = 1.06\,\delta + 5.05 \text{ mm s}^{-1}.$$

Comparing a Au(I) with a Au(III) complex with the same ligands, the isomer shift of the latter is larger and its quadrupole splitting is smaller. The Mössbauer parameters are largest, if the coordinating atoms belong to Group IV, and smallest if they belong to Group VII. An average-environment rule seems to apply, i.e., the Mössbauer parameters of compounds with mixed ligands are averages of those of the pure complexes. Although the $\Delta E_Q/\delta$ correlation lines approach each other for large δ and ΔE_Q values, it still appears possible to determine the oxidation state of gold unambiguously using this plot. Viegers has interpreted his results in the framework of LCAO-MO theory. His studies of the dynamical behaviour of gold in molecular crystals (influence of intra- and intermolecular vibrations on the recoil-free fraction of γ-rays) and his work on small gold particles also deserve particular attention.

7.10.3. Intermetallic Compounds and Alloys of Gold

Various solid-state properties of both intermetallic compounds and alloys of gold, which mostly are strongly influenced by the structure of the valence (conduction) electrons, may be successfully studied using ^{197}Au Mössbauer spectroscopy.

a) Intermetallic Compounds

Dunlap, Darby, and Kimball [7.10.14] have measured the temperature dependence of the ^{197}Au Mössbauer resonance in ferromagnetic Au_4V and found a line broadening below the Curie temperature due to magnetic hyperfine interaction. The magnitude of the effective field indicates that the magnetic moment in Au_4V is localized on the V atoms and little moment, if any, resides on the Au atoms. The temperature dependence of the hyperfine field approximately follows the bulk magnetization. The magnitude of the internal field at the gold sites was later determined using the ^{197}Au Mössbauer effect to be 18.5 ± 2.5 T at 4.2 K in Au_4V and 84.7 ± 2.5 T at 4.2 K in Au_4Mn [7.10.15].

More extensive investigations of the intermetallic compounds Au_4Mn, Au_3Mn, Au_2Mn, $AuMn$, $AuMn_3$ and of dilute alloys of gold in manganese by means of ^{197}Au Mössbauer spectroscopy have been carried out by Patterson, Thomson, Huray, and Roberts [7.10.16]. One of their measured spectra of Au_2Mn is shown in Fig. 7.10.7. It is composed of 8 (M1 + E2)-allowed resonance lines (with $E2/(E2 + M1) \approx 0.1$) arising from magnetic hyperfine interaction, somewhat shifted towards each other by electric quadrupole interaction. The isomer shift, magnetic dipole and electric quadrupole splitting observed in these alloys are discussed in terms of the magnetic and crystalline structure. For Au_2Mn the authors found an effective field at the ^{197}Au nucleus of 159.0 T. The spectrum of Au_2Mn is consistent with the helical magnetic structure assumed for this alloy. Mössbauer measurements under applied pressure revealed a 7% increase in the magnitude of the effective field between 0 and 30 kbar, which is believed to be due to (partial) uncoiling of the Mn spin helix. The observed large quadrupole splitting in Au_2Mn has been attributed to about 0.17 of one 5d hole in the $5d_{x^2-y^2}$ orbital.

Erickson and Roberts [7.10.17] have also carefully investigated the ^{197}Au Mössbauer resonance line shape and recoilless fraction as a function of temperature, absorber thickness, and lattice order in Cu_3Au. The spectra have been interpreted

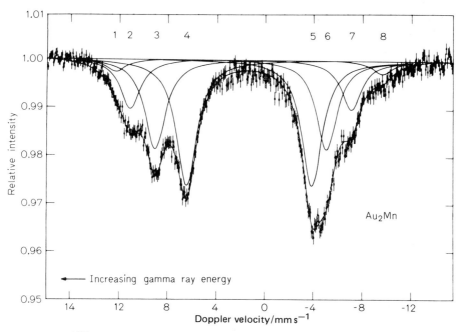

Fig. 7.10.7. ^{197}Au Mössbauer spectrum of Au$_2$Mn at 4.2 K (from [7.10.16])

in terms of a pseudoquadrupole splitting, which has been introduced to describe line broadening due to noncubic local atomic environments of gold atoms, and of the temperature-dependent recoilless fraction for gold. The pseudoquadrupole splitting has been found to be smaller and the recoilless fraction to be larger in the more highly ordered state.

Other ^{197}Au Mössbauer measurements have been carried out on AuAl$_2$, AuGa$_2$, AuIn$_2$, and AuSb$_2$ by Thomson et al. [7.10.18]. The observed isomer shifts with respect to the source of ^{197}Au in Pt are -1.22 (Au metal), $+2.29$ (AuSb$_2$), $+3.49$ (AuIn$_2$), $+4.45$ (AuGa$_2$), and $+5.97$ (AuAl$_2$) mm s^{-1}, and reflect increasing electron density at the Au nucleus in this order of alloys as compared to that of Au metal. The authors concluded that, when volume changes for the alloys with respect to the pure metal are taken into account, approximately one 6s-like electron is transferred to the Au in each of the alloys. From the temperature dependence of the recoil-free fraction, the authors evaluated Debye temperatures for the various intermetallic compounds.

The series of ternary intermetallic gold compounds Li$_2$AuX (X = Ga, In, Tl; Ge, Sn, Pb; Bi) and Li$_2$Au$_{2-x}$In$_x$ ($1.0 \leqslant x \leqslant 1.75$), all of which crystallize in the cubic NaTl structure, have been studied through the use of ^{197}Au Mössbauer spectroscopy at 4.2 K by Gütlich, Odar, and Weiss [7.10.19]. The isomer shifts derived from the single-line spectra have been correlated with the average Allred-Rochow electronegativity [7.10.20], $\overline{EN}$, of the first three coordination spheres around the gold atoms, normalized to the average number of outer electrons $\overline{n}$ and corrected for the distance from the gold atom according to the expression

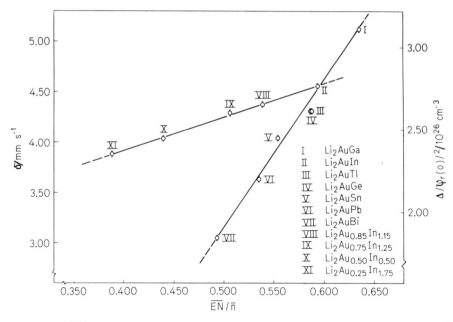

Fig. 7.10.8. ^{197}Au isomer shift and difference in electron density at the gold nucleus, $\Delta|\psi_r(o)|^2$ (relative to metallic gold), in Li_2AuX and $Li_2Au_{2-x}In_x$ as a function of the average electronegativity (Allred-Rochow) normalized to the average number of outer electrons, $\overline{EN}/n$, of the atoms in the first three coordination spheres around a gold atom (from [7.10.19])

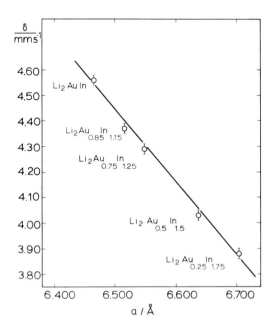

Fig. 7.10.9. ^{197}Au isomer shift in $Li_2Au_{2-x}In_x$ as a function of the lattice constant (from [7.10.19])

$$\overline{EN}/\overline{n} = \frac{EN(Li1) + EN(Li2)\,(r_1^2/r_2^2) + 3\,EN(Au)\,(r_1^2/r_3^2) + EN(X)}{2n(Li) + 3n(Au) + n(X)}$$

(Li1 and Li2 are crystallographically different) for the Li_2AuX series and a similar expression (cf. [7.10.19]) for the series $Li_2Au_{2-x}In_x$. Fig. 7.10.8 shows this correlation of $\delta = f(\overline{EN}/\overline{n})$ for the two series, and also the correlation with changes in electron density at the nucleus, $\Delta|\psi_r(0)|^2$, evaluated from results given in [7.10.7]. The isomer shift in these ternary gold alloys also correlates well with structural data from x-ray diffraction studies as shown in Fig. 7.10.9. From these investigations it is suggested that substantial $5d - 6s$ mixing occurs with nearly matching charge compensation and additional depletion of $5d$ charge through an interaction of the $5d$ band of gold with host orbitals of proper symmetry. The experimental findings appear to be consistent with a net charge-flow off the gold sites in these ternary gold alloys with main-group metals.

b) Alloys

There is a large number of publications dealing with ^{197}Au Mössbauer studies of a great variety of gold alloys. As the primary goal of this review is directed towards applications of Mössbauer spectroscopy to chemical compounds rather than metallic systems, we may be allowed to restrict ourselves here to an enumeration of the most important work by giving in chronological order the reference, the studied systems, and a note concerning the essential points of interest in each case:

Reference	Studied system	Points of interest
[7.10.1]	Au (metal)	First observation of ^{197}Au Mössbauer effect
[7.10.21]	^{197}Au embedded in Au, Pt, stainless steel, Fe, Co, Ni	Recoil-free emission fraction, isomer shift, $|H^{int}|$
[7.10.22]	dilute alloys of Au with Fe, Co, Ni	^{197}Au nuclear data, $|H^{int}|$, line width as function of absorber thickness
[7.10.23]	Au (metal)	Resonance effect as function of absorber thickness, isomer shift, Debye temperature of source and absorber
[7.10.24]	^{197}Au in 19 metals and semiconductors versus gold metal absorber	Electron transfer from isomer shifts, correlation between isomer shift and host electronegativities
[7.10.25]	^{197}Au dissolved in ferromagnetic hosts of Fe, Co, Ni as sources versus Au metal absorber	Nuclear Zeeman effect in Au atoms super-transferred hf fields, $|H^{int}|$ at Au sites
[7.10.26]	Dilute gold alloys with Cu, Ag, Ni, Pd, and Pt as absorbers	Correlation of isomer shift with residual electrical resistivity, wave function at Fermi level, s-band population of gold
[7.10.27]	Au microcrystals	Effect of crystal size on recoil-free fraction
[7.10.28]	Au microcrystals	Debye temperature for microcrystals

Reference	Studied system	Points of interest
[7.10.29]	^{197}Au in iron metal	Magnetic hfs of ^{197}Au, nuclear moment of 77.3 keV state, $|H^{int}|$ of supertransferred field at ^{197}Au
[7.10.30]	^{197}Au and ^{119}Sn in various metallic matrices as absorber	Interpretation of isomer shift
[7.10.31]	Gold alloys with transition metals (Au$_4$V, Au$_4$Mn, Au$_4$Cr), crystallographically ordered and disordered phases	Magnetic behaviour as function of disorder, model for magnetic behaviour of Au-V system
[7.10.32]	Alloys of Pd-Au-Fe (2 at. %)	Mössbauer effect in ^{57}Fe and ^{197}Au, study of band filling, hf fields, isomer shifts
[7.10.33]	Au microcrystals supported in gelatin	Correlation of isomer shift with lattice contraction
[7.10.34]	Neutron irradiated platinum	Debye-Waller factor of Au in Pt after low-temperature neutron irradiation, lattice defect
[7.10.35]	Au$_{0.8}$Cr$_{0.2}$	Hf field and isomer shift in crystallographically ordered and disordered phases
[7.10.36]	Au-Sn alloys	Isomer shift study
[7.10.37]	Au-Hg alloys	Isomer shift as function of Hg concentration
[7.10.38]	Au-Cu and Au-Ag alloys	Isomer shift and electrical resistivity as function of alloy composition and, in Cu$_3$Au, of pressure; model to describe δ in terms of average atomic volume, of short-range parameter and alloy composition; average charge density on Au
[7.10.39]	Au metal	Thick absorber line-shape analysis and interference effects
[7.10.40]	Au-Ni and Cu-Ni-Au alloys	Magnetic hfs at ^{197}Au, $|H^{int}|$ and isomer shift as function of composition, model to describe charge density distribution
[7.10.41]	Various gold alloys	Review of ^{197}Au Mössbauer work on metallic systems of gold
[7.10.42]	Alloys of Au with Ag, Sn, Ni, Pd, Pt, Al	Charge between host and Au atoms from trends of isomer shift and work function data (ESCA) correlated with electronegativity of host elements

References

[7.10.1] Nagle, D., Craig, P. P., Dash, J. G., Reiswig, R. D.: Phys. Rev. Lett. *4*, 237 (1960)

[7.10.2] Stevens, J. G., Stevens, V. E.: *Mössbauer Effect Data Index*. Covering the 1975 literature. New York: IFI Plenum 1976

[7.10.3] Roberts, L. D., Pomerance, H., Thomson, J. O., Dam, C. F.: Bull. Am. Phys. Soc. *7*, 565 (1962)

[7.10.4] Shirley, D. A.: Rev. Mod. Phys. *36*, 339 (1964)

[7.10.5] Shirley, D. A., Grant, R. W., Keller, D. A.: Rev. Mod. Phys. *36*, 352 (1964)

[7.10.6] Charlton, J. S., Nichols, D. I.: J. Chem. Soc. (A), 1484 (1970)
[7.10.7] Faltens, M. O., Shirley, D. A.: J. Chem. Phys. *53*, 4249 (1970)
[7.10.8] Bartunik, H. D., Potzel, W., Mössbauer, R. L., Kaindl, G.: Z. Phys. *240*, 1 (1970)
[7.10.9] Kopfermann, H.: *Kernmomente*, 2nd ed. Frankfurt/Main: Akademische Verlagsgesellschaft 1956
[7.10.10] Leary, K., Bartlett, N.: J. C. S. Chem. Comm., 131 (1973)
[7.10.11] Kaindl, G., Leary, K., Bartlett, N.: J. Chem. Phys. *59*, 5050 (1973)
[7.10.12] Schmidbaur, H., Mandl, J. R., Wagner, F. E., van de Vondel, D. F., van der Kelen, G. P.: J. C. S. Chem. Comm., 170 (1976)
[7.10.13] Viegers, M. P. A.: Ph. D. Thesis, Catholic University of Nijmegen, Netherlands, 1976
[7.10.14] Dunlap, B. D., Darby Jr., J. B., Kimball, C. W.: Phys. Lett. *25 A*, 431 (1967)
[7.10.15] Cohen, R. L., Sherwood, R. C., Wernick, J. H.: Phys. Lett. *26 A*, 462 (1968)
[7.10.16] Patterson, D. O., Thomson, J. O., Huray, P. G., Roberts, L. D.: Phys. Rev. B *2*, 2440 (1970)
[7.10.17] Erickson, D. J., Roberts, L. D.: Phys. Rev. B *9*, 3650 (1974)
[7.10.18] Thomson, J. O., Obernshain, F. E., Huray, P. G., Love, J. C., Burton, J.: Phys. Rev. B *11*, 1835 (1975)
[7.10.19] Gütlich, P., Odar, S., Weiss, A.: J. Phys. Chem. Solids *37*, 1011 (1976)
[7.10.20] Allred, A. L., Rochow, E. G.: J. Inorg. Nucl. Chem. *5*, 264 (1958); Huheey, J. E.: *Inorganic Chemistry*, Principles of Structure and Reactivity. New York, London: Harper and Row 1975, p. 160
[7.10.21] Shirley, D. A., Kaplan, M., Axel, P.: Phys. Rev. *123*, 816 (1961)
[7.10.22] Roberts, L. D., Thomson, J. O.: Phys. Rev. *129*, 664 (1963)
[7.10.23] Andrä, H. J., Hashmi, C. M., Kienle, P., Stanek, F. W.: Z. Naturforsch. *18 a*, 687 (1963)
[7.10.24] Barrett, P. H., Grant, R. W., Kaplan, M., Keller, D. A., Shirley, D. A.: J. Chem. Phys. *39*, 1035 (1963)
[7.10.25] Grant, R. W., Kaplan, M., Keller, D. A., Shirley, D. A.: Phys. Rev. *133*, 1062 (1964)
[7.10.26] Roberts, L. D., Becker, R. L., Obenshain, F. E., Thomson, J. O.: Phys. Rev. *137*, A 895 (1965)
[7.10.27] Marshall, S. W., Wilenzick, R. M.: Phys. Rev. Lett. *16*, 219 (1966)
[7.10.28] Schroeer, D.: Phys. Lett. *21*, 123 (1966)
[7.10.29] Cohen, R. L.: Phys. Rev. *171*, 344 (1968)
[7.10.30] Chekin, V. V.: Sov. Phys.-JETP *27*, 983 (1968)
[7.10.31] Cohen, R. L., Wernick, J. H., West, K. W., Sherwood, R. C., Chin, G. Y.: Phys. Rev. *188*, 684 (1969)
[7.10.32] Longworth, G.: J. Phys. C, Metal Phys. Supl. *1*, S 81 (1970)
[7.10.33] Schroeer, D., Marzke, R. F., Erickson, D. J., Marshall, S. W., Wilenzick, R. M.: Phys. Rev. B *2*, 4414 (1970)
[7.10.34] Mansel, W., Vogl, G., Vogl, W., Wenzl, H., Barb, D.: phys. stat. sol. *40*, 461 (1970)
[7.10.35] Kohgi, M., Yamada, T., Kunitomi, N., Maeda, Y.: J. Phys. Soc. Japan *28*, 793 (1970)
[7.10.36] Charlton, J. S., Harris, I. R.: phys. stat. sol. *39*, K1 (1970)
[7.10.37] Cohen, R. L., Yafet, Y., West, K. W.: Phys. Rev. B *3*, 2872 (1971)
[7.10.38] Huray, P. G., Roberts, L. D., Thomson, J. O.: Phys. Rev. B *4*, 2147 (1971)
[7.10.39] Erickson, D. J., Prince, J. F., Roberts, L. D.: Phys. Rev. C *8*, 1916 (1973)
[7.10.40] Burton, J. W., Thomson, J. O., Huray, P. G., Roberts, L. D.: Phys. Rev. B *7*, 1773 (1973)
[7.10.41] Roberts, L. D., Prince, J. F., Erickson, D. J.: in *Perspectives in Mössbauer Spectroscopy*, S. G. Cohen, M. Pasternak (eds.). New York: Plenum Press 1973
[7.10.42] Chou, T. S., Perlman, M. L., Watson, R. E.: Phys. Rev. B *14*, 3248 (1976)

7.11. Mercury (^{199,201}Hg)

The first Mössbauer measurements involving mercury isotopes were reported by Carlson and Temperley [7.11.1] in 1969. They observed the resonance absorption of the 32.2 keV γ-transition in ^{201}Hg (Fig. 7.11.1). The experiment was performed with zero velocity by comparing the detector counts at 70 K with those registered at 300 K. The short half-life of the excited state (0.2 ns) leads to a natural line width of 43 mm s^{-1}. Furthermore, the internal conversion coefficient is very large ($\alpha = 39$) and the ^{201}Tl precursor populates the 32 keV Mössbauer level very inefficiently ($\approx 10\%$).

In 1971 Walcher [7.11.4] succeeded in observing a resonance effect of about (0.6%) in ^{201}Hg as a function of the Doppler velocity using a Tl$_2$O$_3$ source and an enriched (81% ^{201}Hg) HgO absorber at 4.2 K. The half-width turned out to be $\Gamma_{exp} = 76(10)$ mm s^{-1}, corresponding to a lower limit of the half-life of $t_{1/2} \geqslant 0.1$ ns. It is obvious that the properties of the ^{201}Hg Mössbauer isotope do not render it an interesting isotope from the chemical point of view.

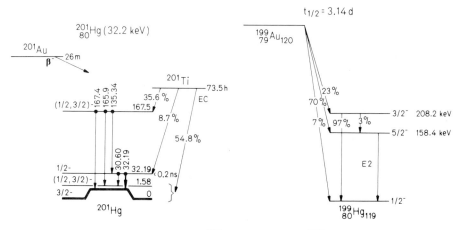

Fig. 7.11.1. Nuclear decay schemes for the ^{201}Hg ([7.11.2]) and ^{199}Hg ([7.11.3]) Mössbauer isotopes

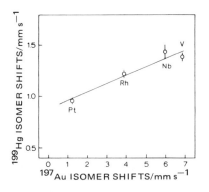

Fig. 7.11.2. Isomer shifts observed for ^{199}Hg in Pt, Rh, Nb, and V matrices plotted versus the corresponding shifts for the 77 keV Mössbauer resonance in ^{197}Au. The sign of the ^{199}Hg shifts has been reversed with respect to the experimental results in order to be consistent with the absorber convention. The shifts are given relative to HgF$_2$ and metallic gold for ^{199}Hg and ^{197}Au, respectively (from [7.11.5])

In 1976 Koch et al. [7.11.5] and Wurtinger [7.11.3] simultaneously succeeded in observing the resonance effect of the high energetic (158.4 keV) $5/2^- - 1/2^-$ E2 transition in ^{199}Hg, which is of the order of some 10^{-5}, using the backscattering geometry [7.11.5] and the transmission geometry with the current integration technique [7.11.3, 5], respectively. As a source a Pt metal foil with ^{198}Pt enriched to about 95% was activated to give ^{199}Pt, making use of the reaction ^{198}Pt(n, γ) ^{199}Pt. ^{199}Pt decays with a half-life of 30 min to the ^{199}Au parent nucleus ($t_{1/2}$ = 3.14 d, cf. Fig. 7.11.1).

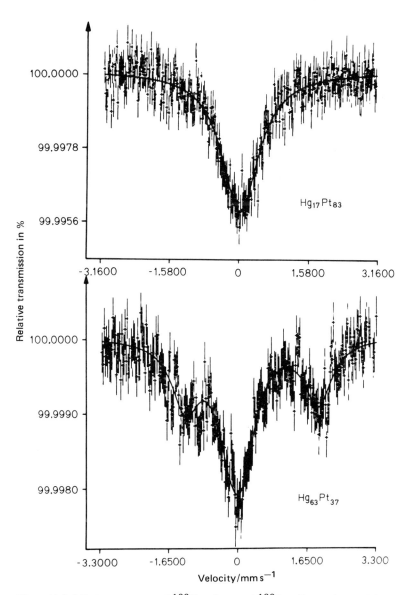

Fig. 7.11.3. Mössbauer spectra of ^{199}Hg$_{17}$Pt$_{83}$ and ^{199}Hg$_{63}$Pt$_{37}$ taken at 4.2 K with a ^{199}Au/Pt source (from [7.11.3])

The line width of $\Gamma_{exp} = 0.74 \pm 0.06$ mm s^{-1} [7.11.3] with a HgPt alloy absorber is in very good agreement with the value of $t_{1/2} = 2.37 \pm 0.07$ ns as measured by γ-γ coincidence technique. Koch et al. [7.11.5] have investigated sources of Pt metal and alloys of the composition $Pt_{0.05}Rh_{0.95}$, $Pt_{0.02}V_{0.98}$ and $Pt_{0.05}Nb_{0.95}$ with HgF_2 as an absorber. In Fig. 7.11.2 the isomer shift of ^{199}Hg is plotted versus the corresponding shifts of the 77 keV resonance in ^{197}Au. A linear relationship with a positive slope is obtained, which implies the sign of $\Delta\langle r^2\rangle$ to be the same for ^{199}Hg as for ^{197}Au, for which a positive value has been established. Comparison with ^{193}Ir, ^{195}Pt and ^{197}Au isomer shifts led the authors of [7.11.5] to an estimation of $\Delta\langle r^2\rangle_{^{199}Hg} \approx 1 \cdot 10^{-3}$ fm^2. Wurtinger [7.11.3] has studied HgPt alloys with a concentration of mercury up to 65%. Two representative spectra are depicted in Fig. 7.11.3. From the evaluation of the Mössbauer spectra the recoil-free fraction, the half-width and the isomer shift could be derived (Fig. 7.11.4). The alloy system with less than 18% mercury content consists of a solid solution with a slightly increasing Debye temperature θ, around 220 K, as a function of the Hg concentration; the Mössbauer spectrum shows a single resonance line. The spectrum of $Hg_{63}Pt_{37}$, however, consists of a quadrupole splitting superimposed by a single line. This has been explained as being due to a mixture of HgPt and Hg_2Pt intermetallic compounds, where the tetragonal structure of Hg_2Pt leads to a quadrupole coupling constant $eQV_{zz}/4 = (1.83 \pm 0.03)$ mm s^{-1}. In HgPt, no observable quadrupole splitting is expected because of its small tetragonal distortion from cubic symmetry. The corresponding Debye temperatures are $\theta_D = (167 \pm 4)$K and $\theta_D = (206 \pm 7)$K for Hg_2Pt and HgPt, respectively.

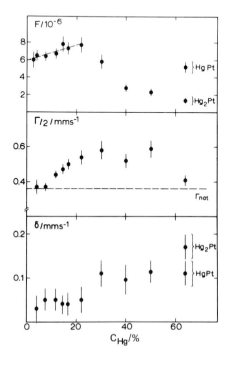

Fig. 7.11.4. $F = f_A f_S \delta_R$ (f_S, f_A Debye-Waller factors of the source and the absorber, respectively; δ_R, experimental correction factor, which is constant as function of the mercury concentration), experimental line width $\Gamma/2$ and isomer shift δ as a function of the Hg content of the PtHg alloy (taken from [7.11.3])

Recently Wurtinger [7.11.6] succeeded in measuring the Mössbauer effect of
^{199}Hg in Hg_2F_2 (Fig. 7.11.5), which shows a large quadrupole interaction with $eQV_{zz}/4$ =
= 3.04 ± 0.36 mm s^{-1}. The difference in the isomer shift between HgF_2 and Hg_2F_2 of
$\delta = -1.77$ mm s^{-1} in connection with the electron density difference derived from
MO calculations, which were calibrated with isoelectronic gold compounds, led to a
$\Delta\langle r^2\rangle$ value in ^{199}Hg of $\delta\langle r^2\rangle_{199_{Hg}} = (3.1 \pm 1.4) \cdot 10^{-3}$. This is in excellent agreement
with muonic data [7.11.7]. Although the ^{199}Hg isotope is not very easy to handle,
recent improvement in Mössbauer technology (current integration technique) allows
one to observe the very small resonance effect and to deduce interesting information
from ^{199}Hg hyperfine interaction and recoilless fraction.

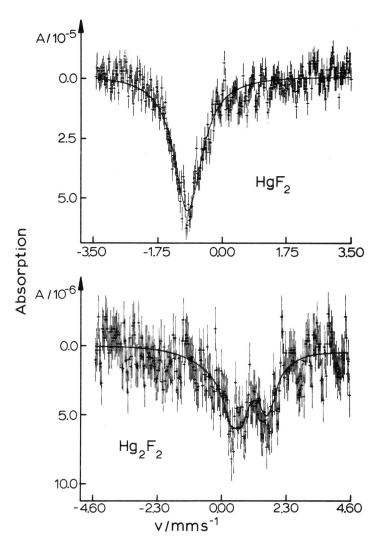

Fig. 7.11.5. Mössbauer spectra of $^{199}HgF_2$ and $^{199}Hg_2F_2$ taken at 4.2 K with a ^{199}Au/Pt source
(from [7.11.6])

References

[7.11.1] Carlson, D. E., Temperley, A. A.: Phys. Lett. *30B*, 322 (1969)

[7.11.2] Stevens, J. G., Stevens, V. E.: *Mössbauer Effect Data Index*, New York: IFI Plenum Press 1969–1975

[7.11.3] Wurtinger, W.: J. Physique *C6*, C6–697 (1976)

[7.11.4] Walcher, D.: Z. Phys. *246*, 123 (1971)

[7.11.5] Koch, W., Wagner, F. E., Flach, D., Kalvius, G. M.: J. Physique *C6*, C6–693 (1976)

[7.11.6] Wurtinger, W.: Dissertation, Technische Hochschule Darmstadt (1977)

[7.11.7] Walter, H. K.: Nucl. Phys. *A 234*, 504 (1974)

8. Some Special Applications

In the preceding chapters we have learned that the chemical and physical state of a Mössbauer atom in any kind of solid material can be characterized by way of the hyperfine interactions which manifest themselves in the Mössbauer spectrum by the isomer shift and, where relevant, electric quadrupole and magnetic dipole splitting of the resonance lines. On the basis of all the parameters obtainable from a Mössbauer spectrum it is in most cases possible to identify unambiguously one or more chemical species of a given Mössbauer atom. This — usually called phase analysis by Mössbauer spectroscopy — is widely used in various kinds of physico-chemical studies, e.g., studies of

— solid-state reactions
— spin and magnetic transitions
— frozen solutions
— corrosion processes
— catalysis
— aftereffects of nuclear transformations
— biological systems
— problems in metallurgy
— minerals
— ancient pottery, etc.

Because of the limited scope of this review it is impossible to give a rigorous account of the work that has been accomplished in these fields. For those, however, who are not working in the field of Mössbauer spectroscopy but wish to gain a feeling for the kind of problems amenable to this spectroscopic technique, we present in the following some selected examples out of the various fields above, together with a choice of references, which the more interested reader may consult for further information.

8.1. Solid-State Reactions

Various kinds of chemical reactions in the solid state have been studied using the Mössbauer effect technique, e.g., thermal decompositions, radiolytical and photolytical reactions, ligand exchange reactions, diffusion, etc.

8.1.1. Thermal Decomposition

An early example for thermal decomposition studies was reported on by Gallagher and Kurkjian [8.1.1], who investigated the decomposition products of some oxalates of iron(III) using the ^{57}Fe Mössbauer effect. Their procedure simply involved taking samples at various temperatures, quenching them and measuring the Mössbauer spectra at room temperature. The changing oxidation state of iron could be clearly followed through the course of the decompositions. For example, the oxidation state of iron in $Sr_3[Fe(C_2O_4)_3]_2 \cdot 2H_2O$ is converted completely to +2 at 300 °C and reverts to +3 at 400 °C. At 700 °C a significant percentage is present as +4 which is gradually reduced to +3 with increasing temperature.

More recent thermal decomposition studies on potassium tris(oxalato)ferrate(III) employing the Mössbauer effect technique have been performed by Bancroft et al. [8.1.2]. They found that, after dehydration at $\sim$ 120 °C, a binuclear complex containing the anion $[(C_2O_4)_2 Fe(Ox)Fe(C_2O_4)_2]^{6-}$ (Ox = quadridentate oxalate group) is formed, which decomposes at temperatures < 380 °C via Fe_3O_4 to either Fe_2O_3 (in air) or Fe (in vacuo).

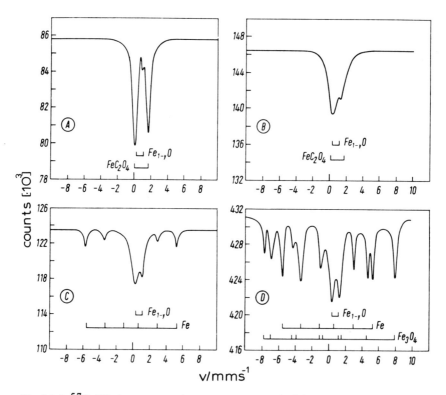

Fig. 8.1.1. ^{57}Fe Mössbauer spectra (at room temperature) of decomposition products of iron(II) oxalate heated to 340 °C for (A) 0.5 hours, (B) 1 hour, (C) 1.5 hours, and (D) after complete decomposition (from [8.1.3])

Förster, Meyer, and Nagorny [8.1.3] carried out similar thermal analyses on the decomposition products of iron(II) oxalate and mixed crystals of iron(II) and magnesium oxalates. By taking ^{57}Fe Mössbauer spectra of samples heated to various temperatures (cf. Fig. 8.1.1), they were able to support the mechanism proposed by Kornijenko for the decomposition of oxalates of iron(II) and similar metals: the primary formation of wüstite, $Fe_{1-y}O$, which reacts with CO to give CO_2 and α-iron, and the formation of magnetite, Fe_3O_4, after complete decomposition at temperatures < 570 °C.

Tominaga et al. [8.1.4] reported on the use of Mössbauer spectroscopy and conventional thermogravimetry to explore the thermal decomposition of iron(II) pyridine and picoline complexes and suggested the following sequences of the decomposition:

$$Fe(py)_4Cl_2 \rightarrow Fe(py)_2Cl_2 \rightarrow Fe(py)Cl_2 \rightarrow Fe(py)_{2/3}Cl_2 \rightarrow FeCl_2$$

$$Fe(\gamma\text{-pic})_4Cl_2 \rightarrow Fe(\gamma\text{-pic})Cl_2 \rightarrow FeCl_2$$

$$Fe(py)_4(NCS)_2 \rightarrow Fe(py)_2(NCS)_2 \rightarrow Fe(NCS)_2.$$

Some of their measured spectra, shown in Fig. 8.1.2, indicate the gradual conversion from $Fe(py)_4Cl_2$ to $Fe(py)_2Cl_2$, which can be followed kinetically. Burger et al. [8.1.5] also analyzed successfully the decomposition products of such complexes $Fe(py)_4X_2$ (X = Cl, Br, I, SCN) using DTA, DTG, and Mössbauer spectroscopy.

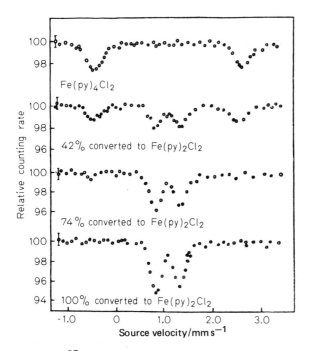

Fig. 8.1.2. ^{57}Fe Mössbauer spectra showing the gradual conversion of $Fe(py)_4Cl_2$ into $Fe(py)_2Cl_2$ (from [8.1.4])

Very useful information has been obtained from thermal dehydration and decomposition studies of hydrated ferrous sulfates and chlorides [8.1.6, 7] and of hydrated ferric sulfates [8.1.8, 9]. ^{57}Fe Mössbauer spectroscopy has also been applied to thermal analyses of $BaFeO_4$ [8.1.10], of various Fe (II)-imidazole compounds, the electronic and chemical structure of which has been determined [8.1.11], of K_2FeO_4 where Fe^{6+} was observed to be directly reduced to the Fe^{3+} state [8.1.10], of dinitrogen tetroxide solvates of iron(III) nitrate [8.1.12], of tris(2,2'-bipyridine)iron(II) chloride [8.1.13], of alkali ferricyanides [8.1.14], of biotites [8.1.15] and of other substances.

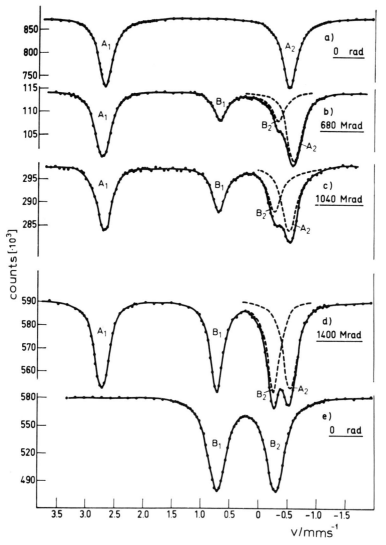

Fig. 8.1.3. ^{57}Fe Mössbauer spectra of γ-irradiated $FeSO_4 \cdot 7H_2O$ as a function of the dose rate; irradiation and Mössbauer measurements were carried out at room temperature. Doublet B_1/B_2 originates from $[Fe(H_2O)_5(OH)]^{2-}$ (from [8.1.16])

8.1.2. Radiolysis Studies

Similarly to the studies of thermal decomposition of polycrystalline substances including both identification of decomposition products and determination of the rate of decomposition by means of Mössbauer spectroscopy (in addition to conventional thermal analysis methods), this new technique has been successfully applied to the investigation of solid-state reactions and decompositions induced by light or irradiation with γ-rays or electrons.

For example, Gütlich et al. [8.1.16] studied the influence of γ-irradiation (^{60}Co source) on hydrated ferrous and ferric sulfates and ferrous oxalate. In ferrous compounds with H_2O ligands coordinated to the iron ion like in $FeSO_4 \cdot 7H_2O$ and $(NH_4)_2Fe(SO_4)_2 \cdot 6H_2O$ with $Fe(H_2O)_6^+$ complexes in the lattice, they observed a dose-dependent conversion into pentaaquohydroxo complexes of iron(III) (cf. Fig. 8.1.3) and suggested the following mechanism: A water ligand is radiolysed in the primary step; the resulting OH radical oxidizes Fe^{2+} to Fe^{3+}.

$$H_2O \underset{\text{Recomb.}}{\overset{\gamma}{\rightleftharpoons}} H \cdot + \cdot OH$$

$$[(H_2O)_5Fe^{II}(\cdot OH)]^{2+} \rightarrow [(H_2O)_5Fe^{III}(OH)]^{2+}$$

$$H \cdot + \cdot H \rightarrow H_2$$

No chemical conversion was observed in γ-irradiated hydrated ferric salts, because the OH radical apparently is not able to oxidize Fe^{3+} to Fe^{4+}. Similar studies, carried out somewhat later by Wertheim and Buchanan [8.1.17] on hydrated ferrous salts, led to full agreement with these findings. More recently Gütlich, Odar and Zyball [8.1.18] irradiated hydrated ferrous salts with 1 MeV electrons (van de Graaff generator) at liquid nitrogen temperature and observed, among other interesting effects, a distinct dependence of the extent of conversion into $[Fe(H_2O)_5(OH)]^{2+}$ complex on the number of coordinated water molecules. Very similar observations were made by Meyers et al. [8.1.19] in studies of the gamma radiolysis of solid iron(II) sulfate and its hydrates.

As early as 1964 a Japanese group [8.1.20] initiated a Mössbauer study of the radiolysis and photolysis of ferric oxalate and found a reduction of Fe^{3+} to Fe^{2+} induced by light as well as by γ-rays. Later on the same laboratory investigated the γ-radiolysis of ferric lactate, ferric citrate, and ferric malate and observed a similar valence change (reduction) of iron (cf. Fig. 8.1.4 [8.1.21]). Bancroft et al. [8.1.22] also looked into γ-radiolysis of potassium trisoxolatoferrate(III) by means of Mössbauer spectroscopy and found, as in the thermal decomposition of this substance, the binuclear anion $[(Ox)_2FeO_2C_2O_2Fe(Ox)_2]^{6-}$ with the quadridentate bridging oxalate group $O_2C_2O_2$ as the final product.

Baggio-Saitovitch, Friedt, and Danon [8.1.23] irradiated some iron(III) chelate perchlorates with acetylacetonate, citrate, EDTA and dipyridyl as chelating ligands with electrons at liquid nitrogen temperature and again observed a reduction of iron(III) to iron(II). They suggested a mechanism involving the formation and fast annihilation of free radicals for explaining the radiation induced reduction process.

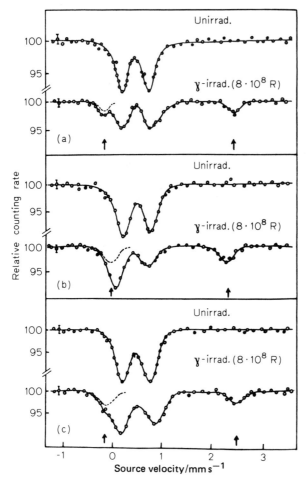

Fig. 8.1.4. Mössbauer spectra of γ-irradiated and unirradiated ferric lactate (a), ferric malate (b), and ferric citrate (c). The quadrupole doublet consisting of the outer two lines (arrows) arises from high-spin ferrous species (from [8.1.21])

8.1.3. Pressure Induced Reactions

The behaviour of solid chemical compounds under pressure up to 200 kbar has been explored using various spectroscopic techniques, including the Mössbauer effect. Most of the contributions have been communicated by the school of Drickamer [8.1.24]. They observed, for instance, that iron(III) in ionic compounds is significantly reduced to iron(II) at high pressure. This is clearly demonstrated by the Mössbauer spectra of ferric phosphate taken at atmospheric pressure and at 140 kbar (cf. Fig. 8.1.5). This phenomenon was found to be reversible. It appears to be associated with a general tendency for the ground state of the ferrous ion to decrease in energy relative to the ligands with increasing compression, which facilitates electron transfer. In other

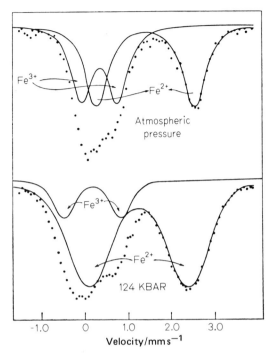

Fig. 8.1.5. Mössbauer spectra of ferric phosphate (from [8.1.25])

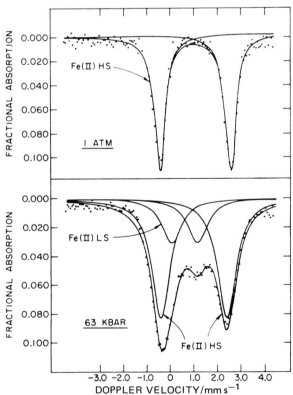

Fig. 8.1.6. Mössbauer spectra of high-spin [Fe(phen)$_2$Cl$_2$] as a function of pressure (from [8.1.26])

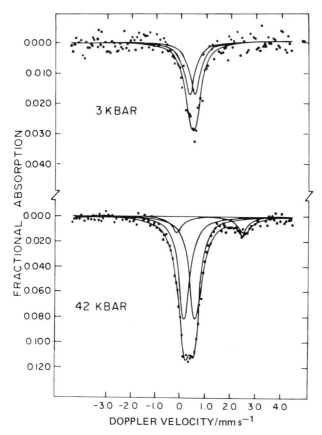

Fig. 8.1.7. Mössbauer spectra of low-spin [Fe(phen)$_3$]Cl$_2$ · 7H$_2$O as a function of pressure. The two outermost lines of low intensity arise from Fe(II) in the high-spin state (from [8.1.26])

experiments Fisher and Drickamer [8.1.26, 27] applied pressure up to 175 kbar to ferrous phenanthroline complexes and followed the change in spin state of the iron by Mössbauer and optical spectroscopy. They observed a significant tendency of the high-spin complexes (bisphenanthroline complexes with halide ligands) to be converted into the low-spin state, and of low-spin complexes (trisphenanthroline complexes) to be converted into the high-spin state at high pressure. A few representative spectra reproduced from their work, shown in Figs. 8.1.6 and 8.1.7, give a clear indication of these effects. The optical absorption measurements under pressure exhibit large red shifts for the ligand π-π^* transition and smaller red shifts for the metal-to-ligand charge transfer peaks. The authors explained these findings in terms of reduced availability of ligand antibonding π^* orbitals for metal-to-ligand back-donation.

8.1.4. Ligand Exchange, Electron Transfer, Isomerism

The Mössbauer effect technique has also been proven very useful in following the gradual ligand exchange in the solid state under various conditions. Gütlich and Hasselbach [8.1.28], for example, have investigated the ligand exchange processes in the polycrystalline system of $FeSO_4 \cdot xH_2O/KCN$ as a function of time, temperature, molar ratio, particle size, and content of crystal water. At room temperature they observed a remarkably fast exchange of coordinated water molecules by CN^- ligands leading to $[Fe(CN)_6]^{4-}$ as the final product. The only transient species observable in the ^{57}Fe Mössbauer spectra is the $[Fe(CN)_5(H_2O)]^{3-}$ complex. Some spectra and essential diagrams of their work are shown in Figs. 8.1.8–10.

An example, demonstrating the successful applicability of Mössbauer spectroscopy to electron transfer reactions in the solid state, has been provided by Tominaga and Sakai [8.1.29]. In mixtures of anhydrous ferric chloride with cobalt(II)acetylacetonate, $Co(acac)_2 \cdot 2H_2O$, they observed a preferred reduction to ferrous chloride rather than the formation of the iron acetylacetonate.

Cyanide linkage isomerism has been studied in polycrystalline iron(II)hexacyanochromate(III) by Brown et al. [8.1.30] employing Mössbauer spectroscopy together with i. r. spectroscopy, x-ray powder diffraction and magnetic susceptibility. Four distinct structures could be revealed at different stages in the isomerization process. The initial face-centered cubic material contains $Fe^{2+}-N\equiv C-Cr^{3+}$ linkages and has interstitial Fe^{2+}. In the first reaction step the interstitial Fe^{2+} displaces Cr^{3+} from

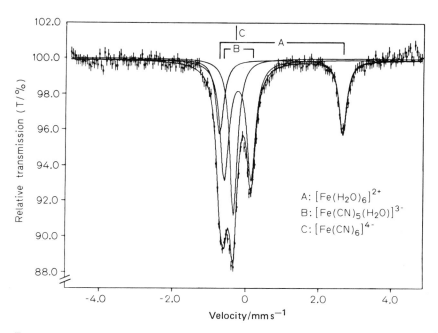

Fig. 8.1.8. Mössbauer spectrum of the polycrystalline mixture $FeSO_4 \cdot 7H_2O/KCN$ (molar ratio 1:6, particle size 0.2–0.3 mm) taken after 720 min of reaction at 5 °C; $^{57}Co/Pt$ source at 291 K, absorber at 175 K (from [8.1.28])

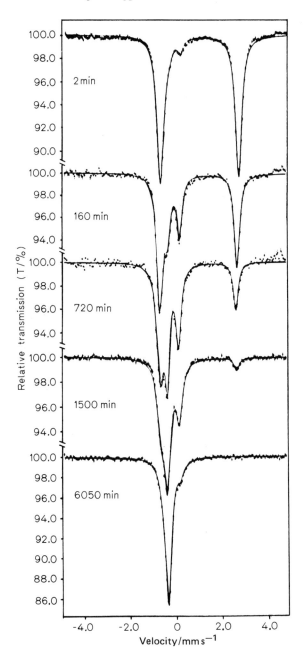

Fig. 8.1.9. Mössbauer spectra of the polycrystalline mixture $FeSO_4 \cdot 7H_2O$/KCN (molar ratio 1:6, particle size 0.2–0.3 mm) as a function of reaction time at 5 °C (from [8.1.28])

the carbon octahedra to give a half-isomerized complex with both Fe^{2+} and Cr^{3+} in the carbon octahedra and Cr^{3+} in interstitial sites. In the presence of air this complex undergoes further rearrangement and partial oxidation. The oxidized compound can then be reduced to give the true linkage isomer containing $Cr^{3+}-N\equiv C-Fe^{2+}$ linkages with Fe^{2+} in interstitial sites.

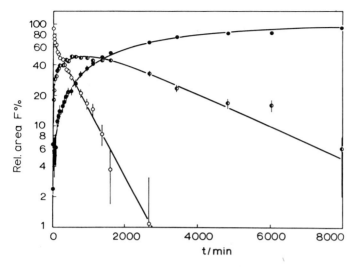

Fig. 8.1.10. Logarithm of relative area of Mössbauer resonance lines of $[Fe(H_2O)_6]^{2+}$ (○), $[Fe(CN)_5(H_2O)]^{3-}$ (◓) and $[Fe(CN)_6]^{4-}$ (●) in the system $FeSO_4 \cdot 7H_2O/KCN$ (molar ratio 1:6, particle size 0.2–0.3 mm) as a function of reaction time at 5 °C (from [8.1.28])

8.1.5. Diffusion in Solids and Liquids

Self-diffusion and impurity atom diffusion in solids and in high-viscosity liquids have been investigated with success using the Mössbauer effect. The papers dealing with diffusion problems, which have appeared so far in the literature, are so numerous that they cannot be reviewed here. We restrict ourselves to giving a few references which may be suitable to inform the reader on how experiments in diffusion studies using the Mössbauer effect are performed and what one can learn from the measured spectra:

References [8.1.31–37] are dealing with diffusion in solids, Refs. [8.1.38–41] with diffusion in viscous liquids, and Refs. [8.1.42–46] particularly with Brownian motion.

References

[8.1.1] Gallagher, P. K., Kurkjian, C. R.: Inorg. Chem. 5, 214 (1966)

[8.1.2] Bancroft, G. M., Dharmawardena, K. G., Maddock, A. G.: Inorg. Chem. 9, 223 (1970)

[8.1.3] Förster, H., Meyer, K., Nagorny, K.: Z. physik. Chem. Neue Folge 87, 64 (1973)

[8.1.4] Tominaga, T., Takeda, M., Morimoto, T., Saito, N.: Bull. Chem. Soc. Japan 43, 1093 (1970)

[8.1.5] Burger, K., Liptay, G., Korecz, L., Kiraly, I., Papp-Molnar, E.: Magy. Kem. Foly. 77, 90 (1971)

[8.1.6] Vértes, A., Zsoldos, B.: Magy. Kem. Foly. 76, 282 (1970); Acta Chim. (Budapest) 65, 261 (1970)

[8.1.7] Dézsi, I., Ouseph, P. J., Thomas, P. M.: Chem. Phys. Lett. 9, 390 (1971)

[8.1.8] Skeff Neto, W., Garg, V. K.: J. Inorg. Nucl. Chem. 37, 2287 (1975)

[8.1.9] Bristoti, A., Viccaro, P. J., Kunrath, J. I., Brandão, D. E.: Inorg. Nucl. Chem. Lett. *11*, 253 (1975)

[8.1.10] Ichida, T.: J. Solid State Chem. *7*, 308 (1973); Bull. Chem. Soc. Japan *46*, 79 (1973)

[8.1.11] Marcolin, H. E., Trautwein, A., Maeda, Y., Seel, F., Wende, P.: Theoret. Chim. Acta *41*, 223 (1976)

[8.1.12] Addison, C. C., Harrison, P. G., Logan, N., Blackwell, L., Jones, D. H.: J. C. S. Dalton, 830 (1975)

[8.1.13] Sato, H., Tominaga, T.: Radiochem. Radioanal. Lett. *22*, 3 (1975)

[8.1.14] Raj, D., Danon, J.: J. Inorg. Nucl. Chem. *37*, 2039 (1975)

[8.1.15] Hogg, C. S., Meads, R. E.: Miner. Mag. *40*, 79 (1975)

[8.1.16] Gütlich, P., Odar, S., Fitzsimmons, B. W., Erickson, N. E.: Radiochim. Acta *10*, 147 (1968)

[8.1.17] Wertheim, G. K., Buchanan, D. N. E.: Chem. Phys. Lett. *3*, 87 (1969)

[8.1.18] Gütlich, P., Odar, S., Zyball, H.: to be published

[8.1.19] Meyers, J., Ladriere, J., Chavee, M., Apers, D.: J. Physique Collq. C 6 *37*, C6−905 (1976)

[8.1.20] Saito, N., Sano, H., Tominaga, T., Ambe, F.: Bull. Chem. Soc. Japan *38*, 681 (1965)

[8.1.21] Saito, N., Tominaga, T., Morimoto, T.: J. Inorg. Nucl. Chem. *32*, 2811 (1970)

[8.1.22] Bancroft, G. M., Dharmawardena, K. G., Maddock, A. G.: J. Chem. Soc. A (19), 2914 (1969)

[8.1.23] Baggio-Saitovitch, E., Friedt, J. M., Danon, J.: J. Chem. Phys. *56*, 1269 (1972)

[8.1.24] Drickamer, H. G., Frank, C. W.: *Electronic Transitions and the High Pressure Chemistry and Physics of Solids*. London: Chapman and Hall 1973

[8.1.25] Champion, A. R., Vaughan, R. W., Drickamer, H. G.: J. Chem. Phys. *47*, 2583 (1967)

[8.1.26] Fisher, D. C., Drickamer, H. G.: J. Chem. Phys. *54*, 4825 (1971)

[8.1.27] Bargeron, C. B., Drickamer, H. G.: J. Chem. Phys. *55*, 3471 (1971)

[8.1.28] Gütlich, P., Hasselbach, K. M.: Ber. Bunsenges. physik. Chem. *78*, 1017 (1974)

[8.1.29] Tominaga, T., Sakai, T.: Radioisotopes *21*, 42 (1972)

[8.1.30] Brown, D. B., Shriver, D. F., Schwartz, L. H.: Inorg. Chem. *7*, 77 (1968)

[8.1.31] Krivoglaz, M. A., Repets'kii, S. P.: Ukr. Fiz. Zh. *11*, 1215 (1966)

[8.1.32] Knauer, R. C., Mullen, J. G.: Phys. Rev. *174*, 711 (1968)

[8.1.33] Valov, P. M., Ya, Vasil'ev, V., Veriovkin, G. V., Kaplin, D. F.: J. Solid State Chem. *1*, 215 (1970)

[8.1.34] Janot, C.: *Mössbauer Spectroscopy. Its Applications*, Proc. Panel 1971. Vienna: IAEA 1972, p. 109

[8.1.35] Helsen, J., Schmidt, K., Chakupurakal, Th., Coussement, R., Langouche, G.: Bull. Groupe Fr. Argiles *24*, 165 (1972)

[8.1.36] Soerensen, K., Trumpy, G.: Phys. Rev. B *7*, 1791 (1973)

[8.1.37] Uskov, V. A., Prudovski, V. J.: Inorg. Mater. *11*, 131 (1975)

[8.1.38] Craig, P. P., Sutin, N.: Phys. Rev. Lett. *11*, 460 (1963)

[8.1.39] Bunbury, D. St. P., Elliott, J. A., Hall, H. E., Williams, J. M.: Phys. Lett. *6*, 34 (1963)

[8.1.40] Elliott, J. A., Hall, H. E., Bunbury, D. St.: Proc. Phys. Soc. *89*, 595 (1966)

[8.1.41] Abras, A., Mullen, J. G.: Phys. Rev. A *6*, 2343 (1972)

[8.1.42] Lisichenko, V. I.: Ukr. Fiz. Zh. *12*, 951 (1967)

[8.1.43] Bhide, V. G., Sundaram, R., Bhasin, H. C., Bonchev, T.: Phys. Rev. B *3*, 673 (1971)

[8.1.44] Gunther, L., Zitkova-Wilcox, J.: J. Physique Collq. C 6 *35*, C6−519 (1974)

[8.1.45] Gunther, L., Zitkova-Wilcox, J.: J. Stat. Phys. *12*, 205 (1975)

[8.1.46] Keller, H., Kündig, W.: Solid State Comm. *16*, 253 (1975)

8.2. Frozen Solution Studies

Mössbauer resonance experiments on frozen solutions incorporating "trapped" Mössbauer isotopes like ^{57}Fe, ^{119}Sn, ^{129}I, and others as microprobes in solidified aqueous and nonaqueous solvents have been carried out widely to gain various kinds of information concerning the crystalline state of the solidified solvent as well as the chemical state of the Mössbauer atom as a function of temperature. Phase transitions between different crystalline states (e.g., cubic to hexagonal ice), glass transition temperature, and transition into supercooled liquid state are some of the crystallographically interesting points extractable from the Mössbauer spectra. The chemical nature of the trapped solvated Mössbauer atom, e.g., with respect to composition and symmetry of the solvation shell, solute-solvent and solute-solute interaction through paramagnetic relaxation, reflected in some cases in the Mössbauer spectra are other points of interests. Finally, usual solution chemistry like electron transfer and ligand exchange reactions, anation, and hydrolysis, is in principle amenable to Mössbauer spectroscopy after rapidly quenching the liquid system to a solid absorber.

The Mössbauer effect technique most often used in frozen solution studies is the conventional transmission experiment with the frozen solution as absorber kept at cryogenic temperatures in an appropriate cryostat. For the detection of phase transitions one often uses the technique of single-channel pulse rate measurements instead of recording the whole velocity spectrum [8.2.1, 2]. This technique is easy to combine with DTA analysis by using an absorber holder as described in [8.2.1]. Fig. 8.2.1 shows the single-channel Mössbauer data and the DTA curve of a frozen ferrous perchlorate (5 mole-%) solution [8.2.1]. Crystallographic phase transitions manifest themselves in general by changes of the percentage resonance effect, the line width, and the quadrupole splitting, whereas changes in the chemical nature of the Mössbauer atom containing species are primarily reflected by changes in the isomer shift and the quadrupole splitting.

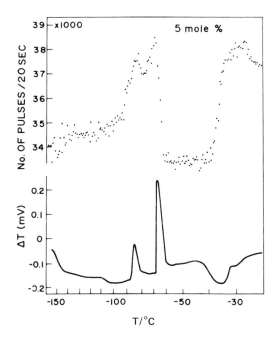

Fig. 8.2.1. Single-channel Mössbauer data and DTA curve of frozen ferrous perchlorate solution (5 mole-%) (from [8.2.1])

Out of the large number of publications on frozen solution work we can name only a reduced list of references; the investigations referred to in the following were carried out using ^{57}Fe as the Mössbauer microprobe.

Frozen aqueous solutions containing iron(II) salts have been widely investigated by various groups [8.2.1–17] using ^{57}Fe Mössbauer spectroscopy, but also differential thermal analysis (DTA) and other methods. The various frozen systems show the same principal behaviour: After rapid quenching to liquid nitrogen temperature they exhibit the properties of an amorphous state; they undergo a glass transition at about 160 K and at somewhat higher temperature they enter a new state with the iron atoms being in a regular environment and DTA revealing an exothermic peak due to recrystallization. These two states (I, II) are found to be independent of the anion in the ferrous compound. In case of ferrous chloride, however, a third state (III) is reached at about 230 K with a reduction of the quadrupole splitting to about half of that in states I and II.

Frozen aqueous solutions containing ferric ions have also been subject to extensive exploration using the ^{57}Fe Mössbauer effect technique [8.2.20–24]. The pattern of the Mössbauer spectra depends strongly on the acidity of the solution, which is certainly due to the replacement of one or more coordinated water molecules (depending on pH) by OH^- ligands [8.2.20]. High-spin ferric ions both in the solid state and in frozen solution tend to exhibit paramagnetic relaxation due to (electronic) spin-spin interaction with relaxation times in the order of the lifetime of the excited Mössbauer level (cf. [8.2.23] and [8.2.21, 22] and references given therein).

Some of the various published ^{57}Fe Mössbauer studies on frozen nonaqueous solutions have been reported in [8.2.25–31].

Various examples demonstrating the usefulness of frozen solution studies by Mössbauer spectroscopy to gain insight into reactions in solution have been communicated in [8.2.32–36]. Maddock et al. [8.2.32] studied the ferric complex species formed upon extraction into nitrobenzene from chloride, bromide, and thiocyanate solutions. Vértes et al. [8.2.33] investigated the hydrolysis of ferric salts in the presence of ClO_4^-, Cl^-, NO_3^-, and SO_4^{2-}, respectively, and found that different species are formed in the solutions during hydrolysis, depending on the nature of the anion. Electron exchange reaction between iron(II) and iron(III) in solution was followed by Komor et al. [8.2.34] via Mössbauer measurements on rapidly cooled solutions. Toma et al. [8.2.35] tried to correlate Mössbauer and visible-u.v. spectra with the aqueous substitution reactivity of several substituted pentacyanoferrate(II) complexes. Birchall and Tun [8.2.36] studied tris(dithiooxalato)ferrate(III) complexes in the solid state and in frozen solution. They found that the preparation of these complexes from iron(III) salts proceeds through a reduction of iron(III) to iron(II), followed by a subsequent reoxidation.

References

[8.2.1] Cameron, J. A., Keszthelyi, L., Nagy, G., Kacsóh, L.: J. Chem. Phys. *58*, 4610 (1973)
[8.2.2] Fröhlich, K., Keszthelyi, L.: Z. Phys. *259*, 301 (1973)
[8.2.3] Dézsi, I., Keszthelyi, L., Pócs, L., Korecz, L.: Phys. Lett. *14*, 14 (1965)
[8.2.4] Dézsi, I., Keszthelyi, L., Molnár, B., Pócs, L.: Phys. Lett. *18*, 28 (1965)

[8.2.5] Cameron, J. A., Keszthelyi, L., Nagy, G., Kacsóh, L.: Chem. Phys. Lett. *8*, 628 (1971)

[8.2.6] Fröhlich, K., Keszthelyi, L.: J. Chem. Phys. *58*, 4614 (1973)

[8.2.7] Sanad, A. M.: Chem. Phys. Lett. *29*, 376 (1974)

[8.2.8] Nozik, A. J., Kaplan, M.: J. Chem. Phys. *47*, 2960 (1967); Chem. Phys. Lett. *1*, 391 (1967); J. Chem. Phys. *49*, 4141 (1968)

[8.2.9] Dilorenzo, J. V., Kaplan, M.: Chem. Phys. Lett. *2*, 509 (1968), Chem. Phys. Lett. *3*, 216 (1969)

[8.2.10] Nundt, W. A., Sonnino, T.: J. Chem. Phys. *50*, 3127 (1969)

[8.2.11] Pelah, I., Ruby, S. L.: J. Chem. Phys. *51*, 383 (1969)

[8.2.12] Ruby, S. L., Zabransky, B. J., Stevens, J. G.: J. Chem. Phys. *54*, 4559 (1971)

[8.2.13] Ruby, S. L., Bernabei, A., Zabransky, B. J.: Chem. Phys. Lett. *13*, 382 (1972)

[8.2.14] Simopoulos, A., Wickman, H., Kostikas, A., Petridis, D.: Chem. Phys. Lett. *7*, 615 (1970)

[8.2.15] Brunot, B., Hauser, U., Neuwirth, W., Bolz, J.: Z. Phys. *249*, 125 (1971)

[8.2.16] Brunot, B., Hauser, U., Neuwirth, W.: Z. Phys. *249*, 134 (1971)

[8.2.17] Ruby, S. L.: in *Perspectives in Mössbauer Spectroscopy*, S. G. Cohen, M. Pasternak (eds.). New York: Plenum Publishing 1973, p. 181

[8.2.18] Oliveira, A. C., Garg, V. K.: Radio-Chem. Radioanal. Lett. *18*, 43 (1974)

[8.2.19] Neuwirth, W., Schröder, H.-J.: Z. Phys. B *23*, 71 (1976)

[8.2.20] Dézsi, I., Vértes, A., Komor, M.: Inorg. Nucl. Chem. Lett. *4*, 649 (1968)

[8.2.21] Mørup, S., Thrane, N.: Chem. Phys. Lett. *21*, 363 (1973)

[8.2.22] Mørup, S.: in *Mössbauer Effect Methodology*. I. J. Gruverman (ed.), Vol. 9. New York: Plenum Press, London 1974, p. 127

[8.2.23] Ohya, T., Ôno, K.: J. Phys. Soc. Japan *34*, 376 (1973)

[8.2.24] Plachinda, A. S., Makarov, E. F.: Chem. Phys. Lett. *25*, 364 (1974)

[8.2.25] Burger, K., Vértes, A., Czakó, I. N.: Acta Chim. Acad. Sci. Hung. *63*, 115 (1970)

[8.2.26] Vértes, A., Burger, K., Suba, M.: Acta Chim. Acad. Sci. Hung. *63*, 123 (1970)

[8.2.27] Champeney, D. C., Woodhams, F. W. D.: J. Phys. B (Proc. Phys. Soc) *1*, Ser. 2, 620 (1968)

[8.2.28] Simopoulos, A., Wickman, H., Kostikas, A., Petridis, D.: Chem. Phys. Lett. *7*, 615 (1970)

[8.2.29] Garbett, K., Williams, R. J. P.: J. Chem. Soc. (A), 3434 (1971)

[8.2.30] Vértes, A., Burger, K., Molnár, B.: J. Inorg. Nucl. Chem. *35*, 691 (1973)

[8.2.31] Afanasov, M. I., Nagy, S., Perfiliev, Yu. D., Vértes, A., Babeshkin, A. M.: Radiochem. Radioanal. Lett. *23*, 181 (1975)

[8.2.32] Maddock, A. G., Medeiros, L. O., Bancroft, G. M.: J. C. S. Chem. Comm., 1067 (1967)

[8.2.33] Vértes, A., Komor, M., Dézsi, I., Burger, K., Suba, M., Gelencser, P.: Proc. Symp. Coord. Chem., 3rd, M. T. Beck (ed.). Budapest: Akad. Kiado 1970, p. 447–93

[8.2.34] Komor, M., Vértes, A., Dézsi, I., Ruff, I.: Acta Chim. Acad. Sci. Hung. *66*, 285 (1971)

[8.2.35] Toma, H. E., Giesbrecht, E., Malin, J. A., Fluck, E.: Inorg. Chim. Acta *14*, 11 (1975)

[8.2.36] Birchall, F., Tun, K. M.: Inorg. Chem. *15*, 376 (1976)

8.3. Surface Studies

The first report on Mössbauer spectra of solid surfaces appeared in 1964. Flinn et al. [8.3.1] measured ^{57}Fe Mössbauer absorption spectra of a sample of η-Al_2O_3 impregnated with ^{57}Fe. Also in 1964 appeared the first paper on the application of Mössbauer emission spectroscopy in surface studies. Brady and Duncan [8.3.2] adsorbed ^{57}Co from solution onto the surface of freshly precipitated oxalates of iron and cobalt and studied the surface effects using these preparations as sources in Mössbauer experiments.

In the following years the number of publications dealing with applications of Mössbauer spectroscopy to investigations of chemical and physical phenomena of surfaces increased rapidly, and the subject was reviewed by Low in 1967 [8.3.3], and by Delgass and Boudart in 1968 [8.3.4]. Two more excellent review articles on surface studies by Mössbauer spectroscopy, covering nearly all the relevant literature up to 1969, were published in 1971 by Goldanskii and Susdalev [8.3.5] and by Hobson [8.3.6].

In a relatively short period of a few years, the application of the Mössbauer effect technique to the study of surface phenomena yielded a number of fundamentally important results on the dynamical properties and the electronic state of surface atoms, on elementary adsorption and catalytic processes, on the influence of dispersion on magnetic and lattice properties, etc., and it became evident that this new spectroscopic method was on its way toward developing to a powerful tool in

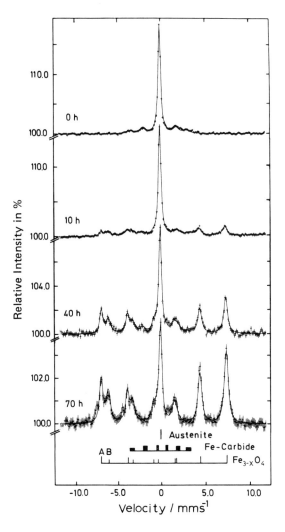

Fig. 8.3.1. CEM spectra of stainless steel 1.4550 (66–70% Fe, 9–11% Ni, 19–23% Cr) oxidized in water at 295 °C and 142 bar for 0, 10, 40, and 70 h. The substrate material contains an austenitic and a non-stochiometric iron carbide phase, which are responsible for the single-line resonance and the magnetic hyperfine pattern, respectively, of the nonoxidized material (t = 0). The oxide layer grown on the steel substrate consists of nonstochiometric magnetite (from [8.3.44])

surface science. A broad variety of chemical and physical phenomena of solid surfaces, like
- electronic state and local symmetry of surface atoms,
- magnetic properties of surface atoms,
- vibrations (intra- and intermolecular) of the surface atoms,
- surface diffusion
- adsorption processes
- chemistry of surface compounds
- catalysis
- topochemical reactions
- boundary effects of a solid on the properties of its internal atoms (variations in the phonon spectra and the nature of atomic vibrations in highly dispersed particles; superparamagnetism)

can be investigated with the help of Mössbauer spectroscopy.

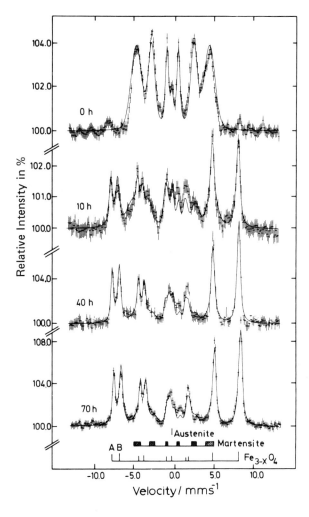

Fig. 8.3.2. CEM spectra of stainless steel 1.4122 (80% Fe, 16.5% Cr) oxidized in water at 295 °C and 142 bar for 0, 10, 40, and 70 h. Beside minor amounts of austenite the nontreated steel sample contains martensite as the dominant phase. A random number of Cr atoms gives rise to a statistical distribution of the magnetic field at the iron nuclei being reflected in the strongly broadened outer resonance lines (top spectrum). The only oxide phase of the corroded steel specimen is nonstoichiometric magnetite (from [8.3.44])

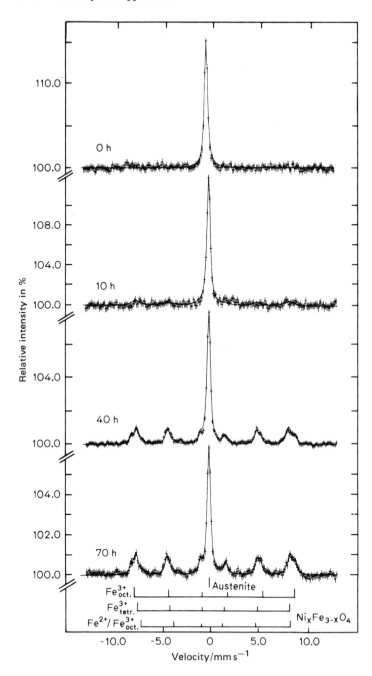

Fig. 8.3.3. CEM spectra of stainless steel Incoloy 800 (40–49% Fe, 30–35 Ni, 19–23% Cr) oxidized in water at 295 °C and 142 bar for 0, 10, 40, and 70 h. Due to the high nickel content of the austenitic steel substrate, the formation of nonstoichiometric nickel ferrite $Ni_xFe_{3-x}O_4$ is observed. The oxide spectrum is a superposition of three subspectra originating from three different iron lattice sites, which are indicated by the stick diagrams (from [8.3.44])

Up to around 1970 one almost exclusively used the transmission geometry, mostly employing a conventional source versus the absorber of interest; in fewer cases the surface effects have been observed by depositing the Mössbauer parent nuclide on the surface of the material of interest and using this as the source versus a suitable single-line absorber. In many instances such as studies of corrosion, oxidation of metals and protecting surface layers, the method of transmission Mössbauer spectroscopy is not sensitive enough to permit surface investigations because the number of Mössbauer atoms in the surface is much smaller than that in the bulk material. In other cases the substrate thickness is too large to be penetrated effectively by the Mössbauer γ-quanta. Corrosion and oxidation studies therefore have often required rather thick surface layers or the removal of corrosion products from the corroded material prior to the Mössbauer measurement; examples are given in [8.3.7–15]. The shortcomings of this technique are evident, particularly in cases where in situ measurements of samples without any physical or chemical pretreatment are highly desirable.

Fortunately, the Mössbauer effect can also be detected by monitoring the radiation which is resonantly reemitted by the absorber after resonance absorption of the incident γ-rays. Useful for ^{57}Fe Mössbauer backscattering measurements are the resonantly scattered γ-rays (energy 14.4 keV; transition probability 0.09), K shell conversion electrons (7.3 keV; 0.81), L shell conversion electrons (13.6 keV; 0.09),

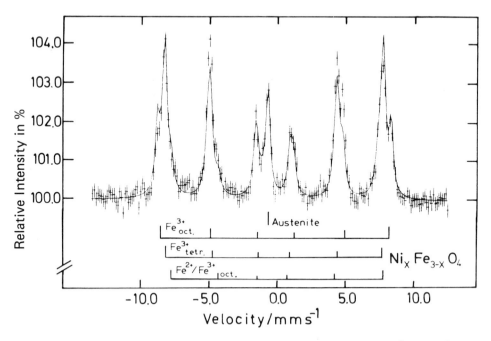

Fig. 8.3.4. CEM spectrum of stainless steel Incoloy 800 oxidized in water at 295 °C and 142 bar for 415 h. The magnetic hyperfine fields and the relative intensities of the three iron sites in $Ni_xFe_{3-x}O_4$ yield a value of x = 0.8. This composition turns out to be constant in the time interval under study (0–415 h, cf. Fig. 8.3.3) (from [8.3.44])

K_α X-rays (6.3 keV; 0.24), and KL Auger electrons (5.4 keV; 0.57). As each of these radiations has a specific penetration range in solid material, it is possible to record depth-selective Mössbauer spectra by changing from one type of backscattered radiation to another. The principle and applications of ^{57}Fe Mössbauer effect measurements by reemitted 14.4 keV γ-rays or 6.3 keV -rays in the backscattering geometry have been reported by many authors; see, for example, [8.3.16–27]. The surface depth, which is analyzed in this technique, ranges from approximately 10–20 μm. In many cases the surface layer (e.g., the corrosion layer) is thinner than this, and overlapping Mössbauer spectra arising from surface layer species and the substrate material will be recorded. If, however, the 7.3 keV conversion electrons or the 5.4 keV electrons are detected in ^{57}Fe Mössbauer backscattering experiments, the depth of penetration will only be $\leqslant 3000$ Å [8.3.28] and the recorded spectra will be representative of the top layer comprising $\lesssim 1\,000$ atoms in thickness. The usefulness of this technique in studies of corrosion, oxidation, and metallurgical problems has been demonstrated by many workers, e.g., in [8.3.29–34]. Theories and ways have been developed for analyzing the resonance line width, the resonance area, and the magnitude of the effect of spectra observed by Mössbauer conversion electron spectroscopy (CEMS) [8.3.35], and for the determination of average thickness of oxide layers from CEM spectra [8.3.36–39]. Special conversion electron detectors have been constructed in various laboratories [8.3.40]. With the presently high standard of development of CEMS, one is able to determine qualitatively and quantitatively various iron oxides and oxyhydroxides, like

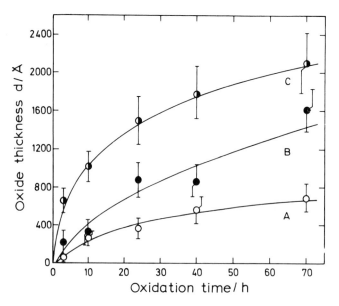

Fig. 8.3.5. Oxide thickness d of the stainless steels (A) Incoloy 800 (○), (B) 1.4550 (●), (C) 1.4122 (◑), oxidized in water at 295 °C and 142 bar as a function of oxidation time. The solid lines represent the theoretical logarithmic rate laws, typical for an oxidation process most probably controlled by short-circuit diffusion (from [8.3.44])

magnetite, haematite, wüstite, and α-, β-, γ-FeOOH, and metallic iron in surface layers, even in cases where several of these species are present simultaneously. This is hardly possible with any other method. Some of the most recent contributions to the study of metal oxidation and corrosion by CEMS [8.3.39, 41−44] may give the reader an impression of the extraordinary power of this technique. Some typical CEM spectra and results from [8.3.44] are shown in Figs. 8.3.1−5.

So far we have paid particular attention only to the study of corrosion and oxidation of iron containing material by means of ^{57}Fe Mössbauer spectroscopy. Of course, other Mössbauer isotopes have also been used for surface studies, although ^{57}Fe undoubtedly has proven to be the most fruitful one for technological and for spectral resolution reasons. Other surface problems besides corrosion and oxidation of metals and alloys have been subject to extensive work using the Mössbauer effect, such as site location and chemical state of Mössbauer atoms on the surface of ion exchange resins, structures of various supports, lattice dynamics, superparamagnetism in small particles, chemisorption, catalytic reactions, etc. Relevant publications that have appeared so far are so numerous that we would not be able to cover them here without going considerably beyond the scope of this book. Instructive examples, however, demonstrating the successful applications of Mössbauer spectroscopy to all these problems are well described in the review articles by Goldanskii and Susdalev [8.3.5] and by Hobson [8.3.6]. Concerning Mössbauer spectroscopy applications to heterogeneous catalysis, the reader may best be informed by consulting the excellent review recently published by Dumesic and Topsøe [8.3.45] and the some 250 references given therein.

References

[8.3.1] Flinn, P. A., Ruby, S. L., Kehl, W. L.: Science *143*, 1434 (1964)
[8.3.2] Brady, P. R., Duncan, J. F.: J. Chem. Soc. (London), 653 (1964)
[8.3.3] Low, M. J.: in *The Gas-Solid Interface*, E. A. Flood (ed.), Vol. 2. New York: Marcel Dekker, Inc. 1967
[8.3.4] Delgass, W. N., Boudart, M.: Catalysis Rev. *2*, 129 (1968)
[8.3.5] Goldanskii, V. I., Susdalev, I. P.: Proc. Int. Conf. Mössbauer Spec., Tihany, 1969, p. 269
[8.3.6] Hobson jr., M. C.: Advanc. Colloid Interface Sci. *3*, 1 (1971)
[8.3.7] Dézsi, I., Vértes, A., Kiss, L.: J. Radioanal. Chem. *2*, 183 (1964)
[8.3.8] Joye, D. D., Axtmann, R. C.: Anal. Chem. *40*, 876 (1968)
[8.3.9] Swartzendruber, L. J., Bennett, L. H.: Scripta Met. *2*, 93 (1968)
[8.3.10] Pritchard, A. M., Dobson, C. M.: Nature *224*, 1295 (1969)
[8.3.11] Channing, D. A., Graham, M. J.: J. electrochem. Soc. *117*, 389 (1970)
[8.3.12] Bancroft, G. M., Mayne, J. E. O., Ridgeway, P.: Brit. Corros. J. *6*, 119 (1971)
[8.3.13] Pritchard, A. M., Haddon, J. R., Walton, G. N.: Corros. Sci. *11*, 11 (1971)
[8.3.14] Pritchard, A. M., Mould, B. T.: Corros. Sci. *11*, 1 (1971)
[8.3.15] Channing, D. A., Graham, M. J.: Corros. Sci. *12*, 271 (1972)
[8.3.16] Major, J. K.: in *Mössbauer Effect Methodology*, I. J. Gruverman (ed.), Vol. 1. New York: Plenum Press 1965, p. 89
[8.3.17] Debrunner, P.: in *Mössbauer Effect Methodology*, I. J. Gruverman (ed.), Vol. 1. New York: Plenum Press 1965, p. 97
[8.3.18] Forsyth, R. H., Terrel, J. H.: Bull. Am. Phys. Soc. *13*, 61 (1968)
[8.3.19] Terrel, J. H., Spijkerman, J. J.: J. appl. Phys. *13*, 11 (1968)

[8.3.20] Collins, R. L.: in *Mössbauer Effect Methodology*, I. J. Gruverman (ed.), Vol. 4. New York: Plenum Press 1968, p. 129

[8.3.21] Ord, R. N.: Appl. Phys. Lett. 15, 279 (1969)

[8.3.22] Chow, H. K., Weise, R. F., Flinn, P. A.: *Mössbauer Spectrometry for Analysis of Iron Compounds*, U. S. Atomic Energy Commission Report No. NSEC-4023-1 (1969)

[8.3.23] Stöckler, H. A.: J. appl. Phys. 41, 825 (1970)

[8.3.24] Meisel, W.: Werkstoffe und Korrosion 21, 249 (1970)

[8.3.25] Ostergaard, P.: Nucl. Instr. Methods 77, 328 (1970)

[8.3.26] Ord, R. N., Christensen, C. L.: Nucl. Instr. Methods 91, 293 (1971)

[8.3.27] Keisch, B.: Nucl. Instr. Methods 104, 237 (1972)

[8.3.28] Swansson, K. R., Spijkerman, J. J.: J. appl. Phys. 41, 3155 (1970)

[8.3.29] Bonchev, Zw., Jordanov, A., Minkova, A.: Proc. Intern. Conf. on Applications of the Mössbauer Effect, Tihany, Hungary, I. Dézsi (ed.). Budapest: Akad. Kiado 1969

[8.3.30] Fenger, J.: Nucl. Instr. Methods 69, 268 (1969); 106, 203 (1973)

[8.3.31] Onodera, H., Yamamoto, H., Watanabe, H., Ebiko, H.: Jap. J. Appl. Phys. 11, 1380 (1972)

[8.3.32] Swartzendruber, L. J., Bennett, L. H.: Scripta Met. 6, 737 (1972)

[8.3.33] Collins, R. L., Mazak, R. A., Yagnik, C. M.: in *Mössbauer Effect Methodology*, I. J. Gruverman (ed.), Vol. 8. New York: Plenum Press 1973, p. 191

[8.3.34] Simmons, G. W., Kellerman, E., Leidheiser jr., H.: Corrosion 29, 227 (1973)

[8.3.35] Krakowski, R. A., Miller, R. B.: Nucl. Instr. Methods 100, 93 (1972)

[8.3.36] Bäverstam, U., Bohm, C., Ekdahl, T., Liljequist, D., Ringström, B.: in *Mössbauer Effect Methodology*, I. J. Gruverman (ed.), Vol. 9. New York: Plenum Press 1974, p. 259

[8.3.37] Huffman, G. P.: Nucl. Instr. Methods. 137, 267–290 (1976)

[8.3.38] Huffman, G. P.: in *Mössbauer Effect Methodology*, C. Seidel, I. J. Gruverman (eds.), Vol. 10. New York: Plenum Press 1976

[8.3.39] Huffman, G. P., Podgurski, H. H.: *Oxidation of Metals*, Vol. 10, No. 6. New York: Plenum Publishing Corp. 1976, p. 377

[8.3.40] Accurate drawings for the construction of a CEMS detector are available upon request from the laboratory of Prof. P. Gütlich, Institut für Anorganische Chemie und Analytische Chemie, University of Mainz, D-6500 Mainz, Germany

[8.3.41] Sette Camara, A., Keune, W.: Corros. Sci. 15, 441 (1975)

[8.3.42] Tricker, M. J., Thomas, J. M., Winterbottom, A. P.: Surface Sci. 45, 601 (1974)

[8.3.43] Thomas, J. M., Tricker, M. J., Winterbottom, A. P.: J. C. S. Faraday Transactions II, 71, 1708 (1975)

[8.3.44] Ensling, J., Fleisch, J., Grimm, R., Grübler, J., Gütlich, P.: "Corrosion Study of Austenitic and Martensitic Steels under Boiler Conditions by Means of ^{57}Fe Conversion Electron Mössbauer Spectroscopy", submitted to Corros. Sci.

[8.3.45] Dumesic, J. A., Topsøe, H.: Adv. Catal. 26, 121–246 (1976)

8.4. Metallurgy

The properties of metals and alloys strongly depend on their preparational history, i. e., on heat treatment, fast or slow cooling, mechanic deformation, corrosion, aging, irradiation of particles, alloying, etc. All these various influences may affect the arrangement of atoms and thus their interaction properties, which consequently influence electric or heat conductivity, magnetism, elasticity, density, reactivity, etc. Since the Mössbauer spectrum is sensitive to changes of electron density, electric field gradient, magnetic field, relative orientation of electric field gradient and magnetic field, with respect to absorbed or emitted γ-rays, and to changes of the mean square displacement (Debye-Waller factors) of the Mössbauer isotope, it is obvious that Mössbauer spectroscopy may serve as useful tool for the investigation of metals and alloys. Hence the application of Mössbauer spectroscopy in physical metallurgy is a rapidly growing field. Several sessions of recent Mössbauer conferences [8.4.1–7] and review articles [8.4.8–13] have been dedicated to this subject. Similar to biological applications, the isotope ^{57}Fe plays the most important role in metallurgical applications – no wonder, because iron is the most important element in metallurgy *and* in Mössbauer spectroscopy. However, the trend to use other Mössbauer isotopes is already noticeable: ^{61}Ni [8.4.14–21], ^{67}Zn [8.4.22], ^{99}Ru [8.4.23], ^{178}Hf [8.4.24], ^{181}Ta [8.4.25–31], ^{182}W, ^{183}W, ^{184}W, ^{186}W [8.4.32–35], ^{189}Os [8.4.36, 37], ^{191}Ir [8.4.38–41], ^{193}Ir [8.4.42–44], ^{195}Pt [8.4.45–47] and ^{197}Au [8.4.48–59]. Among the transition elements Zn and Ta are of particular interest because of their very narrow line widths [8.4.22, 31] (in the μm s^{-1} range), and the noble metals Pt und Au, pure or in alloys, because they may be the ones which can be discussed in terms of available theory with greatest reliability [8.4.57].

8.4.1. Phases and Transitions

8.4.1.1. Interstitial Alloys

Besides the study of phase transitions and magnetic transitions in recent years, many other transitions – displacive, band Jahn-Teller, cooperative Jahn-Teller, ferroelectric, metal-insulator, order-disorder transitions – have been investigated using the Mössbauer effect [8.4.60]. A major advantage of Mössbauer spectroscopy over other techniques such as x-ray diffraction or electron microscopy is the ability to detect the presence of a new phase even when the particles or domains have dimensions of only a few interatomic distances and are therefore too small to be seen in a diffraction pattern or in an electron microscope. Phase transitions may be accompanied by changes in crystal structure, by fluctuations, or they may also occur by a continuous change of composition, possibly leading to significant changes in the Mössbauer spectrum. As an example we discuss the phase transformations in steel, which have been described extensively in the literature [8.4.10, 12, 61–70].

Martensite is a supersaturated solid solution of carbon (or nitrogen) in iron with a body-centered tetragonal lattice, which exhibits ferromagnetic behavior. The impurities C (or N) occupy the octahedral interstices of the iron lattice as indicated in Fig. 8.4.1 [8.4.65], which distort the regular octahedron of next-nearest neighbours of iron. According to Fig. [8.4.1] there are several different types of iron with respect to the central impurity carbon, i.e., the electronic and magnetic behavior of Fe$^{(5)}$, for example, is less influenced by the impurity than Fe$^{(1)}$. Depending on the impurity concentration, the probability of finding Fe$^{(1)}$, ..., Fe$^{(5+\cdots)}$-type iron in steel is different. For a carbon concentration of 4.2 at %, for example, these probabilities, which have been derived from statistical calculations [8.4.12], are 8.2%, 14.5%, 22%, 16% and 39.3%, respectively. This statistical consideration of interstitial impurities in alloys is reflected in the Mössbauer spectrum of martensite, which contains 4.2 at % C (Fig. 8.4.2) [8.4.12]. From inspection of Fig. 8.4.2 we find that the internal magnetic field (H^{int})

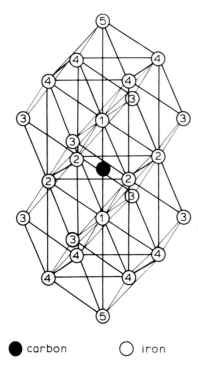

Fig. 8.4.1. Crystal structure of body-centered tetragonal iron-carbon-martensite. The numbers of the iron atoms indicate different neighbour shells of the carbon interstice

● carbon   ○ iron

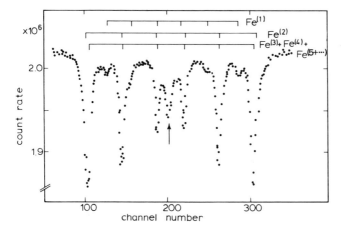

Fig. 8.4.2. Mössbauer spectrum of iron-4.2 at% carbon-martensite. The stick spectra indicate the various contributions to the total spectrum due to iron atoms $Fe^{(i)}$ (i = 1, ...) as defined in Fig. 8.4.1 [8.4.65]

of $Fe^{(3)}$, $Fe^{(4)}$ and $Fe^{(5)}$ corresponds to that of normal α-Fe metal, but H^{int} of $Fe^{(2)}$ is larger and H^{int} of $Fe^{(1)}$ is smaller than H^{int} of α-Fe.

The central "single line" of the spectrum in Fig. 8.4.2 (indicated by an arrow) corresponds to "retained austenite", which appears because of incomplete phase transformation of the high-temperature phase austenite by cooling the sample. Austenite is a solid solution of carbon (or nitrogen) in γ-iron where the C (or N) impurities occupy the interstices of the face-centered-cubic lattice, which exhibits paramagnetism. In the case of relatively low impurity concentration, the probability to find iron atoms with at least one interstitial impurity neighbour is rather low; thus the highest number of iron atoms within the austenite phase is characterized by cubic point symmetry. This high symmetry condition together with the paramagnetic behavior yields a single line absorption pattern. If, however, the impurity concentration of C (or N) increases (for example to 4.2 or 5.8 at%), the probability of finding iron atoms with at least one C (or N) neighbour increases. The presence of a carbon nearest-neighbour destroys the cubic point symmetry of iron. This in consequence leads to a change in electronic structure around the iron nucleus resulting in a quadrupole splitting (A in Fig. 8.4.3) [8.4.63] and also in a small isomer shift of the quadrupole doublet with respect to the single line (B in Fig. 8.4.3). (In Fig. 8.4.2 this quadrupole splitting is not resolved because of the relatively small amount of "retained austenite" in the probe).

Quenching, tempering, and aging considerably influence the phase composition of steel [8.4.71–73]. During tempering or room-temperature aging of Fe-C martensite the carbon atoms tend to cluster to form carbon-rich regions which appear to develop an ordered structure. These carbon-rich phases have been identified from their Mössbauer spectra as ϵ-carbide, χ-carbide (Fe_5C_2) and cementite (Fe_3C) [8.4.62, 64, 66, 70, 74–76]. Similar clustering of impurity atoms has been observed in Fe-N martensite during room-temperature aging [8.4.77].

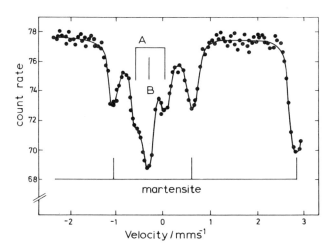

Fig. 8.4.3. Mössbauer spectrum of iron-5.8 at% carbon-austenite plus martensite. A and B represent the austenite absorption lines [8.4.63]

8.4.1.2. Substitutional Alloys

Similar to the investigation of interstitial alloys the qualitative and quantitative analysis of phases is one of the tasks of the investigation of substitutional alloys. The problem of determining the relative amounts of iron in two phases was carefully investigated for the case of mixed austenite and martensite of an iron-nickel alloy [8.4.78]; the Mössbauer results were shown to be significantly more accurate than those obtained by x-ray diffraction, particularly at low austenite concentration.

Steel with composition of about 20 at% Cr, 9 at% Ni, and 71 at% Fe has the property to stabilize the paramagnetic face-centered cubic (γ) phase to low temperatures (stainless steel); correspondingly the Mössbauer spectrum of this material shows a single line which, however, can be considerably broadened [8.4.79, 80]. The broadening has been ascribed to local inhomogenities of ^{57}Fe in this disordered alloy, to the distortions of the local symmetry of ^{57}Fe due to the presence of substitutional impurities Cr and Ni, and also to antiferromagnetic ordering of γ-iron (below about 80 K) [8.4.81]. In other systems the paramagnetic γ-phase has been identified, too, for example in FeCu alloys [8.4.81] or in FeNi alloys [8.4.82].

8.4.1.3. Magnetic Properties

Besides the qualitative and quantitative analysis of phases, the Mössbauer technique has been widely used to study magnetic properties of metals [8.4.60]. In iron-rich alloys the magnetic properties have been found to depend strongly on the composition of the material. Curie temperatures and internal magnetic fields were measured in a series of iron alloys, which contain paramagnetic [8.4.83—98] or diamagnetic [8.4.84, 86, 99—102] substitutional impurities.

Of specific importance have been investigations in which iron was the impurity atom in paramagnetic or diamagnetic host lattices. In this case the ^{57}Fe nucleus can serve as a probe which measures the electronic and magnetic properties of the host, for example in Pd, Pt, Ta, or Ti alloys, which have the interesting property to be able to take up a considerable amount of hydrogen [8.4.29, 103—105]. ^{61}Ni has been utilized as Mössbauer probe in studies of Pd alloys as well [8.4.106, 107]. At very low concentrations (< 100 ppm) of magnetic impurities in metals, local magnetic moments and problems, which are related to the Kondo effect, were studied [8.4.108—110]. The largest moment ($65.2 \pm 1.6\mu_B$) found so far was detected in Ni$_3$Ga (20 ppm Fe) [8.4.111]. Concentrations of magnetic impurities of the order of 0.1 at% in a diamagnetic host lattice (for example Fe or Co in Pt, Au) may yield spontaneous magnetization of the system under study (spin glass behaviour) [8.4.112—117]. Impurity concentrations in the 1 at% range may lead to spin clustering [8.4.103, 118]. In Pt, which contains > 20 at% Fe, even two separated magnetic phases have been measured [8.4.119].

8.4.1.4. Defects

In some cases the Mössbauer effect has helped to identify defects [8.4.120]. If defects are associated with the Mössbauer probe, charge and spin densities, mean-square displacements and lattice symmetry might be strongly affected leading to changes in isomer shift, internal magnetic field, recoilless fraction, and quadrupole splitting. Most of the work in this field has been done with nonmetallic systems; however, there are some examples of defect structure studies in FeAl alloys [8.4.121], in PtFe alloys [8.4.122], and in Ti alloys [8.4.123].

8.4.1.5. Oxidation

The Mössbauer technique is attractive to study oxide systems, in which the sample can be investigated under conditions of varying temperature and oxygen pressure with the possibility of identifying internally oxidized atoms or phases. The internal oxide phases in Cu-rich alloys containing ^{57}Fe were investigated by Gonser [8.4.124]. Knapp et al. have studied the reactions which occur in a blast furnace [8.4.125, 126], and Jones investigated the reduction of metallic oxides [8.4.127]. Several workers have carried out measurements on glassy iron-containing materials related to slags [8.4.128–131], for example, to derive the ferric/ferrous ratio and the distribution of ferric ions between octahedral and tetrahedral sites in a variety of glasses. A series of measurements has been devoted to iron oxides with respect to corrosion studies [8.4.132–143].

8.4.2. Backscatter Technique

The most significant advantage of the application of Mössbauer spectroscopy in metallurgy is that backscatter measurements can be performed on bulk materials, even finished products. This type of application can be used for nondestructive testing [8.4.144], and for the observation of changes in process, i.e., transformations in steels, [8.4.145], changes in magnetic properties in surface layers [8.4.146], and changes in surface properties of corrosive [8.4.133, 147–149] and catalytic processes [8.4.150] (see also Section 8.3). Details of the backscatter technique, especially its depth selectivity, have been described extensively [8.4.145, 151–157].

8.4.3. Further Problems

Further fields in metallurgy which have been approached successfully using Mössbauer spectroscopy are the study of
1. thin film properties, i.e., magnetism [8.4.158–163], multilayer effects [8.4.164], corrosion processes [8.4.163, 165–167], diffusion processes [8.4.168–172], and cluster formation [8.4.173];
2. the invar problem [8.4.12, 174–176];
3. diffusion processes in bulk materials [8.4.123, 168–172, 177–181];

4. texture [8.4.182, 183];

5. amorphous alloys (metallic glasses) [8.4.143, 184–191];

6. intermetallic compounds (Hume-Rothery, Zintl, Laves phases, and Heusler alloys) [8.4.10, 11, 192–194]; and

7. deformation induced transformations [8.4.195–201].

Despite the difficulties which might arise in spectrum analysis due to unresolved broad patterns, the wide field of successful applications made the Mössbauer effect to a standard tool in physical metallurgy.

References

[8.4.1] International Conference on the Application of the Mössbauer Effect. Proceedings, Hungary (1971)

[8.4.2] International Conference on the Applications of the Mössbauer Effect. Proceedings, Israel (1972)

[8.4.3] 5th International Conference on Mössbauer Spectroscopy. Proceedings, Vol. 1 and 2, Bratislava CSSR (1973)

[8.4.4] International Conference on the Applications of the Mössbauer Effect. Proceedings, Bendor France (1974)

[8.4.5] International Conference on Mössbauer Spectroscopy. Proceedings, Cracow Poland (1975)

[8.4.6] International Conference on the Applications of the Mössbauer Effect. Corfu, Greece (1976)

[8.4.7] International Conference on Mössbauer Spectroscopy. Bucharest, Rumania (1977)

[8.4.8] Gonser, U.: Z. Metallkunde 57, 85 (1966)

[8.4.9] Gonser, U.: in An Introduction of Mössbauer Spectroscopy. L. May (ed.). New York: Plenum Press 1971, p. 155

[8.4.10] Keune, W., Trautwein, A.: Metall 25, 27 and 133 (1971)

[8.4.11] Zemčik, T.: see Ref. 5, p. 59

[8.4.12] Fujita, F. E.: in Mössbauer Spectroscopy, U. Gonser (ed.). Topics in Applied Physics 5, 201 (1975)

[8.4.13] Greenwood, N. N., Gibb, T. C.: Mössbauer Spectroscopy, London: Chapman and Hall Ltd. 1971

[8.4.14] Obenshain, F. E., Wegener, H. H. F.: Phys. Rev. 121, 1344 (1961)

[8.4.15] Wegener, H. H. F., Obenshain, F. E.: Z. Phys. 163, 17 (1961)

[8.4.16] Spijkerman, J. J.: Symposia Faraday Soc. 1, 134 (1968)

[8.4.17] Erich, U., Quitmann, D.: in Hyperfine Structure and Nuclear Radiations, E. Matthias, D. A. Shirley (ed.). Amsterdam: North-Holland 1968, p. 130

[8.4.18] Erich, U.: Z. Phys. 227, 25 (1969)

[8.4.19] Erich, U., Kankeleit, E., Prange, H., Hüfner, S.: J. Appl. Phys. 40, 1491 (1969)

[8.4.20] Drijver, J. W., van der Woude, F.: see Ref. 3, p. 33

[8.4.21] Drijver, J. W., van der Woude, F.: see Ref. 6, p. C6–385

[8.4.22] Perlow, G. J.: in Perspectives in Mössbauer Spectroscopy, S. G. Cohen, M. Pasternak (ed.). New York: Plenum Press 1973, p. 221

[8.4.23] Kistner, O. C.: Phys. Rev. 144, 1022 (1966)

[8.4.24] Gerdau, E., Steiner, P., Steenken, P.: see Ref. 17, p. 261

[8.4.25] Sauer, C., Matthias, E., Mössbauer, R. L.: Phys. Rev. Lett. 21, 961 (1968)

[8.4.26] Sauer, C.: Z. Phys. 222, 439 (1969)

[8.4.27] Binder, J., Kaindl, G., Salomon, D., Wortmann, G.: see Ref. 4, p. 184

[8.4.28] Salomon, D., Triplett, B. B., Dixon, N. S., Boolchand, P., Hanna, S.: see Ref. 4, p. 285

[8.4.29] Heidemann, A., Kaindl, G., Salomon, D., Wortmann, G.: see Ref. 4, p. 515

[8.4.30] Salomon, D., Wallner, W., West, P. J.: see Ref. 5, p. 105

[8.4.31] Kaindl, G., Salomon, D.: see Ref. 22, p. 195

[8.4.32] Agresti, D., Kankeleit, E., Persson, B.: Phys. Rev. 155, 1342 (1967)

[8.4.33] Persson, B., Blumberg, H., Agresti, D.: Phys. Lett. *24B*, 522 (1967)

[8.4.34] Persson, B., Blumberg, H., Agresti, D.: Phys. Rev. *170*, 1066 (1968)

[8.4.35] Frankel, R. B., Chow, Y., Grodzins, L., Wulff, J.: Phys. Rev. *186*, 381 (1969)

[8.4.36] Morrison, R. J., Atac, M., Debrunner, P., Frauenfelder, H.: Phys. Lett. *12*, 35 (1964)

[8.4.37] Wagner, F., Kaindl, G., Bohn, H., Biebel, U., Schaller, H., Kienle, P.: Phys. Lett. *28B*, 548 (1969)

[8.4.38] Mössbauer, R. L.: Z. Phys. *151*, 124 (1958)

[8.4.39] Owens, W. R., Robinson, B. L., Iha, S.: Phys. Rev. *185*, 1555 (1969)

[8.4.40] Iha, S., Owens, W. R., Gregory, M. C., Robinson, B. L.: Phys. Lett. *25B*, 115 (1967)

[8.4.41] Wagner, F., Kaindl, G., Kienle, P., Korner, H. J.: Z. Phys. *207*, 500 (1967)

[8.4.42] Wagner, F., Klockner, J., Korner, H. J., Schaller, H., Kienle, P.: Phys. Lett. *25B*, 253 (1967)

[8.4.43] Perlow, G. J., Henning, W., Olson, D., Goodman, G. L.: Phys. Rev. Lett. *23*, 680 (1969)

[8.4.44] Heuberger, A., Pobell, F., Kienle, P.: Z. Phys. *205*, 503 (1967)

[8.4.45] Buyrn, A. B., Grodzins, L.: Phys. Lett. *21*, 389 (1966)

[8.4.46] Buryn, A. B., Grodzins, L., Blum, N. A., Wulff, J.: Phys. Rev. *163*, 286 (1967)

[8.4.47] Atac, M., Debrunner, P., Frauenfelder, H.: Phys. Lett. *21*, 699 (1966)

[8.4.48] Nagle, D., Craig, P. P., Dash, J. G., Reiswig, R. D.: Phys. Rev. Lett. *4*, 237 (1960)

[8.4.49] Andra, H. J., Hashmi, C. M. H., Kienle, P., Stanek, F. W.: Z. Naturforsch. *18A*, 687 (1963)

[8.4.50] Speth, J., Stanek, F. W.: Z. Naturforsch. *20A*, 1175 (1965)

[8.4.51] Roberts, L. D., Thomson, J. O.: Phys. Rev. *129*, 664 (1963)

[8.4.52] Grant, R. W., Kaplan M., Keller, D. A., Shirley, D. A.: Phys. Rev. *133*, A1062 (1964)

[8.4.53] Cohen, R. L.: Phys. Rev. *171*, 343 (1968)

[8.4.54] Thompson, J. O., Hurry, P. G., Patterson, D. O., Roberts, L. D.: see Ref. 17, p. 557

[8.4.55] Barrett, P. H., Grant, R. W., Kaplan, M., Keller, D. A., Shirley, D. A.: J. Chem. Phys. *39*, 1035 (1963)

[8.4.56] Roberts, L. D., Becker, R. L., Obenshain, F. E., Thomson, J. O.: Phys. Rev. *137*, A895 (1965)

[8.4.57] Roberts, L. D., Prince, J. F., Erickson, D. J.: see Ref. 22, p. 127

[8.4.58] Borg, R. J.: see Ref. 4, p. C6–320

[8.4.59] Hurry, P. G., Tung, C. M., Obenshain, F. E., Thomson, J. O.: see Ref. 6, p. C6–371; Viegers, M. P. A., Trooster, J. M.: see Ref. 6, p. C6–293

[8.4.60] Shenoy, G. K.: see Ref. 22, p. 141

[8.4.61] Epstein, L. M.: in *Some Applications of the Mössbauer Effect*. Proceedings of the Symposium on Physics and Non-destructive Testing. San Antonio: Southwest Research Inst. 1963

[8.4.62] Genin, J. M., Flinn, P. A.: Phys. Lett. *22*, 392 (1966)

[8.4.63] Gielen, P. M., Kaplow, R.: Acta Met. *15*, 49 (1967)

[8.4.64] Christ, B. W., Giles, P. M.: Mössbauer Effect Meth. *3*, 37 (1967)

[8.4.65] Moriya, T., Ino, H., Fujita, F. E., Maeda, Y.: J. Phys. Soc. Japan *24*, 60 (1968)

[8.4.66] Genin, J. M., Flinn, P. A.: Trans. Met. Soc. AIME *242*, 1419 (1968)

[8.4.67] Mathalane, Z., Ron, M., Niedzwiedz, S.: J. Mater. Sci. *6*, 957 (1971)

[8.4.68] Lesville, M., Gielen, P. M.: Met. Trans. *3*, 2681 (1972)

[8.4.69] Chow, W. K., Kaplow, R.: Acta Met. *21*, 725 (1973)

[8.4.70] Flinn, P. A.: see Ref. 3, p. 275

[8.4.71] Fujita, F. E.: see Ref. 2, p. F-2

[8.4.72] Genin, J. M., le Caer, G., Simon, A.: see Ref. 3, p. 318

[8.4.73] Williamson, D. L., Keune, W.: see Ref. 5, p. 133

[8.4.74] le Caer, G., Simon, A., Lorenzo, A., Genin, J. M.: Phys. Stat. Sol. *A6*, K97 (1971)

[8.4.75] Borman, H., Campbell, I. A., Fruchart, R.: J. Phys. Chem. Sol. *28*, 17 (1967)

[8.4.76] Ino, H., Moriya, T., Fujita, F. E., Maeda, Y., Cho, Y., Yonokati, Y.: J. Phys. Soc. Jap. *25*, 88 (1968)

[8.4.77] Foyt, J., Genin, J. M.: Compt. Rend. *C270*, 2563 (1970)

[8.4.78] Abe, N.: Ph. D. Thesis, North-Western Univ. (1972)

[8.4.79] Marcus, H., Schwartz, L. H., Fine, M. E.: Tans. ASM *59*, 468 (1966)

[8.4.80] Kocher, C. W.: Phys. Lett. *14*, 187 (1965)

[8.4.81] Gonser, U., Meechan, C. J., Muir, A. H., Wiedersich, H.: J. Appl. Phys. *34*, 2373 (1963); Kenne, W., Halbauer, R., Gonser, U., Lauer, J., Williamson, D.: J. Appl. Phys. *48*, 2976 (1977)

[8.4.82] Crowell, J. M., Walker, J. C.: see Ref. 3, p. 289

[8.4.83] Shull, C. G., Wilkinson, M. E.: Phys. Rev. *97*, 304 (1955)

[8.4.84] Johnson, C. E., Ridout, M. S., Cranshaw, T. E.: Proc. Phys. Soc. *81*, 1079 (1963)

[8.4.85] Ettwig, H.-H., Pepperhoff, W.: Arch. Eisenhüttenwesen *41* (5), 471 (1970)

[8.4.86] Wertheim, G. K., Jaccarino, V., Wernick, J. H., Buchanan, D. N. E.: Phys. Rev. Lett. *12*, 24 (1964)

[8.4.87] Bernas, H., Campbell, I. A.: Sol. Stat. Commun. *4*, 577 (1966)

[8.4.88] Marcus, H. L., Schwartz, L. H.: Phys. Rev. *162*, 259 (1967)

[8.4.89] Collins, M. F., Low, G. G.: Proc. Phys. Soc. *86*, 535 (1965)

[8.4.90] Hüfner, S.: Phys. Rev. *B1*, 2348 (1970)

[8.4.91] Stearns, M. B.: Phys. Rev. *129*, 1136 (1963)

[8.4.92] Johnson, C. E.: Proc. Intern. Conf. Hyperfine Interactions Detected by Nucl. Radiation, Rehovot-Jerusalem, 1970

[8.4.93] Kündig, W.: Nucl. Instr. Meth. *48*, 219 (1967)

[8.4.94] Gabriel, J. R.: in *Mössbauer Effect Methodology*, Vol. 1, I. J. Gruverman (ed.). New York: Plenum Press 1965, p. 121

[8.4.95] Bruch, T. J., Budnik, J. I., Skalski, S.: Phys. Rev. Lett. *22*, 846 (1969)

[8.4.96] Zinn, W., Kalvius, G. M., Kankeleit, E., Kiehle, P., Wiedemann, W.: J. Phys. Chem. Solids *24*, 993 (1963)

[8.4.97] Erich, U., Kankeleit, E., Prange, H., Hüfner, S.: J. Appl. Phys. *40*, 1491 (1969)

[8.4.98] Gros, Y., Paulève, J.: J. Phys. *31*, 459 (1970)

[8.4.99] Shine, G., Chen, C.W., Flinn, P. A., Nathans, T.: Phys. Rev. *131*, 183 (1963)

[8.4.100] Campbell, J. A.: J. Phys. *C2*, 1338 (1969)

[8.4.101] Overhauser, R. W., Stearns, M. B.: Phys. Rev. Lett. *13*, 316 (1964)

[8.4.102] Cohen, R. L., West, K. W.: Proc. IV. Int. Conf. Magnetism, Grenoble, France 1970

[8.4.103] Carlow, J. S., Meads, R. E.: J. Phys. *C2*, 2120 (1969)

[8.4.104] Jannarella, L., Wagner, F. E., Wagner, U., Danon, J.: see Ref. 4, p. C6−517

[8.4.105] Wortmann, G.: see Ref. 6, p. C6−333

[8.4.106] Erich, U., Göring, J., Hüfner, S., Kankeleit, E.: Phys. Lett. *31A*, 492 (1970)

[8.4.107] Fernando, W. H., Segnan, R., Schendler, A. I.: in: Proceedings of 15th Annual Conf. on Magnetism and Magnetic Materials, Philadelphia, 1969

[8.4.108] Hüfner, S., Steiner, P.: see Ref. 22, p. 1

[8.4.109] Steiner, P., Gumprecht, D., Zdrojewski, W. V., Hüfner, S.: see Ref. 4, p. C6−523

[8.4.110] Scherg, M., Seidel, E. R., Litterst, F. J., Gierisch, W., Kalvius, G. M.: see Ref. 4, p. C6 to 527

[8.4.111] Maletta, H., Mössbauer, R. L.: Solid State Comm. *8*, 143 (1970)

[8.4.112] Crangle, J., Scott, W. R.: J. Appl. Phys. *36*, 921 (1965)

[8.4.113] Ericsson, T., Hirvonen, M. T., Katila, T. E., Typpi, V. K.: Sold State Comm. *8*, 765 (1970)

[8.4.114] Maley, M. P., Taylor, R. D., Thompson, J. L.: J. Appl. Phys. *38*, 1249 (1967)

[8.4.115] Kitchens, T. A., Steyert, W. A., Taylor, R. D.: Phys. Rev. *138*, 467 (1965)

[8.4.116] Nussbaum, R. H., Howard, D. G., Ness, W. L., Steen, C. F.: Phys. Rev. *173*, 653 (1968)

[8.4.117] Palaith, D., Kimball, C. W., Preston, R. S., Crangle, J.: Phys. Rev. *178*, 795 (1969)

[8.4.118] Holzapfel, W. B., Cohen, J. A., Drickamer, H. G.: Phys. Rev. *187*, 657 (1969)

[8.4.119] Palaith, D., Kimball, C. W., Preston, R. S., Crangle, J.: Phys. Rev. *178*, 795 (1969)

[8.4.120] see for example Ref. 6, session *Radiation Damage and Defect Structure;* Wertheim, G. K., Hausmann, A., Sander, W.: in *Defects in Crystalline Solids*, S. Amelinckx, R. Gevers, I. Nihoul (eds.), Vol. 4. Amsterdam: North-Holland Publ. Comp. 1971

[8.4.121] Czjzek, G., Berger, W. G.: Phys. Rev. *1*, 957 (1970)

[8.4.122] Dekhtyar, I. Y., Nizin, P. S., Fedchenko, R. G.: Fiz. Metal metallowed *17*, 431 (1969)

[8.4.123] Levis, S. J., Flinn, P. A.: Phil. Mag. *26*, 977 (1972)

[8.4.124] Gonser, U., Grant, R. W., Muir, A. H., Wiedersich, H.: Acta Met. *14*, 259 (1966)

[8.4.125] Knapp, M., Li, K., Philbrook, W. O.: Baikan Institute Intern. Symposium Acad. of Sciences, USSR, Moscow (1971)

[8.4.126] Knapp, M.: Carnegie Technical *37*, 4 (1972)

[8.4.127] Jones, R. D.: Iron and Steel *46*, 33 and 137 (1973)

[8.4.128] Gosselin, J. P., Shimony, U., Grodzins, L., Cooper, A. R.: Phys. Chem. Glasses *8*, 56 (1967)

[8.4.129] Kurkjian, C. R., Sigety, E. A.: Phys. Chem. Glasses *9*, 73 (1968)

[8.4.130] Collins, D. W., Mulay, L. N.: J. Amer. Ceram. Soc. *54*, 69 (1971)

[8.4.131] Pargamin, L., Lupis, C. H. P., Flinn, P. A.: Met. Trans. *3*, 2093 (1972)

[8.4.132] Rossiter, M. J., Hodgson, A. E. M.: J. Inorg. and Nucl. Chem. *27*, 63 (1965)

[8.4.133] Terrell, H. J., Spijkerman, J. J.: Appl. Phys. Lett. *13*, 11 (1968)

[8.4.134] Pritchard, A. M., Mould, B. T.: Corros. Sci. *11*, 1 (1971)

[8.4.135] Pritchard, A. M., Haddon, J. R., Walton, G. N.: Corros. Sci. *11*, 11 (1971)

[8.4.136] Dézsi, I., Vértes, A., Kiss, L.: J. Radioanal. Chem. *2*, 183 (1969)

[8.4.137] Meisel, W.: Z. Chem. *11*, 238 (1971)

[8.4.138] Pritchard, A. M., Dobson, C. M.: Nature *224*, 1295 (1969)

[8.4.139] Johnson, D. P.: Solid State Commun. *7*, 1785 (1969)

[8.4.140] Hrynkiewicz, H. U.: Phys. Stat. Sol. *9*, 611 (1972)

[8.4.141] Romanov, V. P., Checkerskaya, L. F.: Phys. Stat. Sol. *49*, K183 (1972)

[8.4.142] Checkerskaya, L. F.: Phys. Stat. Sol. *19*, K177 (1973)

[8.4.143] Valov, P. M., Vasilev, Ya. V., Veriovkin, G. V., Kaplin, D. F.: J. Solid State Com. *1*, 215 (1970)

[8.4.144] Schwartz, L. H.: Int. J. Nondestr. Test. *1*, 353 (1970)

[8.4.145] Chow, H. K., Welse, R. F., Flinn, P. A.: in *Mössbauer Effect Spectrometry for Analysis of Iron Compounds*, USAEC Report-NSEC-4023-1 (1972)

[8.4.146] Sette Camara, A., Keune, W.: Corros. Sci. *15*, 441 (1975)

[8.4.147] Pritchard, R. M., Polson, C. M.: Nature *224*, 1295 (1969)

[8.4.148] Meisel, W.: Werkstoffe und Korrosion *21*, 249 (1970)

[8.4.149] Terrell, J. H., Forsyth, R. H., Naiman, C. S.: Techn. Rep. of Mithras (1967), Sounders Ass. Inc., 701 Concord Ave., Cambridge, Mass. 02138

[8.4.150] Dumesic, J. A.: see Ref. 6, p. C6–233

[8.4.151] Collins, R. L.: in *Mössbauer Effect Methodology 4*, 129 (1968)

[8.4.152] Swanson, K. R., Spijkerman, J. J.: J. Appl. Phys. *41*, 3155 (1970)

[8.4.153] Keisch, B.: Nucl. Instr. Meth. *104*, 237 (1972)

[8.4.154] Flinn, P. A.: in *Isotopes Development Program Research and Development: 1972*. USAEC Report WASH 1220 (1973), p. 92

[8.4.155] Ord, R. N.: Appl. Phys. Lett. *15*, 279 (1969)

[8.4.156] Stöckler, H. A.: J. Appl. Phys. *41*, 825 (1970)

[8.4.157] Bonchev, Z. N., Jordanov, A., Minkova, A.: Nucl. Instr. Meth. *70*, 36 (1969)

[8.4.158] Lee, E. L., Bolduc, P. E., Violet, C. E.: Phys. Rev. Lett. *13*, 800 (1964)

[8.4.159] Violet, C. E., Lee, E. L.: in *Mössbauer Effect Methodology 2*, 171 (1966)

[8.4.160] Zinn, W., Kalvius, M., Kankeleit, E., Kienle, P., Wiedemann, W.: J. Phys. Chem. Solids *24*, 993 (1963)

[8.4.161] Rabinowitch, K., Shechter, H.: J. Appl. Phys. *39*, 2464 (1968)

[8.4.162] Shimany, V., Rabinowitch, K., Biren, A.: J. Appl. Phys. *41*, 641 (1970)

[8.4.163] Swanson, K. R., Spijkerman, J. J.: J. Appl. Phys. *41*, 3155 (1970)

[8.4.164] Trousdale, W. L., Lindgreen: R. A.: J. Appl. Phys. *36*, 968 (1965)

[8.4.165] Keune, W., Gonser, U.: in preparation

[8.4.166] Elius, D. J., Linnett, J. W.: Transact. Farad. Soc. *65*, 562 (1969)

[8.4.167] Gonser, U., Grant, R. W., Muir, A. H., Wiedersich, H.: Acta Met. *14*, 259 (1966)

[8.4.168] Singwi, K. S., Sjölander, A.: Phys. Rev. *120*, 1093 (1960)

[8.4.169] Knauer, R. C., Mullen, J. G.: Phys. Rev. *174*, 711 (1968)

[8.4.170] Knauer, R. C., Mullen, J. G.: Appl. Phys. Lett. *13*, 4, 150 (1968)

[8.4.171] Mullen, J. G., Knauer, R. C.: in *Mössbauer Effect Methodology*, Vol. 5, I. J. Gruverman (ed.). New York: Plenum Press 1970

[8.4.172] Lewis, S. J., Flinn, P. A.: App. Phys. Lett. *15*, 10, 331 (1969)

[8.4.173] Rubinstein, M.: Solid State Commun. *8*, 919 (1970)

[8.4.174] Window, B.: see Ref. 2, p. F-6

[8.4.175] Frackowiak, J. E., Jankovski, B. M., Panek, T. J.: see Ref. 3, p. 295

[8.4.176] Jankowski, B. M., Frackowiak, J. E., Konsy, J., Moron, J. W., Panek, T. J.: see Ref. 6, p. 137

[8.4.177] Pollak, A., de Coster, H., Amelinckx, S.: Proceedings of the Second International Conference on the Mössbauer Effect, Saclay. France, 1961, D. M. I. Compton, A. H. Schoen, (eds.). New York: John Wiley 1962, p. 298

[8.4.178] Ulrich, D. L., Wilson, J. M., Resch, W. A.: Phys. Rev. Lett. *24*, 8, 355 (1970)

[8.4.179] Krivoglaz, M. A., Repetskiy, S. P.: Fiz. metal. metallowed *32*, 899 (1971)

[8.4.180] Dibar-Ure, M. C.: Ph. D. Thesis, Carnegie-Mellon Univ. (1973)

[8.4.181] Ron, M., Harnstein, F.: see Ref. 4, p. C6−505

[8.4.182] Pfannes, H. D., Gonser, U.: Appl. Phys. *1*, 93 (1973)

[8.4.183] Gonser, U., Ron, M., Ruppersberg, H., Keune, W., Trautwein, A.: Phys. Stat. Sol. *10*, 493 (1972)

[8.4.184] Tsnei, C. C., Longworth, G., Lin, S. C. H.: Phys. Rev. *170*, 603 (1968)

[8.4.185] Lin, S. C. H., Ducvez, P.: Phys. Stat. Sol. *34*, 469 (1969)

[8.4.186] Ura, M.: The Structure and Stability of Amorphous Fe-P-C Alloys, Ph. D.-Thesis, Osaka Univ. (1974)

[8.4.187] Chien, C. L., Hasegawa, R.: see Ref. 6, p. C6−759

[8.4.188] Marchal, G., Mangin, Ph., Piecuch, M., Janot, Chr.: see Ref. 6, p. C6−763

[8.4.189] Schaafsma, A. S., van der Woude, F.: see Ref. 6, p. C6−769

[8.4.190] Arrese-Boggiano, R., Chappert, J., Coey, J. M. D., Liénard, A., Rebouillat, J. P.: see Ref. 6, p. C6−771

[8.4.191] Gonser, U., Ghafari, M., Wagner, H. G.: see Ref. 7

[8.4.192] Valiev, H. H., Daldina, K. A., Kuzmin, R. N.: see Ref. 5, p. 87, Vol. 1

[8.4.193] Wallace, W. E., Ilyushin, A., Lopez, D.: see Ref. 5, p. 97, Vol. 2

[8.4.194] Baggio-Saitovitch, E., Butz, T., Vasquez, A., Vincze, I., Wagner, F. E., Endo, K.: see Ref. 6, p. C6−417

[8.4.195] Window, B.: Phil. Mag. *26*, 681 (1972)

[8.4.196] Nasu, S., Murakami, Y., Nakamura, Y., Shinjo, T.: Scripta Met. *2*, 647 (1968)

[8.4.197] Lagunov, V. A., Polozenko, V. I., Stepanov, V. A.: Sov. Phys. Sol. State *11*, 191 (1969)

[8.4.198] Brümmer, O., Dräger, G., Katzer, D.: see Ref. 3, p. 306

[8.4.199] Fritzsch, E., Schneider, M., Pietzsch, C., Dens, P.: see Ref. 5, p. 119

[8.4.200] Kjeldgaard, J., Trumpy, G., Thrane, N., Mørup, S.: see Ref. 5, p. 127

[8.4.201] Kim, K. J., Schwartz, L. H.: see Ref. 6, p. C6-405

8.5. Aftereffects of Nuclear Transformations

Chemical aftereffects of nuclear transformations in solids have been reviewed by several authors. The reader not familiar with the subject will gain a clear insight into the field by consulting the articles, for example, of [8.5.1−5].

Atoms which are formed in the process of a nuclear transformation take up kinetic energy, the recoil energy imparted to the nucleogenic atom due to particle emission (α, β, protons, neutrons, Auger electrons) and γ deexcitation, and may exist in an electronically excited or ionized state for some length of time which strongly depends on the nature of the surroundings. Energetic nucleogenic atoms are called *"hot atoms"*, and all the changes in properties of the electronic configuration (like the nature and number of bonds, oxidation state, orbital and spin state, energy state, etc.) as compared to those of the decaying parent atom are termed *"chemical aftereffects"* associated with the nuclear transformation; the whole field of research dealing with such effects is called *"hot atom chemistry"*.

For quite a long period, up to the early sixties, research work in hot atom chemistry of the solid state had to rely entirely on wet-analytical methods to find out how much retention and what kind of primary and secondary aftereffect species had been formed in the course of a nuclear transformation process. Undoubtedly, this was too severe a disadvantage, because it would never allow the hot atom chemist, who is interested in solid aftereffects, to determine the nature and lifetime of short-lived transient species due to immediate reactions upon the generally practiced dissolution in a solvent.

Fortunately, with the discovery of the Mössbauer effect, a physical method developed which seemed to be ideally suited to probe the chemical state of a nucleogenic atom in the solid state, at least for a period of the mean lifetime of the excited Mössbauer state (which is generally on the order of some ns to a few hundred ns) without disturbances by chemical treatments. In a "normal" Mössbauer experiment one uses a single-line source (the Mössbauer parent nuclide embedded in a perfectly cubic paramagnetic host lattice to avoid electric quadrupole and magnetic hyperfine splitting) to study the sample of interest as absorber (scatterer) in a transmission (backscattering) experiment; one describes this as *Mössbauer absorption spectroscopy*. The application of the Mössbauer effect to study aftereffects of nuclear transformations, however, means analyzing the hyperfine interactions occurring in the source. This simply involves preparing the systems of interest incorporating the Mössbauer parent nuclide by any of the usual preparative methods (e.g., classical synthesis of a labelled compound, ion exchange, diffusion into a metallic system, implantation using an ion accelerator) and using this as the Mössbauer source versus a single-line absorber; one speaks of *Mössbauer emission spectroscopy* in this case.

In Mössbauer emission spectroscopy only species with lifetimes longer than or on the order of the lifetimes of the excited Mössbauer nuclear state can be detected, and no measurement can be made of events taking place much faster ($\leqslant 10^{-10}$s), e.g., recombination and rearrangement processes in the electronic shells, or of reactions initiated after the Mössbauer nuclear state has decayed. Nevertheless, Mössbauer emission spectroscopy has helped far more than any other physical method (like perturbed angular correlation, positron annihilation, muonic experiments, etc.) to learn about the aftereffects associated with nuclear transformations.

Most of the work in this area has been done on ^{57}Co labelled systems to study the aftereffects of the EC decay of ^{57}Co into ^{57}Fe in various solid surroundings, mostly insulators of the type of ionic and covalent compounds (see, for example, [8.5.7–16] and references cited therein). Many investigations have also been carried out with systems of other Mössbauer isotopes, e.g., of 4d and 5d transition elements like ^{99}Ru, ^{193}Ir, and ^{197}Au by recording the emission spectra of sources of ^{99}Rh, ^{193}Os, ^{197}Pt, and ^{197}Hg incorporated in various compounds [8.5.17, 18]; systems of ^{119}Sn (e.g., [8.5.19–23]); systems incorporating rare earth Mössbauer isotopes (e.g., [8.5.24]); and the actinide isotope ^{241}Am, α decaying into the Mössbauer isotope ^{237}Np (e.g., [8.5.25–27]).

Mössbauer emission spectra have also been recorded of compounds of [129]I decaying into Xe compounds [8.5.28] and of RbX compounds (X = F, Cl, Br, OH) incorporating [83]Rb which decays into [83]Kr [8.5.29], and most recently with [83m]Kr implanted into Al foils to study lattice defects [8.5.30]. A comparison between emission spectra obtained with sources prepared by ion implantation on the one hand with sources prepared by labelling through classical synthesis on the other hand allows additional effects to be deduced, e.g., radiation damage to the lattice caused by the ion beam with some 50–100 keV kinetic energy (see, e.g., [8.5.31]). Chemical and physical effects associated with the neutron activation of [56]Fe to yield the Mössbauer level of [57]Fe via $^{56}Fe(n, \gamma)^{57}Fe* \xrightarrow{\gamma} {}^{57}Fe$ (14.4 keV) have been studied on line in the neutron flux of a nuclear reactor, where the Mössbauer emission spectra have been recorded $\sim 10^{-7}$s after the nuclear reaction, which is eventually short enough to be sensitive to short-lived transient species due to the nuclear recoil (for example, cf. [8.5.32, 33]).

Insulators of the type of ionic and covalent compounds, purely inorganic in nature as well as metal organic compounds, have been by far the most extensively studied systems in hot atom chemistry using Mössbauer emission spectroscopy. Metallic systems have not been as attractive, because the great abundance and mobility of conduction electrons permits fast neutralization and rearrangement processes following the nuclear decay in times much shorter ($\leqslant 10^{-15}$s) than the Mössbauer time scale. Chemically interesting aftereffects are therefore hardly observable under these circumstances.

The kind of information extractable from Mössbauer emission spectra refers to properties like
- oxidation state
- spin state
- lattice site
- local symmetry
- nature of bonding
- dynamical behaviour
of the nucleogenic atom. Perhaps the most frequently observed aftereffect is a change of oxidation state with respect to the parent atom. In hydrates of [57]Co labelled cobalt (II) compounds, for instance, in which water molecules are directly coordinated to the metal atom, a preferred production of ferric species is observed, whose proportion increases with the number of coordinated water molecules (see, e.g., [8.5.34–36]). Fig. 8.5.1 demonstrates clearly this effect. Other examples for the observation of iron species with higher oxidation states than that of the parent [57]Co ions have been described in [8.5.7]. Triftshäuser et al. [8.5.37, 38] have demonstrated by time-delayed coincidence measurements that the relative intensity of the aliovalent charge state did not vary in time from 5×10^{-9}s to 3×10^{-7}s and thus did not exhibit charge relaxation on a time scale comparable to the mean lifetime on the Mössbauer nuclear level. Resonance lines arising from aliovalent iron species were also detected in the [57]Fe Mössbauer emission spectra of [57]Co labelled cobaltic compounds, where iron appeared in the reduced Fe^{2+} state (see for example [8.5.10, 39, 40]). These findings have led the authors to the conclusion that the nature of the iron species formed by EC decay of [57]Co in complex compounds of cobalt might strongly be determined by the redoxproperties of free radicals resulting from the autoradiolysis of the coordi-

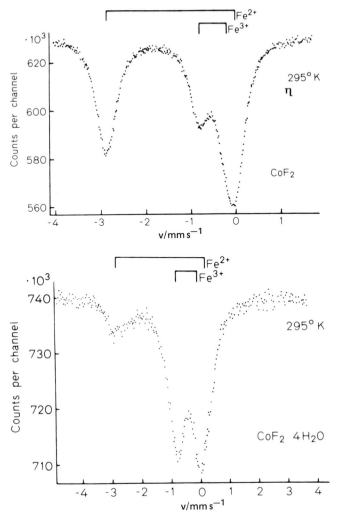

Fig. 8.5.1. ^{57}Fe Mössbauer emission spectra of ^{57}Co laballed CoF_2 and $CoF_2 \cdot 4H_2O$ at 295 K versus $Na_4[Fe(CN)_6] \cdot 10H_2O$ as absorber (from [8.5.34])

nated ligand by Auger electrons or x-rays emitted by the decaying atom. Water molecules as ligands have been suggested to radiolyze into hydrogen atoms and OH radicals, which have a strong oxidizing power against Fe^{2+}. This agrees well with results obtained from ^{57}Fe Mössbauer absorption measurements on hydrated ferrous salts after external irradiation with γ-rays [8.5.41, 42]. Organic ligands such as oxalate, acetylacetonate, citrate, etc. [8.5.39.10, 40], tend to radiolyze upon ^{57}Co decay into radicals with reducing behaviour against Fe^{3+}. In systems with radioresisting organic ligands like phthalocyanine or vitamin B12 no aliovalent iron species was observed in the ^{57}Fe emission spectra [8.5.39, 10].

Recently anomalous spin states of iron(II) have been detected in the emission spectra of various ^{57}Co labelled cobalt(II) diimine complex compounds [8.5.11–15].

"Anomalous" spin state in this context means that the iron species resulting from the parent ^{57}Co atom appears in the same oxidation state as the parent atom, but the spin state is different from that in the corresponding iron(II) compound at the same temperature. It is well accepted that the nucleogenic ^{57}Fe atom occupies the same lattice site as the parent ^{57}Co atom, because the recoil energies resulting from the neutrino emission ($E_R = 3.4$ eV) and from γ-deexcitation processes ($E_R < 1$ eV) are much smaller than the displacement energy (ca. 25 eV) necessary for the nucleogenic ^{57}Fe atom to be removed from the orginal ^{57}Co lattice site. Thus the chemical nature of the ligand sphere is the same around the metal atom in the quasioctahedral source complex compound [^{57}CoN$_{6-x}$N$'_x$] and the corresponding absorber complex compound [^{57}FeN$_{6-x}$N$'_x$]. (N$_{6-x}$N$'_x$ stands for the six coordinating nitrogen atoms belonging either to bifunctional diimine ligand molecules or monofunctional ligands). For example, Fig. 8.5.2 shows variable temperature ^{57}Fe emission spectra of

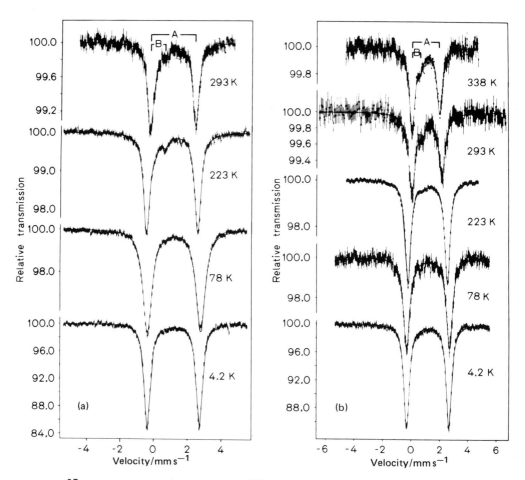

Fig. 8.5.2. ^{57}Fe Mössbauer emission spectra of (a) [^{57}Co(phen)$_2$(NCS)$_2$] and (b) [^{57}Co(bipy)$_2$(NCS)$_2$] at various temperatures versus K$_4$[Fe(CN)$_6$] · 3H$_2$O as absorber (298 K). Quadrupole doublet A is characteristic of high-spin iron(II), quadrupole doublet B arises from a high-spin iron(III) species (from [8.5.12])

[^{57}Co(phen)$_2$(NCS)$_2$] (a) and [^{57}Co(bipy)$_2$(NCS)$_2$] (b) (phen = 1.10-phenanthroline; bipy = α,α′-dipyridyl) versus K$_4$[Fe(CN)$_6$] · 3H$_2$O as absorber (298 K) [8.5.12]. Except for a small fraction of a high-spin iron(III) species (quadrupole doublet B) only observable in the high-temperature region and most probably arising from an

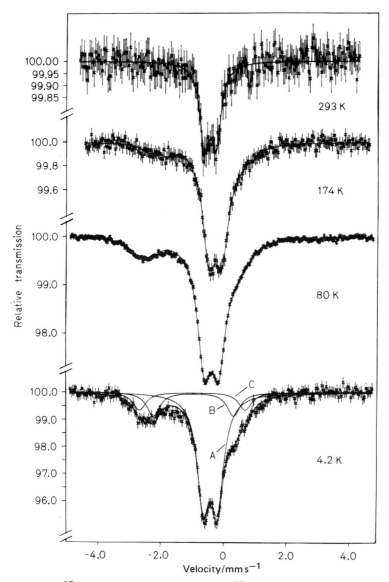

Fig. 8.5.3. ^{57}Fe Mössbauer emission spectra of [^{57}Co(pmi)$_3$] · (ClO$_4$)$_2$ (pmi = 2-pyridinalmethyl-imine) at various temperatures versus K$_4$[Fe(CN)$_6$] · 3H$_2$O (298 K) as absorber. Doublet A arises from the "normal" low-spin (^{1}A$_1$) state of nucleogenic iron(II). Doublets B and C are characteristic of high-spin iron (II) with B being consistent with the ^{5}E state and C with the ^{5}A$_1$ state under D$_3$ local symmetry (from [8.5.14])

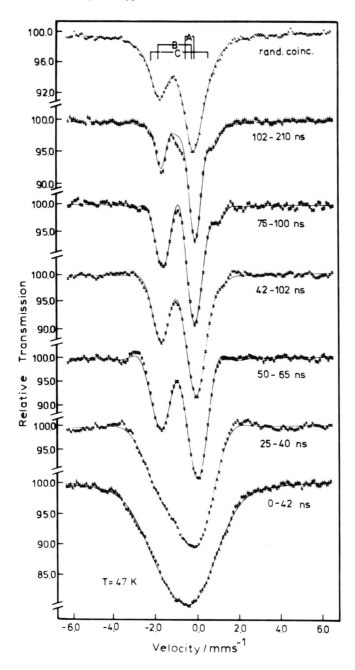

Fig. 8.5.4. Time-differential Mössbauer emission spectra of $[^{57}Co(phen)_3]$ $(ClO_4)_2 \cdot 2H_2O$ versus enriched $K_4[Fe(CN)_6] \cdot 3H_2O$ at $T = 47$ K. The spectra have been corrected for random coincidences. The solid lines are theoretical spectra obtained by fitting time filtering theory to the data points. Finite experimental time resolution and the cosine smearing effect are also included. A: low-spin iron(II); B, C: high-spin iron(II) (taken from [8.5.16])

autoradiolysis effect as mentioned above, the low-temperature spectra exhibit exclusively a high-spin iron(II) quadrupole doublet (A), which is characteristic of the $^5T_{2g}$ substates. On the other hand, the ^{57}Fe Mössbauer absorption spectra of the corresponding iron(II) complex compounds show explicitly the well-established temperature-dependent high-spin ($^5T_2$) $\rightleftharpoons$ low-spin ($^1A_1$) transition as demonstrated in Fig. 6.18. Very similar behaviour was observed for the systems $[^{57}Co(2\text{-}X\text{-}phen)_3] (ClO_4)_2$ (X = CH_3, CH_3O) [8.5.13, 15], whose corresponding iron(II) complex compounds studied in a ^{57}Fe Mössbauer absorption experiment also show a temperature-dependent $^5T_2 \rightleftharpoons {}^1A_1$ transition [8.5.43, 45].

Anomalous spin states have also been seen in the Mössbauer emission spectra of $[^{57}Co(phen)_3] (ClO_4)_2 \cdot 2H_2O$ [8.5.11], $[^{57}Co(phen)_2(CN)_2] \cdot 2H_2O$ [8.5.14] and $[^{57}Co(pmi)_3] (ClO_4)_2$ (pmi = 2-pyridinalmethylimine) [8.5.14], whose iron(II) analogs all are strong-field complexes and show solely low-spin iron(II) quadrupole doublets typical of the $^1A_1$ state throughout the temperature range of interest (< 300 K). The relative intensity of the "anomalous" high-spin state increases with decreasing temperature at the expense of the intensity of the "normal" low-spin state of nucleogenic iron(II); Fig. 8.5.3 demonstrates this. From a theoretical analysis of the temperature dependence of the quadrupole splitting including a ligand-field calculation and considering spin-orbit coupling as well as covalency effects, the "anomalous" high-spin iron(II) quadrupole doublets have been found to be consistent with the $^5E (D_3)$ and $^5A_1 (D_3)$ state of nucleogenic iron(II) [8.5.14]. The mechanism of their formation after the EC decay of ^{57}Co has been discussed in [8.5.16].

The lifetime of such anomalous spin states could be measured recently by Grimm et al. [8.5.16] in temperature-dependent delayed coincidence Mössbauer emission

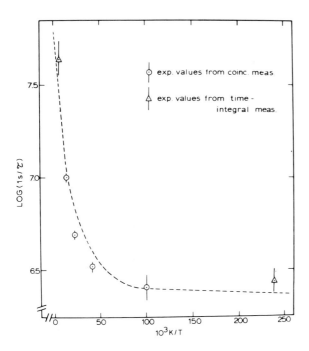

Fig. 8.5.5. Logarithm of $1s/\tau$ (τ = mean lifetime of the anomalous high-spin states) as function of 10^3K/T. The dashed line has been found by fitting the appropriate expression for electron tunneling (as given in [8.5.16]) to the experimental data points (from [8.5.16])

experiments on the system $[^{57}Co(phen)_3](ClO_4)_2 \cdot 2H_2O$ [8.5.16]. Some representative time-differential Mössbauer emission spectra obtained by these authors are shown in Fig. 8.5.4. From these spectra, data for the temperature dependence of the mean lifetime τ were deduced; the results are shown in Fig. 8.5.5 as a plot of log $(1/\tau)$ vs $1/T$. A theoretical function fitting equally well to the data points could be found by assuming two different processes as being rate determining for the decay of the anomalous spin quintet states: (i) electron tunneling to an organic ligand radical formed by autoradiolysis and (ii) radiationless decay via spin-lattice interaction. Both the formation and the decay of the anomalous spin quintet states have been discussed at length in [8.5.16].

References

[8.5.1] Harbottle, G.: Hot Atom Chemistry of the Solid State in *Chemistry Research and Chemical Techniques Based on Research Reactors*. Vienna: IAEA 1963

[8.5.2] Harbottle, G.: Ann. Rev. Nucl. Sci. *15*, 89 (1965)

[8.5.3] Müller, H.: Angew. Chem. *79*, 128 (1967)

[8.5.4] Stchouskoy, T.: Rev. Chim. Min. *5*, 683 (1968)

[8.5.5] Maddock, A. G., Wolfgang, R.: The Chemical Effects of Nuclear Transformations in: *Nuclear Chemistry*, L. Yaffe (ed.), Vol. II. New York: Academic Press 1968, p. 185

[8.5.6] Stöcklin, G.: *Chemische Folgereaktionen von Kernumwandlungen in anorganischen Festkörpern* in: *Chemie heißer Atome*, Ch. 10 (Verlag Chemie, Weinheim 1969)

[8.5.7] Wickman, H. H., Wertheim, G. K.: Spin Relaxation in Solids and Aftereffects of Nuclear Transformations, in *Chemical Applications of Mössbauer Spectroscopy*, V. I. Goldanskii, R. H. Herber (eds.). New York: Academic Press 1968, p. 604

[8.5.8] Friedt, J. M., Danon, J.: Radiochim. Acta *17*, 173 (1972)

[8.5.9] Adloff, J. P., Friedt, J. M.: Mössbauer Studies of the Chemical Effects of Nuclear Transformation, in *Mössbauer Spectroscopy. Its Applications*, Proc. Panel 1971. Vienna: IAEA 1972, p. 301

[8.5.10] Nath, A., Klein, M. P., Kündig, W., Lichtenstein, D.: Radiation Effects, Vol. 2. Glasgow: Gordon and Breach Science Publ. 1970, p. 221

[8.5.11] Ensling, J., Fitzsimmons, B. W., Gütlich, P., Hasselbach, K. M.: Angew. Chem. Intern. Ed. *9*, 637 (1970)

[8.5.12] Ensling, J., Gütlich, P., Hasselbach, K. M., Fitzsimmons, R. W.: Chem. Phys. Lett. *42*, 232 (1976)

[8.5.13] Fleisch, J., Gütlich, P.: Chem. Phys. Lett. *42*, 237 (1976)

[8.5.14] Ensling, J., Fleisch, J., Gütlich, P., Fitzsimmons, B. W.: Chem. Phys. Lett. *45*, 22 (1977)

[8.5.15] Fleisch, J., Gütlich, P.: Chem. Phys. Lett. *45*, 29 (1977)

[8.5.16] Grimm, R., Gütlich, P., Kankeleit, E., Link, R.: J. Chem. Phys. *67*, 5491 (1977)

[8.5.17] Zahn, U., Potzel, W., Wagner, F. E.: Consequences of Nuclear Transformations in Chemical Compounds Studied by the Mössbauer Method, in *Perspectives in Mössbauer Spectroscopy*, S. G. Cohen, M. Pasternak (eds.). New York: Plenum Press 1973, p. 55

[8.5.18] Rother, P., Wagner, F., Zahn, U.: Radiochim. Acta *11*, 203 (1961)

[8.5.19] Sano, H., Kanno, M.: Chem. Commun., 601 (1969)

[8.5.20] Llabador, Y., Friedt, J. M.: Chem. Phys. Lett. *8*, 592 (1971)

[8.5.21] Ambe, S., Ambe, F.: Radiochim. Acta *20*, 141 (1973)

[8.5.22] Ambe, S., Ambe, F., Shoji, H., Saito, N.: J. Chem. Phys. *60*, 3773 (1974)

[8.5.23] Ambe, F., Ambe, S.: Bull. Chem. Soc. Jap. *47*, 2875 (1974)

[8.5.24] Glentworth, P., Nichols, A. L., Large, N. R., Bullock, R. J.: J. C. S. Dalton *21*, 2364 (1973)

[8.5.25] Pillinger, W. J., Stone, A. J.: in *Mössbauer Effect Methodology*, I. J. Gruverman (ed.).,
Vol. 4. New York: Plenum Press 1968, p. 217

[8.5.26] Dunlap, B. D., Kalvius, G. M., Ruby, S. L., Brodsky, M. B., Cohen, D.: Phys. Rev. *171*,
316 (1968)

[8.5.27] Gal, J., Hadari, Z., Yanir, E., Bauminger, E. R., Ofer, S.: J. Inorg. Nucl. Chem. *32*, 2509
(1970)

[8.5.28] Perlow, G. J.: Xenon and Iodine, in *Chemical Applications of Mössbauer Spectroscopy*,
V. I. Goldanskii, R. H. Herber (eds.). New York: Academic Press 1968, p. 378

[8.5.29] Gütlich, P., Odar, S., Walcher, D.: Z. Naturforsch. *256*, 1183 (1970)

[8.5.30] Link, R., Gütlich, P.: Nucl. Instr. Methods, in the press

[8.5.31] Fleisch, J., Gütlich, P., Mohs, E., Wolf, G. K.: Mössbauer Emission Spectroscopy on
[57]Co implanted into Iron Compounds, Proc. Int. Conf. on Mössbauer Spectroscopy,
Cracow (1975), Vol. 1, p. 217

[8.5.32] Berger, W. G., Fink, J., Obenshain, F. E.: Phys. Lett. *25A*, 466 (1967)

[8.5.33] Berger, W. G.: Z. Phys. *225*, 139 (1969)

[8.5.34] Friedt, J. M., Adloff, J. P.: C. R. Acad. Sci. Paris, C. R. Acad. Sci. Paris *264 C*, 1356
(1967)

[8.5.35] Friedt, J. M., Adloff, J. P.: C. R. Acad. Sci. Paris *268* C, 1342 (1969)

[8.5.36] Ingalls, R., Coston, C. J., de Pasquali, G., Drickamer, H. G., Pinajian, J. J.: J. Chem.
Phys. *45*, 1057 (1966)

[8.5.37] Triftshäuser, W., Craig, P. P.: Phys. Rev. *162*, 274 (1967)

[8.5.38] Triftshäuser, W., Schroeer, D.: Phys. Rev. *187*, 491 (1969)

[8.5.39] Jagannathan, R., Mathur, H. B.: Inorg. Nucl. Chem. Lett. *5*, 89 (1969)

[8.5.40] Friedt, J. M., Asch, L.: Radiochim. Acta *12*, 208 (1969)

[8.5.41] Gütlich, P., Odar, S., Fitzsimmons, B. W., Erickson, N. E.: Radiochim. Acta *10*, 147
(1968)

[8.5.42] Wertheim, G. K., Buchanan, D. N. E.: Chem. Phys. Lett. *3*, 87 (1969)

[8.5.43] König, E., Ritter, G., Spiering, H., Kremer, S., Madeja, K., Rosenkranz, A.: J. Chem.
Phys. *56*, 3139 (1972)

[8.5.44] Fleisch, J., Gütlich, P., Hasselbach, K. M., Müller, W.: Inorg. Chem. *15*, 958 (1976)

[8.5.45] Fleisch, J., Gütlich, P., Hasselbach, K. M.: Inorg. Chem. *16*, 1979 (1977)

8.6. Applications to Biology

8.6.1. General Significance of Iron in Biology

Within the transition metal series the biological applications of the Mössbauer effect
are essentially limited to iron-containing compounds. However, due to the general
significance of iron in biology, there is a wide research field open to Mössbauer spec-
troscopists: "Among those elements essential for life, iron enjoys a status of extraordinary importance. It is involved in storage and transport of oxygen, in electron
transport, in the metabolism of N_2 and H_2, in the reduction of ribotides to deoxyri-
botides (precursors of DNA), in oxidation and hydroxylation of a host of inorganic
and organic metabolites and, finally, in the decomposition of utilization of hydrogen
peroxide. In spite of its abundance in the earth's crust, the profound insolubility of
the ferric ion at neutral pH has demanded the evolution of special ligands which can
dissolve, transport and make available the element to aerobic organisms. This affords
yet another, for the most part relatively new, class of iron binding molecules the function of which is the transport of iron itself". (Cited from "Evolution of Biological
Iron Binding Centers" by I. B. Neilands [8.6.1]).

Because of the important role iron plays in Mössbauer spectroscopy as well as in biology it is no wonder that more than 500 contributions have been published in this field since the early days of γ-resonance spectroscopy in 1960. One of the first applications of Mössbauer spectroscopy to biology was the study of hemoglobin reported by Gonser, Grant and Kregdze in Appl. Phys. Lett. *3*, 189 (1963). All further contributions were summarized by L. May in the "Index of Publications in Mössbauer Spectroscopy of Biological Materials" [8.6.2], and many of them were the subject of review articles (for example by Lang, by Oosterhuis et al., and by Debrunner in "Applications of Mössbauer Spectroscopy", Vol. I, ed. by R. L. Cohen, Academic Press, N. Y. 1976, and by Johnson in "Mössbauer Spectroscopy", ed. by U. Gonser, Springer, Berlin 1975).

The limitation to iron receives a bonus in that the Mössbauer effect can selectively probe the electronic and magnetic structure of iron, which to the biologist is important especially under the aspects (i) that other methods such as electron spin resonance and paramagnetic susceptibility often cannot distinguish the different iron spin states as unambiguously as the Mössbauer effect does, and (ii) that these two alternative methods are insensitive to the diamagnetic state of iron.

8.6.2. Heme Proteins

8.6.2.1. Absorption Measurements with Frozen Solutions

From Chapter 2 we know that the key point of the Mössbauer effect is the recoilless emission and absorption of γ-rays. This has the important consequence that the Mössbauer nuclei must be bound in a solid system. The biological sample cannot be studied in aqueous solution; it must be frozen, crystallized, lyophilized or, at least, it has to be embedded in a material with very high viscosity. Because of these unphysiological conditions under which biological compounds are investigated, one may have doubts about the reliability of experimental Mössbauer results with respect to the real physiological state of a molecule. There are, however, arguments from which one can assume, that the molecular or electronic structures under certain circumstances undergo only undetectably small changes upon crystallizing or freezing of the sample. (i) The protein structure of hemoglobin in aqueous solution as inferred from x-ray and neutron small-angle scattering [8.6.3–8.6.5] is in agreement with the atomic coordinates determined by x-ray diffraction analysis of hemoglobin single crystals [8.6.6, 8.6.7]. (ii) The electronic structure of the heme-iron is not noticeably changed in going from the liquid to the solid state as exemplified by susceptibility measurements [8.6.8]. (iii) The quadrupole splittings of an deoxy-myoglobin single crystal at 77 K and of a frozen deoxymyoglobin solution at 77 K (from which the single crystal was originally grown at around 300 K) are equal within the error limits [8.6.9].

In order to conserve the native state of the sample it is very important, however, to freeze solutions as fast as possible. Slow freezing will allow the growth of relatively large crystals of ice, changing the concentration of the solution. Fast freezing probably prevents these changes which have been found to affect the Mössbauer spectra significantly [8.6.10–8.6.12].

Figure 8.6.1 shows Mössbauer spectra of a frozen solution of CO-myoglobin (mbCO) with pH 7.0 measured at 4.2 K [8.6.13], a typical candidate for the ferrous low-spin state. Curve (a) corresponds to a measurement without applied magnetic field. Curve (b) shows a spectrum taken with a magnetic field of $H_0 = 4.7$ Tesla, applied perpendicular to the γ-beam. From the analysis of curve (b) V_{zz} has been derived to be positive [8.6.13]. Both spectra have been found to be nearly temperature independent within the wide range of $4.2\ K \leqslant T \leqslant 200\ K$.

Figure 8.6.2 shows Mössbauer spectra of oxy-hemoglobin (hbO$_2$) at 4.2 K with (a) $H_0 = 0$ and (b) $H_0 = 3.0$ Tesla applied perpendicular to the γ-beam [8.6.14]. ΔE_Q of hbO$_2$ has been found to be temperature dependent, falling to about 1.95 mm s^{-1} at 195 K. From the magnetic perturbation in spectrum (b) one has concluded that V_{zz} is negative.

Spectra of deoxy-myoglobin (mb), characterized by a ferrous high-spin heme-iron are presented [8.6.15] in Fig. 8.6.3. Spectrum (a) refers to 4.2 K and $H_0 = 0$,

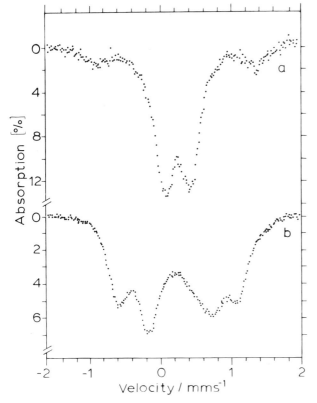

Fig. 8.6.1. Experimental Mössbauer spectra of a frozen solution of CO-myoglobin (mbCO). (a) T = 4.2 K, $H_0 = 0$; (b) T = 4.2 K, $H_0 = 4.7$ Tesla, $\vec{H}_0 \perp \vec{\gamma}$. (The additional resonances at –0.9 mm s^{-1} and +1.33 mm s^{-1} are due to oxy-myoglobin (mbO$_2$) which remained in the sample through the preparation of mbCO (taken from [8.6.13])

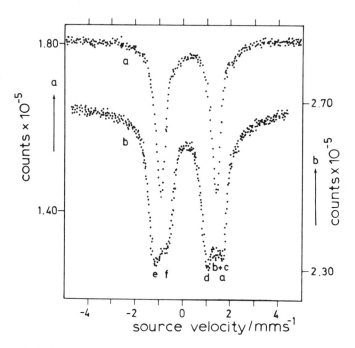

Fig. 8.6.2. Experimental Mössbauer spectra of oxy-hemoglobin (hbO$_2$). (a) T = 4.2 K, H$_o$ = 0; (b) T = 4.2 K, H$_o$ = 3.0 Tesla, $\vec{H}_o \perp \vec{\gamma}$ (taken from [8.6.14])

spectrum (b) to 4.2 K, H$_0$ = 5.0 Tesla and $\vec{H}_0 \perp \vec{\gamma}$, and spectrum (c) to 150 K, H$_0$ = 5.0 Tesla and $\vec{H}_0 \perp \vec{\gamma}$. ΔE_Q decreases to about 1.88 mm s^{-1} at 150 K. Concerning the interpretation of the magnetic perturbation of spectra (b) and (c), there exists considerable controversy in the literature [8.6.13–8.6.18]; the sign of V$_{zz}$ is still questionable. Further measurements – especially with ^{57}Fe enriched mb single crystals under applied magnetic fields – are required to decide whether V$_{zz}$ is positive or negative, and whether the z axis of the principal axes system of the EFG is oriented parallel to the heme-plane or not. An interpretation of the Mössbauer spectra in Fig. 8.6.1–3 on the basis of MO calculations (see Chap. 6) yielded the following essential results:
(i) The bond angle of Fe-C-O and Fe-O-O in CO- and O$_2$-myoglobin (and hemoglobin) appears more likely to be 135° than 180°, i.e., the so-called "Pauling-geometry" is favoured over the "Griffith-geometry" [8.6.13, 20–22].
(ii) In both CO- and O$_2$-myoglobin (and hemoglobin) the iron atom lies *in* the heme-plane [8.6.13, 20–22].
(iii) In deoxy-myoglobin (and hemoglobin) the heme-iron is pentacoordinated and significantly *out* of the heme-plane [8.6.17, 23].
(iv) A pronounced deviation from fourfold symmetry of the heme-normal in deoxy-myoglobin was found [8.6.9, 18].
(v) The electronic structure of the heme-iron in CO-, O$_2$-, and deoxymyoglobin (and hemoglobin) involves *all* iron valence orbitals (3d, 4s, 4p) [8.6.13, 17–23].

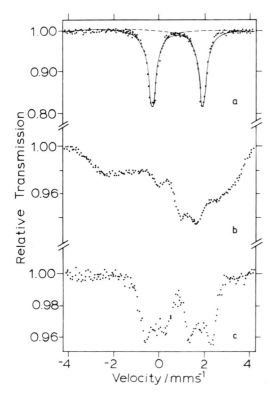

Fig. 8.6.3. Experimental Mössbauer spectra of deoxymyoglobin (mb). (a) T = 4.2 K, $H_o = \underline{0}$; (b) T = 4.2 K, $H_o = 5.0$ Tesla, $\vec{H}_o \perp \vec{\gamma}$; (c) T = 150 K, $H_o = 5.0$ Tesla, $\vec{H}_o \perp \vec{\gamma}$ (taken from [8.6.15])

8.6.2.2. Emission Spectroscopy

Emission spectroscopy uses a single line absorber (for example, stainless steel or $K_4[Fe(CN)_6] \cdot 3H_2O$) to investigate the energy levels of the ^{57}Fe nuclei produced by K-capture of ^{57}Co in the source. Information can thus be gained from an iron complex which may not be accessible by ordinary synthesis, and which may exist only during the lifetime of the 14.4 keV excited nuclear ^{57m}Fe state (ca. 10^{-7}s) after K-capture. Moreover, this technique requires far less material than the corresponding absorption method; less than 1 n mol of ^{57}Co may be sufficient to collect data in a reasonable time limit.

Examples in this field are the studies of vitamin B_{12} [8.6.24], Co-phthalocyanine [8.6.25], and a ^{57}Co labeled protoporphyrin IX complex [8.6.26]. The investigation of the latter sample in its deoxygenated and oxygenated state yielded the interesting result that the isomer shift $\delta(T)$ and the quadrupole splitting $\Delta E_Q(T)$ agree, within the normal uncertainties, with the values of deoxymyoglobin and oxy-myoglobin, respectively. This result is particularly interesting under the aspect that it is possible to handle these compounds as model compounds for heme-proteins in large computers like Univac 1108, IBM 370–168, TR 440, or CDC 3600 without implying size approximation on the molecule under study.

8.6.2.3. Absorption Measurements with Single Crystals

Characteristic of all single crystal absorption spectra is the fact that the intensity ratio of the two absorption lines depends on the orientation of the crystal with respect to the direction of the transmitted γ-beam [8.6.27]. A typical spectrum of a deoxy-myoglobin single crystal [8.6.9] is shown in Fig. 8.6.4.

Recently Zimmermann [8.6.28] has described a method for evaluating single-crystal ^{57}Fe Mössbauer spectra which exhibit quadrupole splitting. There, it is shown that the line intensities can be characterized by an intensity tensor, which is proportional to the macroscopic electric field gradient (EFG) tensor. Applying this method to myoglobin single crystals, we are concerned with the specific symmetry properties of monoclinic crystals, which lead to *manyfold* solutions for the local EFG tensors [8.6.29]. Thus, single-crystal measurements without applied magnetic field do not solve the controversy concerning the sign and the direction of the z axis of the principal axes system of the EFG in deoxymyoglobin, which was already mentioned in connection with Fig. 8.6.3.

8.6.2.4. Scattering Experiments with Single Crystals

For the determination of the electron density map of a molecule the amplitudes and phases of the waves scattered by a single crystal are required for a number of Bragg reflections. Common x-ray techniques yield the product $(A_H e^{i\varphi}H) \cdot (A_H e^{-i\varphi}H) = A_H^2$, where A_H is the amplitude and φ_H the phase of the sum of all waves scattered by the

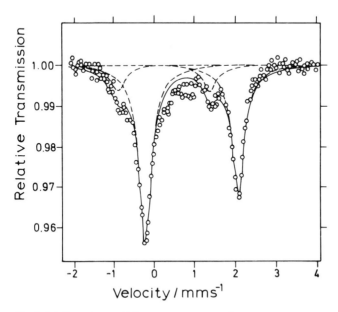

Fig. 8.6.4. Experimental Mössbauer spectrum of a mb single crystal with the b axis of the crystal being perpendicular to the γ-beam, and with $\varphi = 156°$ ($\varphi = 45°$ corresponds to a′‖$\vec{\gamma}$ and $\varphi = 135°$ to c‖$\vec{\gamma}$; a′b and c are mutually orthogonal). Side peaks originate from mbO$_2$ (taken from [8.6.9])

electrons of the molecule into the Bragg reflection H. Thus, the scattering amplitude A_H is obtained, but the phase information is lost. The solution of this phase problem for protein structure determination is based on *Perutz'* and *Kendrew's* isomorphous replacement method [8.6.30–32]. In this procedure Bragg reflections have to be measured at least three times, first on a crystal of native molecules, and then on two crystals, in which reference scatterers (for example Hg atoms) have been substituted at well-defined positions. From the difference of the measured intensities, one can calculate the relative phases without ambiguity.

A supplementary method for the isomorphous replacement method is provided by the Mössbauer scattering technique, which is based on the interference between Rayleigh scattering and γ-resonance scattering. Its application to proteins and its advantages have been described in detail by Parak et al. [8.6.33]. Essential features are as follows:

(i) The ^{57}Fe scatterer is originally built into the protein, i.e., no isomorphous replacement procedure is necessary.

(ii) The nuclear scattering amplitude theoretically amounts to about 6 times the scattering amplitude for the 80 electrons of one mercury atom, i.e., the iron nucleus as resonance scatterer is as powerful as an electronic scatterer with about 500 electrons.

(iii) The wavelength of 0.86 Å of the 14.4 keV γ-radiation of ^{57}Fe is comparable to atomic spacings in the crystal.

(iv) The radiation scattered by the ^{57}Fe nucleus can interfere with the radiation scattered by the electron shell of an atom [8.6.34, 35].

(v) For a unique phase determination, the ^{57}Fe nucleus is sufficient as a single reference scatterer compared to the twofold heavy atom substitution in the isomorphous replacement method.

In principle the necessary phases for the structure determination are derived in the following way: Moving the source with a Doppler velocity $v \gg v_r$ relative to the heme-protein single crystal destroys the nuclear resonance scattering which occurs at resonance velocity v_r by the Doppler effect. The nuclear scattering amplitude is then equal to zero, and the phase shift between the incoming and the nonresonantly scattered wave also becomes zero. Decreasing the Doppler velocity v and approaching v_r increases both the phase shift and the nuclear scattering amplitude. The phase of resonantly scattered γ-radiation ($v = v_r$) is shifted by $\frac{\pi}{2}$ with respect to the incident γ-radiation from the source. By arbitrary choice of two different Doppler velocities close to v_r, two different nuclear scattering amplitudes with known absolute value and phase can be selected as an alternative for the two heavy atom scattering amplitudes of the isomorphous replacement method [8.6.33, 36]. The construction of the phase angle φ_H with the use of the two known nuclear scattering amplitudes n_{Fe} and $\tilde{n}_{Fe}$ and the primarily known scattering amplitude $F_0(H)$, not including nuclear resonance scattering, then is analogous to the procedure for the x-ray structure analysis (Fig. 8.6.5). Several difficulties, however, as for example the extremely long counting time, the enrichment of heme-protein single crystals [8.6.37] with ^{57}Fe and working at low temperatures [8.6.38] (including calibration [8.6.9]) require careful and patient experimentation to solve the phase problem by Mössbauer scattering.

After some test measurements on a single crystal of $K_3[Fe(CN)_6]$ [8.6.36], the first scattering experiments on a CO-myoglobin (mbCO) single crystal have been

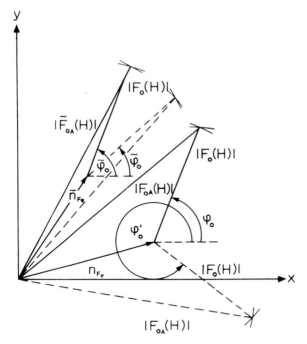

Fig. 8.6.5. Principle of the phase determination (taken from [8.6.36]). $n_{Fe}(\tilde{n}_{Fe})$ is the nuclear resonance scattering amplitude, $F_o(H)$ is the scattering amplitude of the unit cell, not including nuclear resonance scattering. $F_{oA}(H)$ ($\tilde{F}_{oA}(H)$) represents the scattering amplitude of the unit cell including the nuclear resonance scattering. n_{Fe} and $F_{oA}(H)$ correspond to Doppler velocity v_r and $\tilde{n}_{Fe}$, $\tilde{F}_{oA}(H)$ to $\tilde{v}$ ($\tilde{v}$ is very close to v_r) · $\varphi_o(H) = \tilde{\varphi}_o(H)$ is the correct phase, which should be determined

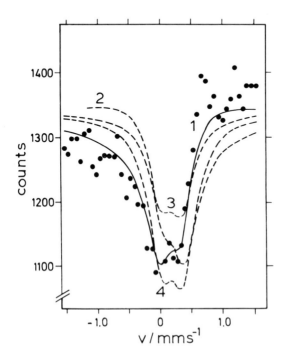

Fig. 8.6.6. Scattering experiment on the $60\bar{3}$ reflection of a mb single crystal. The four calculated curves are explained in the text (taken from [8.6.39])

reported by Parak et al. [8.6.39]. Fig. 8.6.6 shows the result of the scattering experiment on the 603-reflection of mbCO. The counting time in this case was six weeks! From a least-squares-fit of the data (curve 1 in Fig. 8.6.6) a phase value of $\varphi(60\overline{3})$ = $-1°$ was obtained, in good agreement with a value of $\varphi(60\overline{3}) = 0°$, which was calculated from the known structure of mbCO. The other curves in Fig. 8.6.6 were calculated with the phases $\varphi = 180°$ (curve 2), $\varphi = 90°$ (curve 3), and $\varphi = 270°$ (curve 4), indicating that there exists no agreement between these curves and the experimental data. In spite of the large statistical error the accuracy of $\varphi(60\overline{3})$ was estimated to be better than $\pm 45°$, a limit which is certainly sufficient for the purpose of a structure determination. At any rate, the effort to solve the "phase problem" in proteins using the interference between Rayleigh scattering and γ-resonance scattering is promising.

8.6.3. Iron Transport and Storage Compounds

A review on "Biological Iron Transport and Storage Compounds" has recently been written by Oosterhuis et al. [8.6.40]. There the various experimental Mössbauer studies on iron transport proteins (ferrichrome A, deferoxamine, enterobactin, myobactin P, transferrins, and conalbumin) and iron storage proteins (ferritin, phycomyces, hemosiderin, and gastroferrin) are summarized. Here we concentrate on two examples to further illustrate the spin relaxation behaviour (which we have already discussed in Sec. 7.2) of iron compounds.

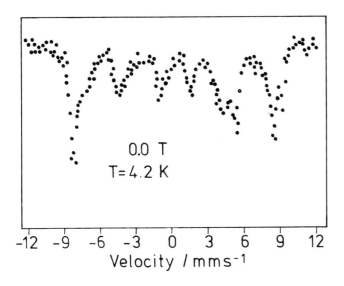

Fig. 8.6.7. Experimental Mössbauer spectrum of human serum transferrin in water, T = 4.2 K, $H_0 = 0$ (taken from [8.6.40])

8.6.3.1. Transferrin

Transferrin is an iron chelating compound found in the blood serum. Its function is the transport of iron from the iron storage areas of the body to areas where iron proteins are synthesized; there is some speculation that iron is taken from ferritin by transferrin to the bone marrow for incorporation into the red blood cells. Transferrin normally has two iron sites, which from the Mössbauer data appear to be in the ferric high-spin state [8.6.41, 42]. Since the spectrum at low temperature (Fig. 8.6.7) shows magnetic hyperfine interaction [8.6.40], it may be concluded that the two iron sites are sufficiently separated from each other to yield spin-spin relaxation within the slow relaxation limit (see Fig. 6.12), thus exhibiting magnetically split spectra.

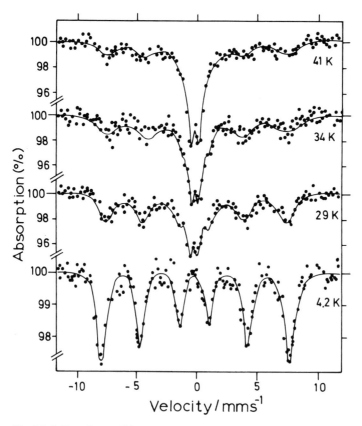

Fig. 8.6.8. Experimental Mössbauer spectra of horse ferritin at various temperatures, $H_0 = 0$ (taken from [8.6.44])

8.6.3.2. Ferritin

Ferritin is found in such diverse organisms as mammals, plants and fungi. The molecule contains a nearly spherical cell of ferric oxyhydroxide (FeOOH) with a diameter of about 70 Å, and an outer protein shell with a thickness of about 25 Å [8.6.43]. The FeOOH core of the molecule becomes antiferromagnetically ordered at about 200 K [8.6.43, 44]. Because of the limit on the volume of iron in the ferritin, it is probable to have a uniform FeOOH core particle size in this case. This is especially interesting under the aspect of the superparamagnetic behaviour of the FeOOH core particle (Fig. 8.6.8), which then can be analyzed along the lines described in Section 6.7.2.4, without necessarily assuming a size-distribution function like that of (6.66).

8.6.4. Enzyme Systems

The chemistry of life is controlled by enzymes, highly specialized proteins catalyzing the reactions in living cells. The number of different enzymes is very large, for example a relatively simple organism such as a bacterium produces about 1,000 different types. Many of them contain transition metal atoms. The importance of transition metal atoms and their complexes such as biocatalysts is based on the same properties which make them useful as industrial catalysts. The partially filled d orbitals of the transition metals are energetically close to s and p orbitals, and as a consequence, electron transfer may take place without consuming much energy.

So far, Mössbauer studies of enzymes have been limited to iron. In an excellent survey, Debrunner [8.6.45] has discussed ^{57}Fe Mössbauer spectra to illustrate three types of iron complexation in bacterial enzymes, namely the cases where iron is bound directly to the protein (oxygenase), the iron-sulphur proteins (nitrogenase), and the cases where iron is incorporated into a prosthetic group such as the heme (cytochrome P450). Here as an example of the third category we shall discuss catalase, which has not been included in Debrunner's review.

Catalase catalyzes the decomposition of hydrogen peroxide into oxygen and water and plays, therefore, an important role in the catalysis of the reactions of peroxide compounds [8.6.46]. Mössbauer experiments [8.6.47–50] on catalase turned out to be very difficult, because of the lack of in vitro ^{57}Fe enrichment in this case. Two representative spectra of native bacterial catalase [8.6.47] (prepared from Micrococcus lysodeikticus) are given in Fig. 8.6.9. Spectrum (a) shows a well resolved quadrupole doublet. The accompanying broad absorption suggests the presence of an unresolved magnetic hyperfine interaction. By applying a small field of 0.1 Tesla, a spectrum was obtained which indicated that the broad part becomes a six-line pattern with an intensity ratio of $3:4:1:1:4:3$ while the quadrupole peak did not change in area and position (Fig. 8.6.9b). It is significant that the areas of quadrupole part (P_Q) and magnetic part (P_M) are equal to the experimental error. P_Q is very similar to the low-temperature spectrum of crystalline hemin, which is characteristic of the ferric high-spin state [8.6.51–53], the symmetrical doublet at 4.2 K is interpreted by fast spin-spin relaxation [8.6.54]. On the other hand, P_M is very similar to that of a magnetically diluted

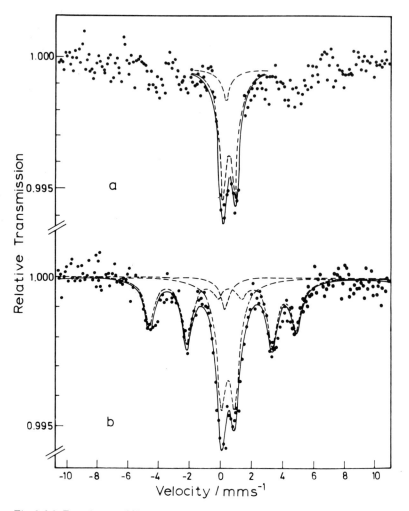

Fig. 8.6.9. Experimental Mössbauer spectra of native bacterial catalase at 4.2 K with (a) $H_0 = 0$ and (b) $H_0 = 0.1$ Tesla, $\overline{H}_0 \perp \vec{\gamma}$. The spectra include the absorption peak (central single line) of iron impurities, contained in the various windows of the Mössbauer cryostat, and identified by a blank run (taken from [8.6.47]

frozen hemin solution [8.6.54]. In the dilute hemin, a similar superposition of P_Q and P_M has been observed.

On the assumption that the catalase molecule consists of four subunits [8.6.55], it was concluded from the above findings [8.6.47] that four prosthetic groups in bacterial catalase show pairwise different interactions; specifically two of the prothetic groups are located at considerably short distance. Therefore, half of the hematins will show such a symmetric quadrupole doublet as crystalline hemin, because the spin-spin relaxation time decreases with decreasing distance between the paramagnetic ions. In view of spin-spin relaxation, however, Fe^{3+} in the P_Q should be separated by approximately 7 Å or less. Such a short separation of subunits has not been observed in hemoglobin for example. As catalase has the larger molecular weight than hemoglobin, the assumption of a short separation between subunits is surprising.

References

[8.6.1] Neilands, J. B.: in *Structure and Bonding*, Vol. 11, Berlin – Heidelberg – New York: Springer 1972, p. 145

[8.6.2] May, L.: Index of Publications in Mössbauer Spectroscopy of Biological Materials, Dept. of Chemistry, The Catholic University of America, Washington, D. C. 20017

[8.6.3] Conrad, H., Mayer, A., Thomas, H. P., Vogel, H.: J. Mol. Biol. *41*, 225 (1969)

[8.6.4] Schneider, R., Mayer, A., Schmatz, W., Kaiser, B., Scherm, R.: J. Mol. Biol. *41*, 231 (1969)

[8.6.5] Schneider, R., Schmatz, W., Mayer, A., Eicher, H., Schelten, J., Franzel, R.: (private communication)

[8.6.6] Perutz, M. F., Muirhead, H., Cox, J. M., Goaman, L. C. G., Mathews, F. S., Mc Gandy, E. L., Webb, C. E.: Nature *219*, 29 (1968)

[8.6.7] Perutz, M. F., Muirhead, H., Cox, J. M., Goaman, L. C. G.: Nature *219*, 131 (1968)

[8.6.8] Grant, R. W., Cape, J. A., Gonser, U., Topol, L. E., Saltman, P.: Biophys. J. *7*, 651 (1967)

[8.6.9] Trautwein, A., Maeda, Y., Gonser, U., Parak, F., Formanek, H.: Proceedings of the 5th International Conference in Mössbauer Spectroscopy, Bratislava (CSSR), Sept. 1973 Gonser, U., Maeda, Y., Trautwein, A., Parak, F., Formanek, H.: Z. Naturforschung *29b*, 241 (1974)

[8.6.10] Cooke, R., Debrunner, P.: J. Chem. Phys. *48*, 4532 (1968)

[8.6.11] Trautwein, A., Eicher, H., Mayer, A., Alfsen, A., Waks, M., Rosa, J., Beuzard, Y.: J. Chem. Phys. *53*, 963 (1970)

[8.6.12] Dézsi, I.: Proceedings of the International Conference on the Application of the Mössbauer Effect, Israel, August 1972

[8.6.13] Trautwein, A., Maeda, Y., Harris, F. E., Formanek, H.: Theoret. Chim. Acta *36*, 67 (1974)

[8.6.14] Lang, G., Marshall, W.: Proc. Roy. Soc. (London) *87*, 3 (1966)

[8.6.15] Maeda, Y., Morita, Y., Marcolin, H. E., Trautwein, A.: (unpublished results)

[8.6.16] Proceedings of the "International Conference on the Applications of the Mössbauer Effect", Corfu (Greece), 1976

[8.6.17] Trautwein, A.: in *Structure and Bonding*, Vol. 20, Berlin – Heidelberg – New York: Springer 1974, p. 101

[8.6.18] Kent, T., Spartalian, K., Lang, G., Yonetani, Y.: (preprint 1976)

[8.6.19] Loew, G. H., Kirchner, R. F.: J. Am. Chem. Soc. *97*, 7388 (1975)

[8.6.20] Halton, M. P.: Theoret. Chim. Acta *23*, 208 (1971); Inorg. Chim. Acta *8*, 131 (1974)

[8.6.21] Dedieu, A., Rohner, M.-M., Veillard, A.: in *Metal-ligand Interactions in Organic Chemistry and Biochemistry*, Proceedings of the 9th Jerusalem Symposium on Quantum Chemistry and Biochemistry, March 1976

[8.6.22] Aronowitz, S., Gouterman, M., Chien, J. C. W.: Theoret. Chim. Acta (in press)

[8.6.23] Trautwein, A., Zimmermann, R., Harris, F. E.: Theoret. Chim. Acta *37*, 89 (1975)

[8.6.24] Nath, A., Klein, M. P., Kündig, W., Lichtenstein, D.: Radiat. Eff. *2*, 211 (1970)

[8.6.25] Mullen, R. T.: in *Mössbauer Effect Methodology*, I. J. Gruverman (ed.), Vol. 5. New York: Plenum Press 1970, p. 95

[8.6.26] Marchant, L., Sharrock, M., Hoffman, B. M., Münck, E.: Proc. Nat. Acad. Sci. USA *69*, 2396 (1972)

[8.6.27] Zory, P.: Phys. Rev. *140*, A 1401 (1965)

[8.6.28] Zimmermann, R.: Nucl. Instr. Methods *128*, 537 (1975); Chem. Phys. Lett. *34*, 416 (1975)

[8.6.29] Maeda, Y., Harami, T., Trautwein, A., Gonser, U.: Z. Naturforschung *31b*, 487 (1976)

[8.6.30] Bragg, W. L., Perutz, M. F.: Proc. Roy. Soc. (London) *A225*, 315 (1954)

[8.6.31] Bodo, G., Dintris, H. M., Kendrew, J. C., Wyckoff, H. W.: Proc. Roy. Soc. (London) *A253*, 70 (1959)

[8.6.32] Perutz, M. F.: in: Scientific American, Nov. 1964, p. 2

[8.6.33] Parak, F., Mössbauer, R. L., Hoppe, W.: Ber. Bunsenges. Phys. Chem. *74*, 1207 (1970)

[8.6.34] Black, D. J., Longworth, G., O'Connor, D. A.: Proc. Phys. Soc. *83*, 925 (1964)

[8.6.35] Bernstein, S., Campell, E. C.: Phys. Rev. *132*, 1625 (1963)

[8.6.36] Parak, F., Mössbauer, R. L., Biebl, U., Formanek, H., Hoppe, W.: Z. Phys. *244*, 456 (1971)

[8.6.37] Parak, F., Formanek, H.: Acta Cryst. *A27*, 573 (1971)

[8.6.38] Thomanek, U. F., Parak, F., Mössbauer, R. L., Formanek, H., Schwager, P., Hoppe, W.: Acta Cryst. *A29*, 263 (1973)

[8.6.39] Parak, F., Mössbauer, R. L., Hoppe, W.: Proceedings of the International Conference on the Applications of the Mössbauer Effect, Corfu (Greece), 1976

[8.6.40] Oosterhuis, W. T., Spartalian, K.: in *Applications of Mössbauer Spectroscopy*, R. L. Cohen (ed.). New York: Academic Press 1976, p. 141

[8.6.41] Spartalian, K., Oosterhuis, W. T.: J. Chem. Phys. *59*, 617 (1973)

[8.6.42] Tsang, C. P., Boyle, A. J. F., Morgan, E. H.: Biochim. Biophys. Acta *328*, 84 (1973)

[8.6.43] Blaise, A., Chappert, J., Giradet, J.: C. R. Acad. Sci. Paris *261*, 2310 (1965)

[8.6.44] Boas, J. F., Window, B.: Aust. J. Phys. *19*, 573 (1966)

[8.6.45] Debrunner, P.: in *Applications of Mössbauer Spectroscopy*, R. L. Cohen (ed.). New York: Academic Press 1976, p. 171

[8.6.46] Theorell, H.: Experientia *4*, 100 (1948)

[8.6.47] Maeda, Y., Trautwein, A., Gonser, U., Yoshida, K., Kikuchi-Torii, K., Homma, T., Orgura, Y.: Biochim. Biophys. Acta *303*, 230 (1973)

[8.6.48] Karger, W.: Ber. Bunsenges. Phys. Chem. *68*, 793 (1964)

[8.6.49] May, L., Hasco, G. M.: Am. Chem. Soc. 156 Natl. Meet., Abst. No. 124 (1968)

[8.6.50] May, L., Arnold, M. H., Kuo, C. T.: in *Perspectives in Mössbauer Effect Spectroscopy*, M. Pasternak (ed.). New York: Plenum Publ. Corp. 1972

[8.6.51] Shulman, R. G., Wertheim, G. K.: Rev. Mod. Phys. *36*, 459 (1964)

[8.6.52] Gonser, U., Grant, R. W.: Biophys. J. *5*, 823 (1965)

[8.6.53] Bearden, A. J., Moss, T. H., Caughey, W. C., Beaudreau, C. A.: Proc. Natl. Acad. Sci. U. S. *53*, 1246 (1965)

[8.6.54] Blum, M.: Phys. Rev. Lett. *18*, 305 (1967)

[8.6.55] Tanford, C., Lovrien, R.: J. Am. Chem. Soc. *84*, 1892 (1962)

Subject Index

Inorganic Chemistry Concepts

Editors:
M. Becke, C. K. Jørgensen, M. F. Lappert,
S. J. Lippard, J. L. Margrave, K. Niedenzu,
R. W. Parry, H. Yamatera

Volume 1
R. Reisfeld, C. K. Jørgensen

Lasers and Excited States of Rare Earths

1977. 9 figures, 26 tables. VIII, 226 pages
ISBN 3-540-08324-3

Contents:
Analogies and Differences Between Mono-
atomic Entities and Condensed Matter. –
Rare-Earth Lasers. – Chemical Bonding and
Lanthanide Spectra. – Energy Transfer. –
Applications and Suggestions.

The authors discuss both numerous spectroscopic questions, including the influence of the
neighbour atoms on positions, and intensities
of transitions in trivalent Pr, Nd, Sm, Eu, Gd,
Tb, Dy, Ho, Er, Tm, and Yb and other effects
of chemical bonding in vitreous and crystalline
solids, as well as the parameters determining
the applications in lasers, such as the radiative
and non-radiative transitions probabilities, the
strategy for suppressing radiationless processes competing with luminescence, and the
energy transfer from trivalent cerium, the
uranyl ion, vanadate, tungstate, ion of
thallium, lead and bismuth isoelectronic with
mercury atom (and from many other species)
to the rare earths. In a final chapter, general
questions of radiation from opaque objects,
communications, geodesy and astrophysics
are discussed, and the use of neodymium glass
lasers for initiating nuclear fusion reviewed.
Many new suggestions are presented in this
novel interdisciplinary field.

Volume 2
R. L. Carlin, A. J. van Duyneveldt

Magnetic Properties of Transition Metal Compounds

1977. 149 figures, 7 tables. XV, 264 pages
ISBN 3-540-08584-X

Contents:
Paramagnetism: The Curie Law. – Thermo-
dynamics and Relaxation. – Paramagnetism:
Zero-Field Splittings. – Dimers and Clusters. –
Long-Range Order. – Short-Range Order. –
Special Topics: Spin-Flop, Metamagnetism,
Ferrimagnetism and Canting. – Selected
Examples.

This book introduces chemists and physicists
to magnetic ordering phenomena. Chemists
have long been interested in magnetic interactions in clusters, but have shied away from the
cooperative phenomena that have always fas-
cinated physicists. Part of the reason for this
is that the most remarkable phenomena occur
at low temperatures, a region where many
chemists have only recently started making
measurements. This work starts with a study
of paramagnetism, and shows that specific
heat results are as valuable a magnetic
measurement as a susceptibility. Emphasized
throughout is the valuable information to
be obtained by measuring single crystals,
rather than powders. The subject of paramagnetic relaxation is also introduced.
Numerous illustrative examples from literature are cited throughout. The book con-
cludes with an extensive literary survey of both
the single-ion properties of the iron-series
ions, and ten interesting examples of magnetic
systems to which the principles introduced in
the beginning of the book have been applied.

Springer-Verlag
Berlin
Heidelberg
New York

Topics in Current Chemistry

Fortschritte der chemischen Forschung
Managing Editor: F. L. Boschke

Volume 65

Theoretical Inorganic Chemistry II

1976. 47 figures, 44 tables. IV, 153 pages
ISBN 3-540-07637-9

Contents/Information:
K. Bernauer: *Diastereoisomerism and Diastereoselectivity in Metal Complexes*
Diastereoselectivity of a single ligand molecule has as yet not been reviewed. Even though some examples of conformational analysis are known and predictions of selectivity can be made, the orientation of a secondary ligand in a chiral mixed ligand complex is still based on empirical data. (134 references)

M. S. Wrighton: *Mechanistic Aspects of the Photochemical Reactions of Coordination Compounds*
Increasing interest in the photochemistry of transition metal containing molecules prompts this survey of the important developments in understanding the chemical transformations resulting from electronic excitation of such molecules. (196 references)

A. Albini, H. Kisch: *Complexation and Activation of Diazenes and Diazo Compounds by Transition Metals*
The interest in the coordinating properties of diazenes stems from the probable intermediacy of diazene complexes in biological nitrogen-fixation as well as from the non-enzymatic conversion of coordinated dinitrogen into diazene derivatives. (119 references)

Volume 56

Theoretical Inorganic Chemistry

1975. 22 figures, 18 tables. IV, 159 pages
ISBN 3-540-07226-8

Contents/Information
C. K. Jørgensen: *Continuum Effects Indicated by Hard and Soft Antibases (Lewis Acids) and Bases*
The origin of the selective affinity of hard central atoms to hard ligands, and of soft atoms to soft ligands is discussed. The softness parameters of Ahrland and of Klopman are defined from atomic spectra and hydration energies of ions. (210 references)

H. Brunner: *Stereochemistry of the Reactions of Optically Active Organometallic Transition-Metal Compounds*
The article describes the stereochemical results obtained with optically active organometallic compounds. The results are discussed according to the stereochemical outcome under retention, inversion, racemization, and epimerization reactions. (69 references)

L. H. Pignolet: *Dynamics of Intramolecular Metal-Centered Rearrangement Reactions of Tris-Chelate Complexes*
NMR-line-broadening techniques have been extensively used to measure rates of reactions in which magnetically nonequivalent environments are averaged or interchanged on the NMR timescale. (85 references)

S. Veprek: *A Theoretical Approach to Heterogeneous Reactions in Non-Isothermal Low Pressure Plasma*
The development of solid-state physics and technology has stimulated a large effort toward the application of plasmas to thin-film preparation and crystal growth. (66 references)

Springer-Verlag
Berlin
Heidelberg
New York